# Ten Questions

# A Sociological Perspective

SEVENTH EDITION

Joel M. Charon
*Professor Emeritus*
*Minnesota State University Moorhead*

Australia • Brazil • Japan • Korea • Mexico • Singapore • Spain • United Kingdom • United States

**Ten Questions: A Sociological Perspective, Seventh Edition**

Joel M. Charon

Publisher/Executive Editor: Linda Schreiber

Senior Acquisitions Editor: Chris Caldeira

Assistant Editor: Erin Parkins

Technology Project Manager: Lauren Keyes

Marketing Manager: Kim Russell

Marketing Assistant: Jillian Myers

Marketing Communications Manager: Martha Pfeiffer

Project Manager, Editorial Production: Jared Sterzer

Creative Director: Rob Hugel

Art Director: Caryl Gorska

Print Buyer: Paula Vang

Permissions Editor: Roberta Broyer

Production Service: Pre-Press PMG

Copy Editor: Pre-Press PMG

Cover Designer: Yvo Riezebos

Cover Image: Réunion des Musées Nationaux/Art resource, NY

Compositor: Pre-Press PMG



Library of Congress Control Number: 2008942965

Student Edition:
ISBN-13: 978-0-495-60130-2
ISBN-10: 0-495-60130-6

**Wadsworth, Cengage Learning**
10 Davis Drive
Belmont, CA 94002-3098
USA



Printed in Canada
1 2 3 4 5 6 7 8 9 13 12 11 10 09

# Contents

# Preface

Sociology is a perspective—that is, one way of thinking, or one way of looking at and investigating the universe. It focuses on the human being as a member of society, so its questions should have importance to all of us who seek to understand who we are.

*Ten Questions: A Sociological Perspective,* Seventh Edition, is written for students in introductory sociology courses and for those who do not have enough time to take an entire sociology course but still wish to understand how sociologists think. It is written for sociologists, who sometimes forget the excitement of sociology as they become involved in the tasks of teaching and research; and for critics of sociology, whose criticisms are too often without foundation. It is for English teachers, physicists, psychologists, artists, poets, and many other scholars whose lives are filled with the same questions but whose approaches differ; and it is for all people who value education and believe, like the Greeks, that "the unexamined life is not worth living."

This book introduces the perspective of sociology by posing twelve questions and then answering them, thereby trying to describe the sociological approach. Sociologists wonder about these questions regularly, and most debate them with colleagues, students, or, at the very least, themselves.

*How do sociologists study society?* How do we "observe" society? Will our personal biases prevent our understanding? Can sociology be scientific? What does it mean to be scientific?

*What does it mean to be human?* What is human nature? What are the key qualities that make humans *human?* What is the role of society, language, culture, and socialization in what humans become?

*How is society possible?* Is it force or is it a willingness to cooperate that makes society possible? How is it that a number of unique individuals are able to give up their self-interests and wishes to some extent in order for a community to exist?

*Why are people unequal in society?* Is it human nature? Is inequality built into the nature of society? What are the consequences for society if there is excessive inequality? Is it possible to create a society in which people are basically equal?

*Are human beings free?* Are our ideas really our own? Are we really responsible for our own actions? What are the ways in which social forces impact our lives? Is freedom real or an illusion?

*Why can't everyone be just like us?* Why is it that people have a difficult time respecting people who are different from them? What is ethnocentrism, and why does it exist? What, after all, creates the many differences among people?

*Why is there misery in the world?* What causes human problems? Why is life so difficult for so many people? How does society create its own problems? Is it possible to build a better society? Is misery inevitable?

*Does the individual really make a difference?* Is this simply wishful thinking? When can the individual really make a difference? What works against it? What actually causes society to change?

*Is organized religion necessary for society?* What is religion? How has religion traditionally contributed to society? Can religion be harmful in society? Is religion still important in modern society?

*Is the world becoming one society?* What is globalization? To what extent is the world becoming a single unit? Are we becoming a world that promises order, democracy, and equality? Or are we becoming a capitalist world that is controlled by a few corporations? What qualities must be developed before we call ourselves a single society?

*Why study sociology?* How does sociology help students understand society? How does sociology play a role in liberal arts? How is sociology the study of democracy? What are some of the important ways sociology helps understanding, questioning, and caring about society and the individual?

*Should we generalize about people?* Is it better to understand people only through treating each person as unique? How should one generalize? What is stereotyping? What do social scientists try to do when they study people?

These are the twelve questions that make up the chapters in this book. They are the most important questions that sociology helps my understanding of society and my life.

The order of these chapters should not determine how the book should be read. Instructors have found many ways to order the chapters in their classes. I am sure that those who have read this book on their own have jumped around considerably according to their purpose. One of the attractions to instructors is that the chapter order can easily adjust to their own course outline.

In this Seventh Edition I made five changes suggested by the reviewers of the Sixth Edition.

1. I added a new chapter that examines the issues related to globalization. I found it very difficult to describe, explain, predict, and evaluate globalization, but that still was my goal. Throughout the chapter I discussed capitalism, world economy, technology, nations, societies, inequality, and democracy.

   I think this question is increasingly relevant to understanding social life. The problem, of course, is that the book now has twelve questions rather than ten. As you probably know, I included 10 questions in the first edition of this book. However, the book has evolved without changing the title. I added the question "Should We Generalize about People" in the third edition, and now I feel the need to add "Is the World Becoming One Society?"

   I still need to keep the title Ten Questions for purpose of our advertising, your ordering, and my fear that everyone will become confused by the change of the book title.

2. I made the chapter on generalization, social science, and stereotypes as an "Afterword." It is an important chapter, but I still do not know where it should be in the book. Some would like me to integrate the chapter into Chapter 1, but that makes Chapter 1 much too long and not really workable. I know that some instructors use it before all the other chapters, some use it right after the first chapter, others use it right before the last chapter, and others use it as

the very last chapter. I am certain that there is no consensus, and that is why I made it an "Afterword."

3. I brought together the material of the "Introduction" and Chapter 1. Thus, there is no introduction, and Chapter 1 is revised.

4. I spent a lot of effort making the chapter on freedom better organized and more clear. Reviewers made suggestions and I realized I needed a better presentation.

5. I had the good fortune to have helpful reviews by professors who have used the earlier editions of the book. I carefully evaluated their suggestions, and I changed what I thought would make the various chapters more interesting, more clear, and more up to date.

Feedback from students and faculty is welcome. In fact, I welcome comments and questions from anyone. My e-mail address is **charonj@mnstate.edu.** I will try my best to reply. When the next edition is written, such input will make a difference in what I write.

A basic assumption underlies this book: that students will enjoy discussing and wondering about these questions. They will recognize education to be more than accumulating facts, and if challenged to debate issues that shed light on the human being, students will discover a fervor in learning that is too often ignored.

This was an exciting project for me. It forced me to make explicit my assumptions about the nature of sociology. The encouragement I received from reviewers was gratifying, and their suggestions for improving the manuscript were invaluable. Especially important were Andrea Bertotti Metoyer, Gonzaga University; Tim Pippert, Augsburg College; Dennis C. Shaw, Lower Columbia College; Rebecca Plante, Ithaca College; and Daniel Sarabia, Roanoke College. Also, Professor Lee Vigilant and Professor Joan Ferrante reviewed and made suggestions for the chapter on globalization.

I would like to thank all the various editors who have supported this project: Serina Beauparlant, Eve Howard, Bob Jucha, and Chris Caldeira. All four have had faith in what I was trying to do in this book. It remains a labor of love for me. The questions and debates included are really a result of many discussions with colleagues and students. I was very fortunate to live my life in wonderful schools and universities where I found so many people interested in issues that encouraged me to be concerned with the kinds of questions presented in this book. I owe a lot to many people. I thank my friends and teachers from Minneapolis North High School and the University of Minnesota, colleagues and students from St. Paul Harding High School and St. Paul Highland Park High School, and, of course, Minnesota State University Moorhead. I especially the kindness and inspiration of David Cooperman, my advisor. I cannot forget to thank those who wrote great books and articles that caused me to wonder, understand, question, and think, and, of course, all the people outside of the academic world who questioned me, learned from me, and taught me ideas that did not simply come out of the academic world. Whatever I know is clearly the result of social interaction with people who inspired me to think critically and explained many of the understandings that allowed me to write this book.

I would like to dedicate this book to my wonderful family: my wife Susan, who helps make my life worth living; and my sons, Andrew and Daniel, whose intelligence, creativity, thoughtfulness, and individuality make me proud.

## Supplements

The following supplements are available from Thomson Wadsworth.

**Online Instructor's Manual.** This manual is available for download at the book companion website, **http://sociology.wadsworth. com/charon7e/**. It offers chapter outlines, in-class discussion questions, and testing suggestions for both multiple-choice and essay questions.

**Extension: Wadsworth's Sociology Readings Collection.** Create your own customized reader for your sociology class, drawing from dozens of classic and contemporary articles found on the exclusive Cengage Wadsworth TextChoice database. Using the TextChoice website **(http://www.TextChoice.com),** you can preview articles, select your content, and add your own original material. TextChoice will then produce your materials as a printed supplementary reader for your class.

## Classroom Presentation Tools for the Instructor

**JoinIn™ on TurningPoint®.** Transform your lecture into an interactive student experience with *JoinIn*. Combined with your choice of keypad systems, *JoinIn* turns your PowerPoint® application into audience-response software. With a click on a handheld device, students can respond to multiple-choice questions, short polls, interactive exercises, and peer-review questions. You can also take attendance, check student comprehension of concepts, collect student demographics to better assess student needs, and even administer quizzes. In addition, there are interactive text-specific slide sets that you can modify and merge with any of your own PowerPoint lecture slides. This tool is available to qualified adopters at **http://turning point.cengage learningconnections.com.**

**Wadsworth's Lecture Launchers for Introductory Sociology.** An exclusive offering jointly created by Wadsworth/ Cengage Learning and Dallas TeleLearning, this video contains a collection of video highlights taken from the "Exploring Society: An Introduction to Sociology" Telecourse (formerly "The Sociological Imagination"). Each 3- to 6-minute-long video segment has been especially chosen to enhance and enliven class lectures and discussion of 20 key topics covered in any introductory sociology text. Accompanying the video is a brief written description of each clip, along with suggested discussion questions to help effectively incorporate the material into the classroom. Available on VHS or DVD.

**ABC Videos (Introduction to Sociology Volumes I–IV).** ABC Videos feature short, high-interest clips from current news events as well as historic raw footage going back 40 years. Perfect for discussion starters or to enrich your lectures and spark interest in the material in the text, these brief videos provide students with a new lens through which to view the past and present, one that will greatly enhance their knowledge and understanding of significant events and open up to them new dimensions in learning. Clips are drawn from such programs as World News Tonight, Good Morning America, This Week, PrimeTime Live, 20/20, and Nightline, as well as numerous ABC News specials and material from the Associated Press Television News and British Movietone News collections.

**Wadsworth Sociology Video Library.** Bring sociological concepts to life with videos from Wadsworth's Sociology Video Library, which includes thought-provoking offerings from Films for Humanities, as well as other excellent educational video sources. This extensive collection illustrates important sociological concepts covered in many sociology courses.

**InfoTrac® College Edition with InfoMarks®.** Available as a free option with newly purchased texts, InfoTrac College Edition gives instructors and students four months of free access to an extensive online database of reliable, full-length articles (not just abstracts) from thousands of scholarly and popular publications going back as far as twenty-two years. Among the journals available 24/7 are *American Journal of Sociology, Social Forces, Social Research,* and *Sociology.* InfoTrac College Edition now also comes with InfoMarks, a tool that allows you to save your search parameters, as well as save links to specific articles. (Available to North American college and university students only; journals are subject to change.)

**Wadsworth's Sociology Home page at http://sociology.wadsworth.com.** Combine this text with the exciting range of web resources on Wadsworth's Sociology Home Page, and you will have truly integrated technology into your learning system. Wadsworth's Sociology Home Page provides instructors and students with a wealth of *free* information and resources, such as Sociology in Action, Census 2000: A Student Guide for Sociology, Research Online, a Sociology Timeline, a Spanish glossary of key sociological terms and concepts, and more.

# How Do Sociologists Study Society?

## Researching the Social World

### Concepts, Themes, and Key Individuals

- ❑ Academic discipline
- ❑ Five thinkers: Marx, Weber, Durkheim, Mead, and Berger
- ❑ Science, social science, and sociology
- ❑ Philosophy and critical thinking
- ❑ Proof, logic, and empiricism
- ❑ Empirical studies in sociology
- ❑ Objectivity, natural law, and natural cause

### Introduction

In the 1950s I believed that men and women were different. I honestly believed that women were not capable of playing football, baseball, boxing, or hockey, and that women were not capable of being managers, mathematicians, soldiers, or engineers. I grew up believing that to be a man was to control women, that virginity was important for women but not men. I believed that women who did not marry were "old maids," and women who did not have children could not have a meaningful life. I believed that it was more important for men to have a college education than for women, and that women were good for nurses, elementary school teachers, secretaries, and airplane stewardesses. Of course, not everyone believed such ideas, but I remember believing them. Slowly I questioned my beliefs. The evidence around me was overwhelming—and questioning, observation, reading, discussion, and formal study made it impossible for me to continue accepting these earlier beliefs. Psychologists, sociologists, and anthropologists challenged what I had learned, and respected teachers and friends questioned my beliefs. These individuals became important models to me. I eventually decided there was really no difference between men and women. I argued, I listened, I thought a lot, and I recognized that some diffferences may actually exist, but always the differences were probabilities, not complete and distinct differences. I asked myself what really is the importance of sex differences, and I have come to believe that whatever natural differences exist are much less important than differences based on social influence.

I have many more beliefs. Each has a similar narrative as to how I arrived at them. As I look back, I wonder why—why did I choose one belief rather than another. More importantly, I ask how do I really know that my beliefs are true?

My student experience, my twelve years of teaching high school history and the thirty years of lecturing, research, and writing at a university have taught me how difficult it is to understand anything related to the human being. I recognized that there are many ways that people search for the truth. I learned that *how* one arrives at his or her truth is central to whether I am willing to consider what others tell me. I try to find out, if someone is highly biased, what that bias is, what its background is, what kind of evidence is presented, whether the arguments are logical and clear, and whether there is arrogance in the presentation. I realize that if I am going to understand something about the human being, I need to be open-minded, careful, critical of those who are influencing me and critical of my own set of beliefs. Most of the time my education has been a consistent conversation with others and myself—evaluating, questioning, and seeking more understanding. I respect people who recognize the difficulties in understanding others. Nevertheless, I am suspicious of anyone who argues that he or she knows the "truth" when it comes to the human being.

## Sociology

I came to sociology late in my formal schooling, but I was prepared for it because of my love of history, psychology, and philosophy. Sociology is an *academic discipline,* an attempt to seek knowledge and understanding through painstaking and critical investigation. Like other academic perspectives—from physics and chemistry, art and mathematics, or philosophy and psychology—it answers questions with care, debate, and uncertainty. This book, *Ten Questions,* is *my* understanding of sociology. For many sociologists it is far too simple, and for others my analysis will not be satisfactory. Over time I realized that *Ten Questions* is not meant to be a book of answers: It is much more a *book about thinking.* Although all sociologists might not always agree with the answers in this book, most would probably agree that it describes *how sociologists think,* how they would deal with the questions posed.

This book does not emphasize many of the specific specialties, research studies, and conclusions that all sociologists believe. It is not meant to be a textbook, nor an encyclopedic list of "facts." The questions I consider are among the most important ones that sociologists study. Indeed, they are fascinating questions that most thinking people investigate for much of their lives. They form the basis for what a serious education *should* investigate.

As you will shortly see, the sociological perspective is different from the way most people—maybe even you—see reality. We live in a society that emphasizes the individual and tends to look for the reasons for action within individuals. Our religious and political heritage and our tendency to focus on psychology too often cause us to overlook the importance of society in understanding human life. Whereas most people emphasize personality, character, heredity, and individual choice when they discuss human beings, the sociologist keeps crying out to us: "Don't forget society! Remember, human beings are social, and that makes a difference in what we all are."

I cannot escape the power of sociology to affect the way I think. Like almost everyone, I am repulsed by violent crimes. Injustice and inhumanity upset me. War and murder, exploitation and physical abuse, racism and sexism, theft and the destruction of property, feeding other people's addictions, and refusing to help the poor—all anger me and, frankly, that anger caused me to become a sociologist. But sociologists approach such problems differently from those of other people. "In what kind of society," we ask, "does this happen? What *social conditions* cause individuals to lose their humanity? What are the *social causes* of poverty, crime, and destructive violence?"

Whenever I read or hear about a horrible crime, my first reaction is "What a horrible thing to happen!" My second reaction is "How can people do that? What's wrong with them?" But then I get a third reaction, one that takes more self-discipline and care: "What are the underlying reasons for such acts? From what kind of world does inhumanity such as this arise?" As a sociologist, I am driven to understand the nature of society (including my own), and I appreciate all the different ways in which society affects the human being. Of course, I know that this is not the only way to understand human action, but I believe that it goes a long way.

## Five Thinkers

Throughout this book I draw from the works of several important sociologists. These writers have had the greatest influence on my own thinking. Their ideas are the most exciting and meaningful to me, so I will briefly introduce them now.

Sociology owes much to the work of Karl Marx (1818–1883). Marx, of course, is best known for *The Communist Manifesto* (1848) and *Das Kapital* (1867), both of which are critiques of capitalism and the society as he knew it. Marx was dissatisfied with how his society functioned; out of that dissatisfaction (which really amounted to great anger), he developed a theory of society that focuses on social class, social power, and social conflict. Marx's analysis is challenging to what most Americans believe, and he brought to sociology a critical and sophisticated approach to understanding society. Underlying all that he wrote was the idea that social inequality is the key to understanding society.

No one has influenced the development of the sociological perspective more than Max Weber (1864–1920), a German social thinker best known for *The Protestant Ethic and the Spirit of Capitalism* (1905). In this work he shows us that Protestant religious thinking was a central contributing factor to the development of capitalism in the West. Like almost everything he wrote, that book exhibits Weber's interest in describing the importance of *culture* in influencing how people act. People behave the way they do, he argues, because of this *shared belief system,* and the only way in which social scientists can understand people's actions is to understand their culture. That is why Weber is so important for the study of religion,

Modernization, legitimate authority, bureaucratization, science, and tradition—bureaucratization, science, and tradition—all particular ways of *thinking* that characterize people living together. If we think of Marx as the *critical sociologist,* then we should think of Weber as the *cultural sociologist.* This view is slightly misleading, however, because Weber was broader than that; like Marx, he also was deeply interested in social class, social power, and social conflict.

When the name of Émile Durkheim comes up in discussion, my thinking immediately shifts to "social order." Durkheim (1858–1917) was driven to understand all the various ways in which society is able to work as a unity. Society, he maintains, is not simply a bunch of individuals; it constitutes a larger whole, a reality that is more than the sum of the individuals who make it up. What keeps it together? How is this unity maintained? Durkheim documents the important contributions of religion, law, morals, education, ritual, the division of labor, and even crime in maintaining this unity. Every one of his major works examines it. For example, his most famous work, *Suicide* (1897), shows how the lowest and the highest levels of social solidarity result in high suicide rates. His last important work, *The Elementary Forms of Religious Life* (1915), documents the importance of religion, ritual, sacred objects, and other elements of the sacred world for social solidarity. Durkheim also contributed greatly to our appreciation of the influence of social forces on the individual, from suicide to knowledge of right and wrong.

In many ways sociology owes its perspective to the work of Marx, Weber, and Durkheim. Two other sociologists, both from the United States, appear now and then in the following chapters. Both have taught me much about the social nature of the human being, and especially about the power our social life has over the way we think. George Herbert Mead (1863–1931), a social psychologist who taught at the University of Chicago, has been extremely important in helping me understand the many complex links between society and the human being. His most important contribution to sociology is *Mind, Self and Society* (1934), which was written from his lecture notes by devoted students after his death. Throughout his work, certain questions are addressed over and over: What is human nature? That is, what characterizes the human being as a species in nature? How does society shape the human being? How does the individual in turn shape society? Mead persuasively shows that human beings are unique because they use symbols to communicate and they can think about their own acts and the acts of others. To Mead, *symbol use, selfhood,* and *mind* are qualities that create a being who can change society and not simply be passively shaped by it. The individual's relationship with society is complex, however, because symbols, self, and mind are socially created qualities, possible only *because* we are social beings.

The other American sociologist is Peter Berger (b. 1929). Along with Mead, Berger has tremendously influenced my thinking about the meaning and importance of sociology. His *Invitation to Sociology* (1963) and (with Thomas Luckmann) *The Social Construction of Reality* (1966) describe sociology as a special type of consciousness, a perspective that is profound, unusual, critical, and humanistic in its concerns. To Berger, sociology is liberating because it helps to reveal our taken-for-granted realities for what they are: social creations that appear true on the surface, but on closer inspection are usually found to be partially true or even untrue. In all of his work, Berger shows the power of society to shape human action and thought. Society socializes the human being to accept its ways. For Berger, to understand the power of society is the first step toward understanding who we are and what we can do to control our lives. It is also important to recognize the strong influence Berger has made in understanding the importance of and changes in religion that are addressed in this book.

What I write here is hardly ever my own idea because it is so heavily influenced by people such as Marx, Weber, Durkheim, Mead, and Berger. This book is inspired by all of these thinkers—and others—so I hope that if you like it, you will turn your attention to their works. I also hope you find in them the inspiration that I have found.

## The Importance of Rational Proof

We are all indebted to the ancient Greeks (whose civilization reached its peak about 300 BCE) for the great number of contributions they made to the world. They left us masterpieces in sculpture, pottery, architecture, and drama. They influenced mathematics, science, literature, and democracy as we have come to know them. But their special approach to understanding reality is what interests us here. Their culture began and encouraged the study of *philosophy*. To this day, the contributions of Socrates, Plato, and Aristotle are unparalleled in the history of thought. It is not the ideas alone that stand out, however, but their critical approach, their questioning attitude, that are central to Western thought.

In their search for understanding, the Greek philosophers developed a *critical* approach to ideas—a questioning of the ideas that people believed at the time. In their constant questioning, the Greek philosophers were teaching people who would listen to them—and people who would eventually read their works—to reject authority alone as the basis for truth, and to be suspicious of what the culture in Greek society claimed to be true. Instead, they argued, the truth and falsehood of any idea must be measured against a neutral standard, some acceptable measuring stick that prevents people from simply believing something because it "feels good" or it agrees with what they already believe. This measuring stick was "the rational proof," or what we today call *logic.* Just as we can determine if one line is longer than another by applying a ruler to both, so we can determine if one idea is truer than another by applying the rules of logic to both.

The ancient Greeks developed the *logical proof,* and since then these rules of logic, altered over time as new rules were discovered, became the basis for understanding in much of the academic world.

Honesty is at the heart of sound proof: Do not twist your thinking to prove what you want to prove; do not exaggerate, jump to unwarranted conclusions, or scream to the sky that an idea is wrong simply because it disagrees with what you already believe. Be suspicious of those people who are not logical; be careful with those who trick you to accept something on a basis other than logic. Instead, accept ideas because they are arrived at through a careful, organized, and honest examination of the idea. Take it apart: Look at its assumptions, search for its contradictions, examine it according to what else has already been proved, dissect it into its components, and examine them. If an idea can be supported through such activities, then you can be relatively sure that something approximating truth has been arrived at. If not, then it should be rejected, no matter who says it, no matter how much you want to believe it, even if it is what you have always been taught.

Greek civilization eventually declined, and its influence in the Western world was gradually replaced by an all-powerful Christian church that became the source of all truth. From around 300 CE until 1400, the most important approach to truth was a spirit of faith. People were expected to accept what their church taught rather than to seek wisdom on their own. The spirit of critical, rational analysis of established ideas was defined as heresy and punished.

Eventually, however, the spirit of Greek philosophy became acceptable again in the intellectual community. With the rise of a critical philosophy and science around the fifteenth century, Greek philosophy became a powerful force in the Western world. The development of social science in the eighteenth century and the founding of sociology as a social science in the nineteenth century were consequences of these developments and became part of this critical tradition.

## Proof, Science, and Sociology

### *The Need for Scientific Sociology*

Whenever I try to explain sociology to a mathematician friend, he gives up and declares, "I think higher math is easy compared with understanding human beings." When we puzzle over personal, social, and worldly problems, it becomes obvious that human beings are difficult to make generalizations about simply because of their complexity. This complexity is not the only thing that makes sociology such a difficult discipline. People walk into sociology, unlike mathematics, already believing that they know a great deal about the subject at hand: human beings and society. What could be more familiar? It is difficult to get people to question what they already "know." It is even more difficult to get them to recognize that what they know is not necessarily true.

The heart of rational proof is the recognition that the basis for truth must be found in reason, in a careful appraisal of ideas. Socrates, one of the greatest of Greek philosophers, best represented this recognition. The Socratic method of investigation is a continuous set of questions posed to someone. Through this method he revealed to the individual the assumptions people made without careful thought, the illogical conclusions they reached, and the poor evidence they relied on for their set of beliefs. Socrates, through his questioning, did not discover truth so much as he uncovered untruth, and in so doing he caused others to seek truth in a more careful, thoughtful way.

### *Empirical Proof*

This questioning is the purpose of sociology. Auguste Comte (1798–1857), the French thinker who coined the term *sociology,* argues that the critical methods of the ancient Greeks can and should be applied to society. In fact, these methods go even further: They should rely on a strictly *scientific approach,* a measuring stick even more demanding than rational proof. Science, too, demands that a neutral measuring stick be applied to the truth or falsehood of an idea, but this must not only conform to sound logic, but also be ultimately founded on *careful observation.* Comte believed that the purpose of the discipline he was founding, sociology (the "science of society"), would be to analyze the nature of society carefully and objectively through careful observation rather than to accept what has been handed down to us from the past. From Comte to the present, this has been sociology's central goal—to understand what society actually is irrespective of what people want it to be. This is the reason for sociology, and this has been its strength.

This emphasis on rational proof laid the foundation for modern science. Philosophical questioning and the idea that conclusions must be carefully arrived at through a measuring stick of rational thinking developed into a proof—known as *empirical proof*—that became the basis for science. Empirical proof trusts careful observation in measuring the truth or falsehood of an idea. An idea is rationally developed, but then it must be empirically tested—that is, tested against what we can see in the universe around us. Does the idea conform to careful observation in a laboratory or through a microscope or through a telescope? Does the idea conform to what we see in natural settings or in experimental settings, in everyday life, in historical records, or in election results?

Even the Greeks occasionally used empirical proof. In fact, some people trace empirical proof to Archimedes, a renowned Greek thinker and one of history's first scientists. Archimedes wanted to know how he could measure the volume of a mass (such as a king's crown). After all, one could not measure the mass of a crown with a simple ruler, because it was such an impossible shape. One day, while taking a bath, he noticed that the water rose in the tub when he got in and lowered when he got out. That, in a flash of insight, was the answer to his problem: To measure the volume of anything, all one must do is measure how much water is displaced by the object. The story is told that Archimedes ran naked into the streets crying out, "Eureka! Eureka! I have found it!"

How did Archimedes find his answer? Simply put, he observed it. This example illustrates well what empirical proof is: It is proof that is observed. Most important, once an idea has some empirical support, it can be shared with others, who then are able to observe the idea in a different situation in order to check out the first observer. This type of proof eventually became the basis for all the sciences from physics and biology to psychology and sociology.

Rational proof and empirical proof are both ways to test whether or not an idea is accurate. Philosophers and mathematicians rely heavily on rational proof; scientists rely on the empirical. Yet it is important to realize these are still different approaches to proof; one relies on careful thinking, whereas the other requires careful observation. An example might help. Once in my life I was interested in the game of roulette; I thought I had a system that could beat it. Because I understood statistics, including probability theory, I decided to figure out whether I could win by applying probability theory to my system. I discovered that there was no way I could win over the long run. In the end, the chances were that I would eventually lose everything. But, after all, I did not want to believe something as wonderful as my own system unless I saw with my own eyes that I would eventually lose. By looking at actual numbers on a random table to see how often eight odd numbers in a row came up, I realized that probability theory was correct: I would eventually lose all my money with my system. Probability theory is rational proof; counting how often odd numbers actually came up in a row was empirical proof. Of course, I recognized that if luck entered my game I might win, but most people with a system like mine are forever doomed in the long run to pay for the casino palaces where roulette is played.

*Empirical proof is the basis for sociology* and for many of the ideas in this book. It is the basis for the conclusions in the various specialties of sociology, from the study of family and religion to the study of revolution and culture. Observation is the basis for science, and so it is the basis for sociology. Acceptable evidence is that which can be *observed* by one individual and *shared* with others so that they can observe it, check it, criticize it, build on it, or disprove it. Observation can take place in a laboratory or in the natural environment. We can observe prominent people, items checked on a questionnaire, or diaries and letters. We can observe people in a gang, a corporation, a religious group, a football team, or an army. We can observe the speeches of political leaders and articles written in newspapers and magazines. Sometimes observation is a relatively easy matter—for example, seeing how many people marked item 7 on a questionnaire, or how many people under twenty-one committed suicide in Minnesota during 1993. Sometimes observation is far more difficult, and researchers must be especially imaginative and careful in their observation. How do poor people go about finding work? How do wealthy people exert power in government? What things do little boys say to little girls that reflect their understanding of gender roles? No matter what the question, if sociology is being done, then researchers must use empirical evidence to support their ideas, evidence that can somehow be observed and therefore be checked by others.

Two examples of empirical work in sociology will give you some idea how sociologists "observe." Durkheim's study of the causes of suicide is the first example. He was concerned about the high suicide rates in many communities and societies in Europe at the end of the nineteenth century. He could "see" those rates by simply looking at the number of suicides per 100,000 people in a given population. He found a remarkable consistency in the rate of suicide in France over time; in comparison with other societies, France's rate was higher than some and lower than others. He wanted to know why rates differed among societies and why they differed among communities within a society. He theorized that these rates were heavily influenced by the level of social solidarity in the community—that is, by how integrated the community was. He could not "see" the level of solidarity, so he applied what he logically thought out about solidarity in various communities (and he expresses exactly what he is thinking to the reader of the study). He argues, for example, that Catholic communities will have higher levels of social solidarity than Protestant communities because Catholics are more embedded in their church, whereas Protestants emphasize the individual's relationship to God. He further argues that the Jewish community is more integrated than either the Catholic or Protestant one (because Jews in nineteenth-century Europe were more separated from the larger community and their religion permeated every aspect of their daily living).

Then Durkheim was ready to "see" the evidence concerning suicide rates. He found what he expected: Protestant communities had the highest suicide rates, and Jewish communities had the lowest. He contrasted other communities: urban versus rural and college-educated versus noncollege-educated, for example. In every case, the more individualistic communities had higher suicide rates. Durkheim observed data collected by the government—imperfect, incomplete, and perhaps even biased data. Of course, one need not believe his theory about social solidarity and suicide rates, but the beauty of science is that one knows how he thought and how he observed. One can go and observe the same data to check him out or show that the same data can be understood in another way. One can now examine data in the United States or in any other part of the world. One does not have to take Durkheim's word for anything.

The second example is a study published in 1977 by Rosabeth Kanter. Kanter was interested in how, by its very nature, a large organizational system worked against equality between men and women. She knew that there were many ways to "observe" men and women in the corporation: She could take a nationwide survey, she could interview presidents of corporations, or she could examine existing data on how many women were in managerial positions or on boards of directors. What Kanter decided to do, however, was an in-depth study of one corporation. She sent out a mail survey to a sample of sales workers and sales managers. She interviewed many employees about their work and their positions in the corporation. She systematically examined 100 appraisal forms filled out on secretarial performance. She attended group discussions, observed training programs, and examined many documents within the corporation. She informally visited with employees at lunch, in hallways, and wherever else she could meet them. The success of the study, of course, depended on how well she could convince other social scientists that her methods had been careful, objective, and thorough. The strengths of her study were how deeply she was able to investigate in one corporation and the diverse ways in which she observed. Of course, the weakness was that she had only enough time and money to study one corporation. We need not believe what she found, however: We can do another case study. We can compare what she found in her observations with what others found in theirs. We can check out national surveys or do our own national survey, or we can examine already collected government data on all corporations.

These are not necessarily the best studies in sociology, but they are good examples of what is meant by empirical evidence. In 1918, W. I. Thomas and Florian Znaniecki published a study of the Polish peasants who settled in the United States; they had examined the immigrants' diaries and letters. In the 1920s, Frederic Thrasher and his associates studied gangs in Chicago by observing their actions on the street and interviewing members; and Robert and Helen Lynd studied the community of Muncie, Indiana, by asking people questions door-to-door in order to understand class and power in that community. In 1944, Gunnar Myrdal published a landmark study of race relations that identified "The American Dilemma"—the American conflict between the democratic creed and the treatment of African Americans. His study relied on observing documents, museums, and everyday social interaction and social events. In the 1940s, Samuel Stouffer and his colleagues published a massive questionnaire study of the American soldier in World War II; and, in 1950, Theodor Adorno and his associates produced a study called *The Authoritarian Personality* based on what they heard and observed in interviews and questionnaires of people's attitudes toward minorities. In the 1950s, Robert Bales studied leadership in small groups by carefully observing conversations in small laboratory groups, and C. Wright Mills studied power in the United States by observing people's names on a variety of lists and in newspapers. William H. Sewell in the 1960s and 1970s wanted to understand how social class affected success in the American system of education, so he observed students through the system by giving them questionnaires over several years. William Julius Wilson in the 1980s studied the lack of opportunity in the inner city by observing population and employment data gathered for the cities throughout the United States; and Gary Alan Fine studied preadolescent socialization through observing how boys acted on Little League baseball teams. In the 1990s, through intensive investigations, Richard Gelles and Murray Straus were able to better understand the nature and causes of family violence. Jonathan Kozol observed schools throughout the United States and interviewed principals and teachers in order to describe and explain the "savage inequalities" among and within school districts. And Candace Clark took on an interesting and complex study of "sympathy in everyday life," using a mixture of research techniques including (1) an analysis of fiction, Hallmark cards, and descriptions of the *New York Times*'s "Neediest Cases"; (2) "intensive eavesdropping"; (3) "focused discussions concerning sympathy"; (4) "freewriting" by students on issues related to their feelings; (5) a questionnaire given to more than 1,000 adults and sixty schoolchildren; and (6) more than sixty intensive interviews. The number of empirical studies in sociology is too many to mention, but this brief list gives us some indication of how observation was used in a wide diversity of research techniques in the United States.

## *Observation in Sociology*

Sociology is a particular type of science. It is not easy to observe groups, societies, power, interaction, or social class because they are not physical entities like leaves, skin, rock, or stars. Nor is it easy to observe people's ideas or values, their morals, or their hopes. Thus, the scientist must watch how human beings present themselves to others—what they do, what they say, and what they write—and then look beyond and infer the existence of a more abstract social reality. Thus, when people act together, we infer the existence of a group and then draw from our observations the qualities of groups, the ways in which they form and function, and their effects on individual action. When Durkheim tried to understand society, for example, he focused on people's rituals, the moral outrage they exhibited toward certain individuals, and the objects they worshiped. He showed how these acts and beliefs revealed the power of society over human action, and he showed their necessity for the continuation of society.

Sociologists do not have a narrow, rigid view of science. On the contrary, their view is that science must be open and that its techniques must be varied. They recognize that certainty is almost impossible and that their ideas must thus remain tentative. Max Weber contends that all scientists must be prepared to see their own ideas overturned with new evidence in their own lifetimes, especially scientists who study the human being. He emphasizes that there are exceptions to almost every conclusion we make, but they do not negate the conclusion. They do make it more tentative and complex. For example, people generally end up in the social class in which they are born—but *not everyone* does. We must ask why birth is so important to class placement and why there are exceptions. With each conclusion there are new questions and new directions for investigation. And with each conclusion there are some who are skeptical and decide to test it in a slightly different way. It is not final truth that characterizes sociology (and other sciences); it is a constant debate in which scientists, through published writings, put forth their ideas and evidence and wait for others to agree or disagree.

The refusal of sociologists to take a narrow view of science is also seen in their diverse and creative research. Sociologists will not get very far by setting up laboratory experiments, using microscopes, running rats through mazes, or mixing chemicals in test tubes. They use laboratory experiments when they can, but such experiments are less common in sociology than in certain other sciences. Although it is desirable to have complete control over your environment in order to have faith in your results, this is usually impossible. Systematic observation of people is difficult, and we must develop creative techniques that respect the complexity of the subject—human society—that is being investigated. Thus human beings are often motivated by ideas, values, attitudes, and moral concerns that cannot be observed but can only be understood through questionnaires, interviews, and analyses of their speech and writing. To understand conflict, cooperation, inequality, agreement, and power, sociologists observe people in groups wherever they can be found, they enter into and study a community for one or two years, or they feed data from newspapers and magazines into computers. They study crime rates, suicide rates, divorce rates, and unemployment rates to understand what qualities characterize a certain society. They observe everyday rituals and more formal religious rituals to understand how people think about the universe and their own lives. Every single act of the human being is something for sociologists to study because it can help us make sense out of a larger picture that is not easily seen. We take whatever seems reliable from our observations, preserving data in which we have confidence and discarding ideas that the evidence does not seem to support.

### *Objectivity in Science*

Science is not merely observation; it is *careful observation.* The purpose of science is to exercise control over our observations, to help us determine that what we say we have observed actually "is there" and not just what we want to see. Weber describes science—and sociology as a science—as "value-free" investigation: an attempt to carefully and objectively observe the world "as it is" rather than as we would like it to be. He means that our only commitment must be to scientific investigation itself; our conclusions always remain open to further investigation.

To be *objective* means literally to see the world as an "object" apart from ourselves, to separate it as much as possible from our subjective perception. This is more easily said than done, however. It is the reason for the many rules that scientists agree to follow in posing a question, setting up a hypothesis, testing that hypothesis, arriving at a conclusion, and relating that conclusion to the original question posed. Strict rules tell scientists how to create good theory, how to sample, how to observe accurately, how to control the study so that it focuses only on what they want to study, how to interpret data carefully, and how to refine theory on the basis of the evidence. Strict guidelines tell scientists how to report to other researchers the way in which an idea was formed, how a test was developed, what was observed, and how the results were interpreted. All of the rules that are there for the scientist to follow have one basic purpose: to ensure, as much as possible, that the work done is *objective* and that the personal bias of the scientist is minimized.

Complete objectivity, of course, is impossible, especially when the subject matter is society and the human being. Total objectivity is probably more difficult to achieve in social science than in natural science and is probably more elusive in sociology than in any other social science. All scientists have many biases that they can never be fully aware of, and such biases always enter into their studies, sometimes a little bit, sometimes a lot. Focusing on the social world rather than the biological is an inherent bias in sociology, just as psychology's focus on individual development is a bias. The questions we ask, the way we do our studies, our methods of our interpretation of our studies, and our conclusions are all possible biases. Good scientists can only be honest in their work. If they are alert, they can continue to uncover biases by critically evaluating one another's work. The fact that total objectivity is impossible does not change the fact that objectivity must still be our goal.

## Two Assumptions of Science

Religions make several assumptions about the universe. Most assume that a God exists and that this God has given humankind a set of moral laws to live by. Most of them also assume that people's souls live after death, and they assume that a body of truth has been given to humankind by God.

Science, too, makes assumptions about the universe. The first is that nature is lawful. The second is that natural events are caused by other natural events.

### *Natural Law in Science and Sociology*

To believe that nature is lawful is to hold that nature is governed by predictable regularities. Scientists believe that it is possible to explain the past and, on that basis, predict the future because events happen according to natural law. It is regularity that governs nature rather than haphazard, unpredictable chaos. We can therefore generalize about events in nature rather than simply believe that each event is unique. We can generalize about illness, gravity, the composition of matter, energy, plants, and all living things because we assume that each is understandable and, to some extent, predictable. The purpose of science is to understand these natural laws. Scientists are driven to solve the puzzles that are assumed to exist in nature.

Before science, people explained nature by the acts of supernatural forces, God or gods intervening and determining who would live and who would die, which wars would be fought and who would win, what progress would be made, and what losses would be suffered. It may be that the universe is, in fact, controlled by supernatural forces and that natural law is not important. Science—if it is to understand anything—must proceed on the assumption that events occur because of regularities in nature. Events occur in nature the way they do because underlying natural laws exist.

Sociology is a science, and thus, like other sciences, it assumes natural law. Human beings and human society are part of nature, and they are subject to regularities that can be isolated, understood, and predicted. When people interact, for example, they almost always develop a system of inequality, a set of expectations for each actor (called *roles*), and a shared view of reality (called *culture*). When a society industrializes, there is a strong trend away from tradition and toward individualism, and because the individual is less firmly embedded in social groups, a higher suicide rate results. When an oppressed group's expectations race ahead of what the dominant group in society is able or willing to deliver, there will be violent rebellion, even widespread revolution. Evidence seems to show that such patterns have occurred in the past and will probably occur in the future. And we can point to evidence to help explain what other conditions aid their occurrence and why exceptions or different patterns arise.

It is not difficult to see how far we have gone in our knowledge if we simply compare what we know now with what people used to believe about poverty, social change, suicide, alcoholism, racial inequality, gender inequality, crime, social power, and social class. As such issues have been studied as part of a natural order rather than simply random events or the result of supernatural forces, we have been able to explain why such societal qualities arise and how they affect human behavior.

### *Natural Cause in Science and Sociology*

The second assumption of science is that events in nature are caused by other natural events. Actually, this assumption of natural cause is the most important aspect of natural law. It is the basis for explanation and prediction; it is the essence of the order we call *nature*. Natural law assumes order; natural cause assumes that the order is governed by natural cause. For example, if the conditions are the same, then objects will fall to the ground because a natural law is at work. They fall to the ground because of a natural force we call *gravity,* which pulls objects toward the center of the earth (not quite a perfect sphere) and which results from the earth turns. This is the natural cause. Microbes cause some diseases. Biological inheritance causes certain forms of cancer. Poverty causes some crime. Exploitation of disadvantaged groups helps to create and perpetuate their inferior position in society. The whole purpose of experiments in science is to link independent variables (influences, causes) to dependent variables (results, effects) to show that when variable *X* occurs, it produces variable *Y.*

Cause is not easy to establish. Scientists must go through a long and difficult process to uncover it. Note how difficult it has been to establish that smoking is a cause of lung cancer. One must show that smokers are at a higher risk than nonsmokers. One must show that it is not only smoking combined with air pollution or eating red meat that causes cancer, but also smoking by itself that will be causal. One must be able to show that the more smoking one does, the more likely one is going to develop lung cancer. One must show that it is not personality, gender, class, or place of residence that causes both smoking and lung cancer, but rather that smoking itself is directly linked to lung cancer. And one must try to go further: What exactly is it about smoking that causes cancer? If one stops smoking, does that reduce the risk? Do other activities or human characteristics increase the risk for smokers?

Sociology applies this principle of cause to the human being. Human belief and action have causes. Other social sciences share this assumption with sociology. Psychology shows us how environment and heredity interact to shape the person and how the qualities of the person, in turn, shape what he or she thinks or does. Economics isolates economic forces in society, political science looks at political forces, and anthropology examines biological and cultural forces. Social psychology shows us how other people around us cause what we do.

Because of the complexity of the human being, this underlying assumption of cause is more difficult to apply with certainty in social science than it is in natural science (that is, the physical and biological sciences). It is rare when we can isolate clear, inevitable causes for what people think and do. We tend to uncover tendencies and probabilities. We are more likely to call a cause in social science an "influence" or a "contributing factor." Such information is highly valuable, even as imperfect as it is. Being abused as a child is an important influence on whether one abuses children as a parent. The size of an organization is an important contributing factor to its developing a bureaucratic structure. Being born into poverty affects one's chances of becoming rich, of being successful in school, and of becoming president of the United States.

Sociology has come very far since the nineteenth century toward convincing the general population that social circumstances make an important difference for individual action. Consider, for example, the almost total absence of a social explanation for suicide before 1900. The work of Durkheim was an important breakthrough: Societies have different suicide rates, and those rates are a result of *social forces* such as the degree of social solidarity or the degree of social change. Most people today recognize that suicide is not simply an isolated decision, that society in many ways influences that decision. Individual acts of crime, individual acts of divorce, and individual choices to have children are partly a result of the social world one lives in. Segregated schools create conditions for unequal education, and sex discrimination in hiring, paying, and promoting creates a segregated and unequal labor force. To understand human life, many people have been influenced to examine social forces.

## Summary and Conclusion

Sociology, by its very nature, is a questioning perspective, a "critical point of view." All science must be suspicious of what people know from their everyday experience, but sociology especially must be suspicious, and this suspicion leads to a questioning, probing, doubting, analytical approach to understanding society and the human being. It questions what many people take for granted.

Earlier in this chapter we described the contributions of the ancient Greeks in developing the rational proof. Probably the most important Greek was Socrates, a philosopher who was never satisfied with the answers people gave him. He questioned what people thought, forcing them to be critical: "What is goodness?" "What is virtue?" "What is the good society?" The replies people gave were stock answers, which had been learned but rarely thought out. No matter what answer people gave, Socrates had another question that caused further thought. To him, this is what education must be: a continual search for understanding through asking questions and exposing superficial answers, causing the student to grasp an idea through careful examination rather than simply reciting what was taught.

This is also the mission of sociologists: to probe the answers people give, uncover what they believe, examine reality through controlling personal and social bias, and see the human being in society as clearly and carefully as possible. Sociologists study the assumptions and problems of capitalism when many people today are claiming it to be humankind's salvation. We study the causes of crime and the problems associated with increasing the prison population when many people care only about the evils of crime and the necessity for isolating dangerous individuals from legitimate society. We study the functions of religion in society when many people see religion as part of the sacred world, not to be studied but to be accepted and used for guidance. We study the world of the violent youth gang to understand culture, stigma, and adolescent reactions to society when many people simply want such groups to disappear. We see dimensions of inequality that others ignore, we examine the meaning and contributions of deviance to society as well as its causes, and we try to uncover the purpose of social ritual rather than simply perform it. Sociology goes beyond the obvious; it asks questions when most people do not. It recognizes, as Peter Berger claims, that the first wisdom is that things are almost never what they seem.

Putting together the puzzle of society involves a critical approach to understanding, and that is the essence of science. Throughout this chapter I have attempted to show that it is important to study society scientifically because science encourages care and objectivity. This chapter has emphasized three points concerning how sociologists try to understand society:

1. Ideas must be supported by empirical research. Such research must be careful, creative, and diverse.
2. As a science, sociology must constantly attempt to be objective. It must critically investigate ideas that most people have come to accept as part of their culture. This makes sociology a very difficult science.

3. The human being and human society are part of nature, and thus they are governed by regularities or patterns, by a set of natural laws. Human events, it is assumed, are caused by identifiably natural—including social—causes.

If it is important to understand our life honestly, then we must look at ourselves, examine our ways, and question what we know and see. This is what a liberal arts education should be. This, in the end, is the purpose of sociology as a science.

### Questions to Consider

1. Many instructors believe that it is important for students to become critical thinkers. What does this mean? Do you agree that it is easier to be critical of other people's ideas than one's own ideas? Explain.
2. Based on the short descriptions of the five thinkers—Marx, Weber, Durkheim, Mead, and Berger—is there any one that stands out as the most interesting to you?
3. If sociology cannot be a science of society, then what? Is there a better alternative for understanding society?
4. If all perspectives are biased to some extent, and if science is a perspective, then what is its bias? What limits science in its attempt to understand reality? What does it ignore? What are its assumptions?
5. If we cannot be certain about our truths, then why seek truth? Isn't it better to leave it alone?
6. How would a physicist or chemist answer the question: "Are sociologists really scientists?"

## References

The following works contain introductions to science, issues related to sociology as a science, or examples of good empirical studies in sociology.

**Adorno, Theodor W., Else Frenkel-Brunswick, D. J. Levinson, and R. N. Sanford**. 1950. *The Authoritarian Personality*. New York: Harper & Row.

**Ammerman, Nancy T.** 2005. *Pillars of Faith: American Congregations and Their Partners*. Berkeley, CA: University of California Press.

**Anderson, Eric.** 2005. *In the Game: Athletes and the Cult of Masculinity*. New York: State University of New York.

**Babbie, Earl R.** 2001. *The Practice of Social Research*. 9the ed. Belmont, CA: Wadsworth/Thomson Learning.

———. 2002. *The Basics of Social Research*. 2nd ed. Belmont, CA: Wadsworth/Thomson Learning.

**Bailey, Kenneth D.** 1999. *Methods of Social Research*. 5th ed. New York: Free Press.

**Baker, Therese L.** 1998. *Doing Social Research*. 3rd ed. New York: McGraw-Hill.

**Bales, Robert F.** 1950. *Interaction Process Analysis*. Reading, MA: Addison-Wesley.

**Barnes, Sandra L.** 2005. *The Cost of Being Poor: A Comparative Study of Life in Poor Urban Neighborhoods in Gary Indiana*. New York: State University of New York Press.

**Becker, Howard S.** 1976. *Boys in White: Student Culture in Medical School*. Rev. ed. Chicago: University of Chicago Press.

**Berger, Bennett.** 1990. *Authors of Their Own Lives: Intellectual Autobiographies of Twenty American Sociologists*. Berkeley: University of California Press.

**Berger, Peter L.** 1963. *Invitation to Sociology*. Garden City, NY: Doubleday.

———. 1969. *The Sacred Canopy: Elements of a Sociological Theory of Religion*. Garden City, NY: Doubleday.

**Berger, Peter L., and Thomas Luckmann**. 1966. *The Social Construction of Reality*. Garden City, NY: Doubleday.

**Best, Joel.** 2001. *Damned Lies and Statistics: Untangling Numbers from the Media, Politicians, and Activists*. Berkeley: University of California Press.

———. 2004. *More Damned Lies and Statistics: How Numbers Confuse Public Issues*. Berkeley: University of California Press.

**Brown, Michael K., Martin Carnoy, Ellioitt Currie, Troy Duster, David B. Oppenheimer, Marjory M. Schultz, and David Wellman.** 2003. *Whitewashing Race: The Myth of a Color-Blind Society*. Berkeley: University of California Press.

**Charon, Joel M.** 2002. *The Meaning of Sociology*. 7th ed. Upper SaddleRiver, NJ: Prentice Hall.

**Clark, Candace.** 1997. *Misery and Company: Sympathy in Everyday Life*. Chicago: University of Chicago Press.

**Clarke-Stewart, Alison, and Cornelia Brentano**. 2006. *Divorce: Causes and Consequences*. New Haven, CT; Yale University Press.

**Cohen, Morris R., and Ernest Nagel**. 1934. *An Introduction to Logic and Scientific Method*. New York: Harcourt Brace Jovanovich.

**Collins, Randall, and Michael Makowsky**. 2004. *The Discovery of Society*, 7th ed. New York: McGraw-Hill.

**Coser, Lewis A.** 1977. *Masters of Sociological Thought*. 2nd ed. New York: Basic Books.

**Cuzzort, R. P., and E. W. King.** 1994. *Twentieth Century Social Thought*. 5th ed. New York: Harcourt.

**Denzin, Norman K., and Yvonna S. Lincoln**, eds. 2002. *The Qualitative Inquiry Reader*. Thousand Oaks, CA: Sage.

**DeWalt, Kathleen Musante, and Billie R. DeWalt**. 2002. *Participant Observation: A Guide for Fieldworkers*. Walnut Creek, CA: AltaMira Press.

**Dowdy, Thomas E., and Patrick H. McNamara**, eds. 1997. *Religion: North American Style*. 3rd ed. New Brunswick, NJ: Rutgers University Press.

**Durkheim, Émile.** [1893] 1964. *The Division of Labor in Society*, translated by George Simpson. New York: Free Press.

———. [1895] 1964. *The Rules of the Sociological Method*, translated by Sarah A. Solovay and John H. Mueller. New York: Free Press.

———. [1897] 1951. *Suicide*, translated and edited by John A. Spaulding and George Simpson. New York: Free Press.

———. [1915] 1954. *The Elementary Forms of Religious Life*, translated by Joseph Swain. New York: Free Press.

———. 1972. *Selected Writings*. Ed. and trans. Anthony Giddens. Cambridge, England: Cambridge University Press.

**Edin, Kathryn, and Maria Kefalas**. 2007. *Promises I Can Keep: Why Poor Women Put Motherhood before Marriage*. Berkeley: University of California Press.

**Ehrenreich, Barbara.** 2006. *Bait and Switch: The (Futile) Pursuit of the American Dream*. New York: Metropolitan Books

**Erikson, Kai T.** 1976. *Everything in Its Path*. New York: Simon & Schuster.

**Farley, John E., and Gregory D. Squires**. 2005. "Fences and Neighbors: Segregation in 21st-Century America." *Contexts* 4, no. 1: 33–39.

**Fernandez, Ronald.** 2003. *Mappers of Society: The Lives, Times and Legacies of the Great Sociologists*. Westport, CT: Greenwood Publishing Group.

**Fine, Gary Alan.** 1987. *With the Boys: Little League Baseball and Preadolescent Culture*. Chicago: University of Chicago Press.

**Fonow, Mary Margaret, and Judith A. Cook**, eds. 1991. *Beyond Methodology: Feminist Scholarship as Lived Research*. Bloomington: Indiana University Press.

**Fenton, Steve.** 1984. *Durkheim and Modern Sociology*. Cambridge, England: Cambridge University Press.

**Fernandez, Ronald.** 2003. *Mappers of Society: The Lives, Times and Legacies of the Great Sociologists.* Westport, CT: Greenwood Publishing Group.

**Gelles, Richard J., and Murray A. Straus**. 1990. *Physical Violence in American Families: Risk Factors and Adaptations to Violence in 8,145 Families.* New Brunswick, NJ: Transaction.

**Goldblatt, David**, ed. 2000. *Knowledge and the Social Sciences: Theory, Method, Practice.* New York: Routledge/Open University.

**Iceland, John.** 2003. *Poverty in America: A Handbook.* Berkeley, CA: University of California Press.

**Inkeles, Alex.** 1964. *What Is Sociology? An Introduction to the Discipline and Profession.* Englewood Cliffs, NJ: Prentice Hall.

**Johnson, Allan G.** 1997. *The Forest and the Trees: Sociology as Life, Practice, and Promise.* Philadelphia: Temple University Press.

**Jones, Robert Alun** 1986. *Émile Durkheim: An Introduction to Four Major Works.* Beverly Hills: Sage.

**Kanter, Rosabeth.** 1977. *Men and Women of the Corporation.* New York: Basic Books.

**Kincaid, Harold.** 1996 *Philosophical Foundations of the Social Sciences.* Cambridge, England: Cambridge University Press.

**Kivisto, Peter**, ed. 2000. *Social Theory: Roots and Branches.* Los Angeles: Roxbury.

**Kozol, Jonathan.** 1991. *Savage Inequalities.* New York: Crown.

———. 2005. *The Shame of the Nation: The Restoration of Apartheid Schooling in America.* New York: Random House.

**Kuhn, Thomas S.** 1962. *The Structure of Scientific Revolutions.* Chicago: University of Chicago Press.

**Lemert, Charles.** 2007. *Thinking the Unthinkable: The Riddles of Classical Social Theories.* Boulder, CO: Paradigm Press.

**Liebow, Elliot.** 1967. *Tally's Corner.* Boston: Little, Brown.

**Lofland, John.** 1966. *Doomsday Cult.* Englewood Cliffs, NJ: Prentice Hall.

**Lukes, Steven.** 1977. *Émile Durkheim: His Life and Work: A Historical and Critical Study.* Harmondsworth, England: Penguin.

**Lynd, Robert S., and Helen Merell Lynd**. 1929. *Middletown: A Study in American Culture.* New York: Harcourt Brace.

———. 1937. *Middletown in Transition: A Study in Cultural Conflicts.* New York: Harcourt Brace.

**Macrae, Donald G.** 1974. *Max Weber.* New York: Viking Press.

**Madge, John.** 1965. *The Tools of Social Science.* Garden City, NY: Doubleday.

**Marx, Karl.** [1845–1886] 1956. *Selected Writings,* edited by T. B. Bottomore. New York: McGraw-Hill.

——— [1867] 1967. *Das Kapital.* Vol. 1. New York: International Publishers.

**Marx, Karl and Friedrich Engels.** [1848] 1955. *The Communist Manifesto.* 1955 ed. New York: Appleton-Century-Crofts.

**Mead, George Herbert.** 1934. *Mind, Self and Society.* Chicago: University of Chicago Press.

**Mills, C. Wright.** 1956. *The Power Elite.* New York: Oxford University Press.

———. 1959. *The Sociological Imagination.* New York: Oxford University Press.

**Morrison, Ken.** 1995. *Marx, Durkheim, Weber: Formations of Modern Social Thought.* Thousand Oaks, CA: Sage.

**Myrdal, Gunnar.** 1944. *An American Dilemma.* New York: Harper & Row.

———. 1969. *Objectivity in Social Research.* New York: Pantheon.

**Neuendorf, Kimberly A.** 2002. *The Content Analysis Guidebook.* Thousand Oaks, CA: Sage.

**Neuman, W. Lawrence.** 2004. *Basics of Social Research: Qualitative and Quantitative Approaches.* Boston: Pearson.

**Nisbet, Robert.** 1974. *The Sociology of Émile Durkheim.* New York: Oxford University Press.

———. 2002 *Sociology as an Art Form.* New Brunswick, NJ: Transaction.

**Perrucci, Robert, and Earl Wysong.** 2003. *The New Class Society: Goodbye American Dream?* 2nd ed. Landham, MD: Rowman & Littlefield Publishers.

**Phillips, Bernard S.** 2001. *Beyond Sociology's Tower of Babel: Reconstructing the Scientific Method.* New York: Aldine de Gruyter.

**Riley, Gresham**, ed. 1974. *Values, Objectivity, and the Social Sciences.* Reading, MA: Addison-Wesley.

**Ritzer, George.** 1995. *Sociological Theory.* 4th ed. New York: McGraw-Hill.

———. ed. 1999. *Classical Social Theory.* 3rd ed. New York: McGraw-Hill.

———. 2003a. *The Blackwell Companion to Major Classical Social Theorists.* Malden, MA: Blackwell Publishing.

———. 2003b. *The McDonaldization of Society.* 3rd ed. Boston: Pine Forge Press.

**Roth, Louise Marie.** 2006. *Selling Women Short: Gender and Money on Wall Street.* Brunswick, NJ: Princeton University Press.

**Rousseau, Nathan.** 2002. *Self, Symbols, and Society: Classic Readings in Social Psychology.* New York: Rowman & Littlefield Publishers.

**Scaff, Lawrence A.** 2000. "Weber on the Cultural Situation of the Modern Age." Chapter 5 in *The Cambridge Companion to Weber*, edited by Stephen Turner. New York: Cambridge University Press.

**Schwab, William A.** 2005. *Deciphering the City*. Upper Saddle River, NJ: Pearson Education, Inc.

**Sewell, W. H., R. M. Hauser, and D. L. Featherman**, 1976, *Schooling and Achievement in American Society*. New York: Academic Press.

**Skolnick, Arlene S. and Jerome Skolnick**. 2007. *Family in Transition*. 14th ed. Boston: Allyn & Bacon.

**Stouffer, Samuel A., Arthur A. Lumsdaine, Marion Harper Lumsdaine, Robin M. Williams, Jr., M. Brewster Smith, Irving L. Janis, Shirley A. Star, and Leonard S. Cottrell, Jr.** 1949. *The American Soldier: Combat and Its Aftermath*. Vol. 2. New York: John Wiley & Sons.

**Swingewood, Alan**. 2000. *A Short History of Sociological Thought*. 3rd ed. New York: St. Martin's Press.

**Thomas, William I., and Florian Znaniecki**. [1918] 1958. *The Polish Peasant in Europe and America*. New York: Dover.

**Thompson, Kenneth.** 1982. *Émile Durkheim*. London: Tavistock.

**Thrasher, Frederic.** 1927. *The Gang*. Chicago: University of Chicago Press.

**Tonry, Michael.** 2004. *Thinking About Crime: Sense and Sensibility in American Penal Culture*. New York: Oxford University Press, Inc.

**Tucker, Kenneth H.** 2002. *Classical Social Theory: A Contemporary Approach*. Malden, MA: Blackwell.

**Tucker, Robert C.**, ed. 1972. *The Marx-Engels Reader*. New York: Norton.

**Turner, Jonathan H.** 1991. *The Structure of Sociological Theory*. 5th ed. Belmont, CA: Wadsworth.

**Turner, Stephen Park, and Jonathan H. Turner**. 1990. *The Impossible Science: An Institutional Analysis of American Sociology*. Newbury Park, CA: Sage.

**Weber, Max.** [1905] 1958. *The Protestant Ethic and the Spirit of Capitalism*, translated and edited by Talcott Parsons. New York: Scribner's.

———. [1919] 1969. "Science as a Vocation." In *Max Weber: Essays in Sociology*, translated and edited by H. H. Gerth and C. Wright Mills. New York: Oxford University Press.

———. [1924] 1964. *The Theory of Social and Economic Organization*, edited by A. M. Henderson and Talcott Parsons. New York: Free Press.

**Western, Bruce.** 2006. *Punishment and Inequality*. New York: Russell Sage Foundation.

**Whyte, William Foote.** 1955. *Street Corner Society*. Chicago: University of Chicago Press.

**Wilson, William Julius.** 1987. *The Truly Disadvantaged: The Inner City, the Underclass, and Public Policy*. Chicago: University of Chicago Press.

**Zetterberg, Hans L.** 2002. *Social Theory and Social Practice*. New Brunswick, NJ: Transaction.

**Zimbardo, Philip.** 1972. "Pathology of Imprisonment." *Society*, 9: 4–8.

**Zirakzadeh, Cyrus.** 2006. *Social Movements in Politics: A Comparative Study*. New York: Palgrave Macmillan.

# What Does It Mean to Be Human?

## Human Nature, Society, and Culture

### Concepts, Themes, and Key Individuals

- ❑ Human nature
- ❑ Social essence of human beings
- ❑ Cultural essence of human beings

*The Twilight Zone,* the popular 1960s television series, was exciting and sometimes eerie. We seemed to know that a surprise—scary, wondrous, or both—awaited us if we patiently followed the story. One episode has stayed with me. It concerned a journey by American astronauts. They landed on a distant planet and befriended the inhabitants (who looked human) and were pleased to find themselves in a luxurious house, much like one they might have had on earth. However, they gradually became aware that they could not leave their new home, that they had become prisoners. Then a wall opened and revealed a large pane of glass with spectators peering in. The astronauts were on display under the label "*Homo sapiens* from the Planet Earth."

Since then, I have been bothered by a question that probably few people asked after seeing that episode: What would those creatures from earth, that we call "human beings," have to do in the cage for those outside to understand what human beings are really like? Phrasing the question differently: What is the human being? What makes us "human" and not something else? In what ways are we like all other living creatures? What do we have in common with other animals? How are we different? Of course, these questions have probably teased the thinking person from the beginning of human existence. Look around. We see worms, dogs, cats, bees, ants, and maybe fish. Are we unique? All species of animals are unique. But how are we unique? What is our essence as a species? What would the astronauts in the cage have to do to reveal the essence of the species they represent? We might begin by recognizing that we share many qualities with other animals. Human beings are mammals. This means we are warm-blooded, we give birth to live young, females nurse their young, and we have hair covering parts of our body. We are also primates; therefore, we are mammals who are part of an order within nature that is characterized by increasing manual dexterity, intelligence, and the probability of some social organization. But what makes us different?

Philosophers have made various claims about what is our outstanding characteristic, our key quality. They have pointed to our ability to make and use tools, to love, to know right from wrong, to feel, to think, and to use language. Religious leaders emphasize that we each have a soul and a conscience. They may also stress that we are created in God's image (thus, we are closest to God) or that we are selfish and sinful (thus, we are similar to other animals). The more cynical critic maintains that we are the only animal that makes war on its own kind (even though other animals are clearly aggressive toward members of their own species).

Psychologists may focus on the fact that humans are instinctive, that they are driven by their nonconscious personality, that they are conditioned like many other animals, or that, unlike other animals, they act in the world according to the ideas and perceptions they learn. Most will maintain that human beings develop traits early in life out of an interplay of heredity and environment. Often, for the psychologist, the essence of behavior is to be found in the brain of the person.

Sociologists, too, have much to say about the nature of the human being. They maintain that our unique qualities include that we are:

1. *social,* in that our lives are linked to others and to society in many complex ways; and

2. *cultural,* in that what we become is not a result of instinct, but of the ideas, values, and rules developed in our society.

Without these two core qualities, we would not be what we are. Put us in a zoo, take away either quality, and visitors to the zoo would see something quite different. To understand human beings as a species, therefore, it is important to understand how these two core qualities enter into our lives. It is also important to recognize the complex interrelationship between the social and the cultural: Our culture arises from our social life, and the continuation of our social life depends on our culture.

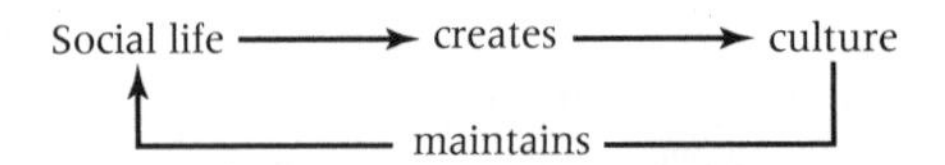


## Human Beings Are Social Beings

Many animals are social beings in a general sense. Fish are social in that they swim in schools, probably for protection. Bees and ants are better organized than any human society. Our closest relatives, apes and monkeys, are social, and their social lives are similar to ours.

To claim that the first human beings were social is simply to recognize that our social life was always important to us and that humans never existed without this quality. The first humans were not isolated individuals but beings who interacted, were socialized, depended on one another, and lived their whole lives around others. Of course, some may have chosen an existence apart from others as they reached adulthood, as do a few individuals today, but all were social in their early lives, and the vast majority were social throughout their adult lives.

### *Survival*

What does it mean to be "social"? On the simplest level it means that *humans need others for their very survival.* Infants need adults for their physical survival: for food, shelter, and protection. A great deal of evidence suggests that infants also need adults for emotional support, affection, and love. Normal growth—even life itself—seems to depend on this support. Studies of infants brought up in nurseries with little interaction with adults show us that these babies suffer physical, intellectual, and emotional harm and that this harm is lasting (Spitz 1945). Of course, the horrible discovery in 1990 of infants brought up in Romanian government nurseries attests to the same problems: Neglecting the basic emotional needs of children brings severe retardation of growth and often death.

Adults also need other people. We depend on others for our physical survival: to grow and transport our food, to provide shelter and clothing, to provide protection from enemies, and almost all the things we take for granted. As adults we also depend on others for love, support, meaning, and happiness. Human survival, therefore, is a social affair. Almost all of our needs—physical and emotional—are met through interaction with others.

It is also clear that our closest relatives in nature also depend on their social life for survival, and the closer they come to humans in nature, the more important are both physical and emotional dependence throughout life.

### *Learning How to Survive*

To be social also means that much of what we become depends on socialization. *Socialization* is the process by which the various representatives of society—parents, teachers, political leaders, religious leaders, the news media—teach people the ways of society and, in so doing, form their basic qualities. Through socialization people learn the ways of society and internalize those ways—that is, make them their own. No other animal depends on socialization for survival as much as the human being.

Almost all other animals depend primarily on biological instinct rather than socialization for survival. Those closest to humans are least dependent on instinct and most able to learn from experience; some rely also on imitation. Humans know how to do little through instinct; we are not born knowing how to deal with our environment. Because we do not survive through instinct, our social nature becomes essential. We do not have to learn that we need to eat, but we do have to learn how to get food (to gather, hunt, fish, farm, or buy it). In most societies, we must also learn how to build a shelter, use weapons, make clothing, and handle other people, to name only a few of the things that matter. In fact, we must learn thousands of things if we are to survive in the particular society we live in, from learning the ABCs to learning how to discourage others from robbing us to learning how best to dress and talk so we can be popular.

In short, human beings live in a world *where socialization is necessary for survival; that socialization is ongoing, lifelong, and broad in scope.*

### *Individual Qualities*

Besides showing us how to survive, *socialization is also necessary for creating our individual qualities.* Our talents, tastes, interests, values, personality traits, ideas, and morals are not qualities we have at birth but qualities we develop through socialization in the context of the family, the school, our peers, the community, and even the media.

We become what we do because of a complex mixture of heredity and socialization. We may have certain biological predispositions, but how others act toward us, what they teach us, and the opportunities they provide for us are all important for what we become. As we interact with others, we choose the directions we will take in life: crime or legitimate business, school or on-the-job training, the single life or the married life, life on the farm or life in the city. Some of us may have all kinds of talent, but whether we direct it toward making money through selling illegal drugs or helping people solve their problems through psychoanalysis depends on our interactions and resulting socialization.

The treatment of women in our society highlights this point. White women born in the United States were denied many opportunities and rights reserved for men. Family, religion, political leaders, and schools—together with discrimination in the economic order—told both men and women what was to be expected of them. Women became the property of men. Eventually, the relationship was altered as women increasingly became partners and were socialized to take care of the household in return for male economic support. In the twentieth century, and especially after World War II, this relationship moved toward a more equal one. As economic opportunities opened up, white women joined the paid labor force in real numbers. Their success in the political, educational, and economic worlds altered the expectations in society for women, and it increasingly altered the female role. After the war, our view of the differences between women and men continued to blur. By the 1990s, an acceptance of the idea that women can do almost anything traditionally reserved for men had clearly evolved, even though opportunities remained limited. Such an idea influences the socialization of children, and that socialization affects choices made in life. Opportunity and socialization have influenced each other, and the result is a society less differentiated and stratified on the basis of gender. Although barriers will continue to exist in society for a long time, we are clearly living within a real-life experiment that offers clear support for the idea that socialization is extremely powerful for what people become!

It is important to see that socialization is exceptionally complex. It involves not only learning things but also modeling one's behavior on that of individuals whom one respects, being affected by perceived opportunities "for people like us," and being influenced by one's successes and failures. When we see socialization this way, we can better understand the harmful effects of discrimination, segregation, and persecution. To be put down by others directly has an impact; to see others like oneself in a deprived existence has an effect on the value one places on oneself as well as the expectations that one develops for oneself. Of course, some individuals overcome such conditions, but these exceptions do not disprove the power of socialization. Indeed, they help clarify the importance of socialization as we try to identify the conditions that encourage individuals to be different from those around them. Socialization helps explain why poverty is so powerful a force on what children "choose" to do with their adult lives.

We can also turn this explanation around. The opportunities that wealthy and privileged children have in society socialize them to seek directions closed to most other people in society. Prestigious high schools and colleges that provide professional training help ensure high placement in society and a life of affluence. Robert Coles (1977) describes the final result of socialization into the wealthy class to be *entitlement:* The children of the affluent learn that they are entitled to certain things in their lives that other children cannot take for granted and often do not even know exist. "The child has much, but wants and expects more, all assumed to be his or hers by right—at once a psychological and material inheritance that the world will provide" (p. 55). In what their parents give and teach, affluent children learn what they have a right to expect from life, what is their *due* because of who they are.

Socialization may not determine all that we are, but its influence cannot be easily denied. Much of what each of us has become can be traced to our interaction with others, and thus our individual qualities are in this sense really social ones. The sociologist emphasizes how socialization influences our choices, abilities, interests, values, ideas, and perspectives—in short, the directions we take in our lives. And, as we will see in later chapters, socialization is not something that happens to us in childhood alone; instead, it continues throughout our lives. At every stage we are being taught or shown by others how we should act, what we should think, and who we are. Early socialization may be the most important, but later socialization may reinforce these early directions or lead us in new ones. Socialization forms the individual actor and is the third way we are social beings.

### *Basic Human Qualities*

We have looked at three ways in which we are social: Our survival depends on others, we learn how to survive through watching and learning from others, and we develop our individual qualities largely through socialization with others. A fourth quality of the human being attests to the importance of our social life: our very humanity.

At what point does the human being *become* human? Religious leaders differ: Some argue that it is at the point of conception, while others say that it occurs when the fetus can survive on its own or at birth or after one year of survival. Indeed, in some religious perspectives children are not really fully human; and for some religious perspectives, women are less than fully human. Leaders in every society have joined in defining certain immoral or different people as less than human and thus nondeserving of human rights. Political leaders also define what constitutes a human with full human rights (this is sometimes based on citizenship, ethnic-group membership, religion, gender, and even correct political beliefs). Philosophers, psychologists, biologists, and artists also have their views. Although this is a highly emotional topic, it is a critically important one. It revolves around the question of human essence. If we believe in a soul, then we will use that as the defining quality. If we believe in God-given human rights at the point of conception, then we will use that as the defining quality. Philosophers might focus on mind as the defining quality, psychologists human intelligence, and biologists the fertilization of the egg or the birth or development of the mature fetus.

It is, in fact, a religious, political, and scientific question, and there is little agreement. Scientists typically attempt to identify certain attributes that make human beings human: intelligence, problem-solving ability, language use, or culture, for example. *Sociologists typically focus on three interrelated qualities: the use of symbols, the development of self, and thinking.* It is only when these three qualities are in evidence that human beings are able to act like the animal we call human. Perhaps the sociologist exaggerates, but there is something profoundly important here: These three qualities, central to the human being, are socially created. In this sense, our very humanity is developed only through social interaction. We are unfinished beings at birth, potentially able to act as other humans do, but that potential is realized only through our social life. Let us examine briefly each quality.

**The Use of Symbols** The more we understand about human beings, the more centrally important becomes their use of *symbols.* A symbol is something that stands for something else and is used in its place for purposes of communication. Although we communicate through the use of nonintentional body language, unconscious facial expressions, and so on, symbols have the additional quality of being understood by the user. Symbolic communication is meaningful: It represents something to the one who communicates as well as to the one receiving the communication. It is an act of intentional communication.

Words are the best example of symbols. They stand for whatever we decide they do. We use words intentionally to communicate something to others, and we use words to think with. Besides words, however, we also decide that certain acts are symbolic (shaking hands, kissing, raising a hand). And humans also designate certain objects to be symbolic: flags, rings, crosses, and hairstyles, for example. Such objects are not meaningful in themselves, but they are designated to be.

Where do such representations come from? It is true that many other animals communicate with one another: wagging tails, making gestures, giving off smells, and growling, for example. The vast majority of these behaviors, however, are instinctive. They are not learned, and they are universal to the species. They are performed by the organism automatically and usually do not appear to have any meaning to the user. (The bee, for example, will do its "dance" communicating to other bees where nectar is located even when the hive has been emptied and no other bees around to see the dance.) The closer we get to the human being in the animal kingdom, however, the more the forms of communication take on a different quality: The acts represent something else only because they are agreed on in social interaction. In other words, *the tools of communication are socially based.* Because the meanings of symbols are socially based, what something represents is pointed out—intentionally taught—to the organism. Thus, the animal learns and *understands* that something stands for something else. When the act is performed, the animal does not simply give off communication but understands the meaning of it. It is clear that human beings depend on socially derived representations for almost everything they do and are; even if other organisms use symbols in this sense, the use is quite limited.

This ability *to create and use symbols that are understood by the user is part of our social essence.* And this ability is so important to us that it undoubtedly qualifies as a central human quality alongside our social essence. Consider what we do with symbols: We use them *to communicate* ideas, feelings, intentions, identities; *to teach others* what we know; *to communicate* to others and *to cooperate* with others in organization; and *to learn* roles, ideas, values, rules, and morals. We can hand down to future generations what we have learned, and they are able to build on what others have taught; symbols make *the accumulation of knowledge possible.* We use symbols *to think* with: to contemplate the future, apply the past, figure out solutions to problems, consider how our acts might be moral or immoral, generalize (about anything, such as all living things, all animals, or all human beings), and make subtle distinctions between smart and not-so-smart candidates for office. Our whole lives are saturated with the use of symbols. *And, far from being created for us by nature, symbols are created by human beings in social interaction.* It is through social interactions that our representations are developed, communicated, and understood by us.

**Selfhood** In a similar way, humans develop self-awareness only through interaction with others, and self-awareness, too, qualifies as a central human quality. Humans develop a realization that they exist as objects in the environment. "This is me." "I exist." "I live, and I will die." "I think, I act, and I am the object of other people's actions." This self-realization should not be taken for granted. It arises through the acts of others. We see ourselves through the eyes, words, and action of others; it is clearly through socialization that we come to see ourselves as objects in the environment. Selfhood develops in stages, and each stage depends on a social context. Through interaction with significant others, we first come to be aware of the self, and we see it through the eyes of one other person at a time. (Children may see themselves through the eyes of their mother, then their father, then their nursery school teacher, then Mister Rogers—all in the same day.) Over time, our significant others merge into a whole, into "them," "society," "other people," or what George Herbert Mead calls a "generalized other," and we begin to use the generalized other to see and direct ourselves. We then see ourselves in relation to a group or society, in relation to many people simultaneously. We thus guide our own acts in line with an organized whole: our family, our elementary school, the United States, all people in our church, or all humanity. We see and understand a relationship between our acts and these other organized wholes. Selfhood makes possible many human qualities.

Specifically, we are able to do three things because *we have a self. First, we can see and understand the effects of our own actions, and we are able to see and understand the effects of the acts of others on us*. We are thus able to plan strategy, alter our directions, and interpret situations as we act. For example, in choosing a major, students can examine themselves: their abilities, interests, values, and past achievements. They can evaluate their experiences, future chances, and possible occupational opportunities. They will probably try to imagine what they would look like in a certain occupation and whether the work would be enjoyable.

Second, *selfhood also brings us the ability to judge ourselves:* to like or dislike who we are or what we do, to feel proud or mortified. We develop a self-concept, an identity, and self-love or self-hatred.

Third, *self also means self-control,* our ability to direct our own actions. We can hold back; we can let go at will; we can go one direction, and, upon evaluation, decide to tell ourselves to go quite another. We are not simply subject to our environment—we are able to alter our own acts as we make decisions, and we are able to do something other than what we have been taught to do.

The more we investigate the meaning and importance of having a self, the more obviously it can be recognized as one of our central qualities. And it is *a socially developed quality:* Without our dependence on social interaction, selfhood would certainly not exist.

**Mind** George Herbert Mead made sociologists aware that the ability to think is intimately related to selfhood and symbol use. Mead called this ability *mind.* Humans, like all other animals, are born with a brain, but the mind—the ability to think about our environment—is a socially created quality. Symbols are agreed-on representations that we use for communication. When we use them to communicate to our *self,* we call this thinking; and all this *communication* that we call thinking, Mead called *mind.* Humans do not simply respond to their environment; they point things out to themselves, manipulate the environment in their heads, imagine things that do not even exist in the physical world, consider options, rehearse their actions, and consider how others will act. (In other words, they figure out their world; they decide how to act in situations; they do not simply respond to their environment.) This ability, so central to what humans are, is made possible through symbols and self, which (as we saw above) are possible *only through social interaction.*

To be social, therefore, means that humans need others for survival and socialization to learn to survive. Socialization also creates our individual qualities. *And social interaction is important for developing our essence: It creates our central qualities of symbol use, selfhood, and mind.*

### *A Life of Interaction within Society*

Humans are social in a fifth sense, however. For whatever reason, *we live our entire lives interacting with others, and find ourselves with others, and find ourselves a part of many groups, organizations, communities, and society.*

We live an organized existence, not an existence apart from others. Almost everyone spends his or her life in a world of *social rules* (morals, laws, customs) and *social patterns* (established systems of inequality, types of families, schools, and religious worship, for example), a world that directs much of what he or she does. As we try to understand what human beings are objectively, we inevitably see them as animals who are born into a society they did not create, who are likely to live their entire existence in that society, and who belong to a host of groups, formal organizations, and one or a few communities there. To observe humans in an environment without social organization is not to observe them as they actually live their lives. We are not solitary beings, but social ones. Some would argue that this is simply in our biology. Others would argue that nature sets us up: we are with little or no instinct, we learn very early that survival depends on others, we develop symbols, self, and mind from others, and the importance of organization becomes a necessary aspect of our existence.

### *Our Dependence on the Social: A Summary*

To emphasize the idea that human beings are social by their very nature is to see something very profound about what we are. Take away our social life and there is nothing left that we might call human. *Our very survival depends on society because of its protection and its socialization; much of what we become as individuals and as a species depends on socialization; and almost everything we do is based on and includes a strong element of social interaction and social organization.*

## Human Beings Are Cultural Beings

To say that human beings are *cultural* is to maintain that we are characterized by certain other qualities not yet described. Many animals are social, but what makes some animals cultural? The answer to this question entails determining what the foundation of a society is. Most social animals live together out of *instinct.* Nature commands that they cooperate, and it directs exactly how that cooperation should take place. Worker bees, queen bees, and drones do not understand what they are doing nor do they figure out how to play their various roles. Instead, they are born with instincts that control their behavior, making cooperation possible.

Some animals learn how to act in society, but much of that learning is *imitative.* They watch and do what others do. In this way they learn their place in the organization. In still other animal societies, adults actually teach the young what to do. This teaching is instinctive; that is, nature commands the organism how the young are to be trained. It is difficult to determine how close to culture some animals come, but it is clear that human beings are cultural, and their social organization is founded on culture, not on *instinct, simple imitation, or species-based teaching.*

Humans are cultural. Culture is here defined as the ideas, values, and rules that are socially created and are understood. Culture is abstract. Instead of physically responding to our environment, we bring a socially constructed perspective of the environment and ourselves that influences our actions. We discuss our world, we have to think about our world, we use abstractions to understand and act in situations. The knowledge we learn in our lives is not lost when we die but is passed down to others. Because of this cultural quality, societies differ considerably from one another. Each has a somewhat unique approach to living. Culture distinguishes organizations of people.

Even our internal world is cultural, not simply physical. Our physical internal state may change as something happens to us (as someone points a gun at us or surprises us or tells us he or she loves us). But a change in our internal state does not automatically produce a response. Responses are defined, controlled, and directed by us, and they are guided by what our culture teaches. Between the internal physical response and what we do lies culture. Although many animals cry out toward their environment in what we might call "anger," human beings have the ability to understand that quality in themselves. They are taught by other people to distinguish anger from love, jealousy, pride, hatred, and fear. The culture that we learn tells us when it is appropriate to get angry and when it is appropriate to show it. We learn how to control anger, how to express anger, and how to feel sorry, guilty, or happy about our anger. It also teaches us many ideas about anger ("anger is natural," "anger is one important cause of prejudice," "the extent of anger is related to frustration"), and we apply these ideas to understanding our internal responses. Even the word *anger*—the label we give our internal state—is cultural. Experts are able to show us different types of anger and different levels of it. We can even learn when anger is "healthy" and "unhealthy," and we can learn how and when it can be "useful" or "harmful" to our goals.

We also label and act toward other people culturally, not "naturally." We see middle-class people and working-class people, conformists and nonconformists, nice people and nasty people. These labels are cultural. They help us divide up reality, and behavior that we perceive as deviant at one time or in one society may not be perceived that way in another (for example, polygamy, homosexuality, cocaine use, and divorce).

Through all of his work, Max Weber emphasizes the important point that we all live in a world of meaning. To understand human action, he argues, we must understand how people define their world, how they think about it. That thinking is anchored in a socially created *culture*. Weber focuses his attention on the influence of religious culture. He shows, for example, that in the seventeenth century, Protestantism was an important influence on the way people acted in the work world. In his view, Protestantism fostered a strong work ethic in society, encouraging individuals to strive for economic success. We are not isolated beings, and we are not simply trained to respond. Through our social and cultural lives, we learn, understand, and think about situations, and we are influenced by those with whom we interact.

## The Importance of It All

What difference does it really make that we are social and cultural beings? *First, to be social and cultural means that we are not set at birth but can become many different things and can go in many different directions.* Because we are social and cultural, we are capable of becoming a saint or sinner, a warrior or business executive, a farmer or nurse. One can become only what one knows, and that depends on what one learns. Although biology may have something to do with differentiating us from one another, making it possible for some of us to excel in various spheres rather than others, our flexibility is still great, and thus society, culture, and socialization play an important role in what we all become.

*Second, distinct societies arise.* Societies based on culture rather than instinct, imitation, or universal-species teaching will vary greatly in what they emphasize, and thus what they socialize their populations to become. We can become a peaceful people or a people that worships militarism. As a people, we can come to believe that the most important goal in life is to make money, or we can believe that the good life is one of unselfish giving. We can emphasize past, present, or future; people or things; competition or cooperation; this life or an afterlife; rock music or opera. Nature does not command what a society becomes, just as it does not command what an individual becomes. Social interaction and culture do, and thus we have evolved a wide diversity of societies. This also means that as new circumstances and problems arise, people can reach new understandings and change their ways. It means that, in contrast to other primates, humans are much better able to evaluate their ways and improve their cooperative endeavors. How a society comes to define reality changes and this, in turn, changes the direction of society. Agricultural societies become industrial societies; peaceful societies turn their attention to war or architecture; tastes in food and music, technology, and employment possibilities change over time.

*Third, to be social and cultural also means that to a great extent each of us is controlled by other people.* We are located within a set of social forces that shape and control what we do, what we are, and what we think. The culture that we learn becomes a part of our very being and comes to influence every aspect of our lives. Unlike other animals, it is not nature that commands us. Nor, unlike what most of us may think, it is not free choice that characterizes many of our decisions. We are social and cultural beings, and it is impossible to escape the many complex influences that this fact has on us.

*Fourth, we become active beings in relation to our environment.* As we depend on the social and cultural, we are no longer passive organisms who must respond according to instinct or conditioning. Instead, socialization into a society with culture allows us to understand what is around us. The word *understanding* might be defined as the ability to stand apart from our environment and describe it with words to ourselves and to others. It also means that we are able to apply our own knowledge to many different situations. To be social and cultural creates a being who is active, problem-solving, and creative. Society and culture may control us, but society and culture also give us a more active relationship to our environment by allowing us to rise above a simple response to it in a fixed way.

## Summary and Conclusion

Look around you. Look in your classroom, on the campus mall, in your dorm, home, or apartment. Watch football games, symphony concerts, and serious drama. Look at the e-mails you send to others, and what others send to you. Look at Google, Yahoo, Facebook. Look at your family, your friends, your classmates, your university. What is it that you see? What is the real essence of that being you see that we call human? The sociological answer is that you see a:

1. being who is *social* in nature, who survives through a dependence on others, who learns how to survive from others, who develops both human qualities and individual qualities through socialization, and who lives life embedded in society; and a
2. being who is *cultural* in nature, who interprets the world according to what he or she learns in society, and, therefore, a being whose nature is not fixed by biology but who is tremendously diverse.

## Questions to Consider

1. What would the individual be like if he or she never interacted with others?
2. How important is society in influencing the differences between men and women—or is it biology that rules?
3. For some people, the most important difference between humans and other animals is that humans have "culture." Is

this a real difference, or do other animals also have culture? What other animals besides humans exhibit behavior that indicates the use of culture?

4. If humans are social and cultural, is it part of their "nature" or is it something learned through socialization?

5. How would someone who holds a religious perspective answer the question? What does it mean to be human?

## References

The following works attempt to explain the link between society and culture on the one hand, and the nature of the human being on the other.

**Berger, Bennett M.** 1995. *An Essay on Culture: Symbolic Structure and Social Structure.* Berkeley: University of California Press.

**Berger, Peter L., and Thomas Luckmann.** 1966. *The Social Construction of Reality.* Garden City, NY: Doubleday.

**Blumer, Herbert.** 1969. *Symbolic Interactionism: Perspective and Method.* Englewood Cliffs, NJ: Prentice Hall.

**Charon, Joel M.** 2003. *Symbolic Interactionism: An Introduction, an Interpretation, an Integration.* 8th ed. Upper Saddle River, NJ: Prentice Hall.

**Coles, Robert.** 1977. "Entitlement." *Atlantic Monthly,* September, .

**Collins, Randall, and Michael Makowsky.** 2004. *The Discovery of Society,* 7th ed. New York: McGraw-Hill.

**Cooley, Charles Horton.** [1902] 1964. *Human Nature and the Social Order.* New York: Schocken Books.

———. [1909] 1962. *Social Organization.* New York: Schocken Books.

**Davis, Kingsley.** 1947. "Final Note on a Case of Extreme Isolation." *American Journal of Sociology* 52: 432–437.

**Durkheim, Émile.** 1893 [1964]. *The Division of Labor in Society.* Translated by George Simpson. New York: Free Press.

———. [1895] 1964. *The Rules of the Sociological Method.* Translated by Sarah A. Solovay and John H. Mueller. New York: Free Press.

———. [1915] 1954. *The Elementary Forms of Religious Life.* Translated by Joseph Swain. New York: Free Press.

**Elkin, Frederick, and Gerald Handel.** 1995. *The Child and Society.* 5th ed. New York: Random House.

**Erikson, Kai T.** 1976. *Everything in Its Path.* New York: Simon & Schuster.

**Forschi, Martha, and Edward J. Lawler**, eds. 1994. *Group Processes: Sociological Analysis.* New York: Nelson-Hall.

**Freie, John F.** 1998. *Counterfeit Community: The Exploitation of Our Longings for Connectedness.* Lanham, MD: Rowman & Littlefield.

**Freud, Sigmund** [1930] 1953. *Civilization and Its Discontents.* London: Hogarth.

**Geertz, Clifford.** 1965. "The Impact of the Concept of Culture on the Concept of Man," pp. 93–118, in *New Views of the Nature of Man,* edited by John R. Platt. Chicago: University of Chicago Press.

**Glendinning, Tony, and Steve Bruce**. 2006. "New Ways of Believing or Belonging: Is Religion Giving Way to Spirituality." *British Journal of Sociology,* 57:(3), 399–414.

**Gordon, Milton M.** 1978. *Human Nature, Class, and Ethnicity.* New York: Oxford University Press.

**Hall, John R., and Mary Jo Neitz**. 1993. *Culture: Sociological Perspectives.* Englewood Cliffs, NJ: Prentice Hall.

**Hallinan, Maureen T.** 2005. *The Socialization of Schooling.* New York: Russell Sage Foundation.

**Heelas, Paul, Linda Woodhead, Steel Benjamin, Karin Tusting, and Baron Szerszynski**. 2004. *The Spiritual Revolution: Why Religion Is Giving Way to Spirtuality.* Oxford, UK: Blackwell.

**Hertzler, Joyce O.** 1965. *A Sociology of Language.* New York: Random House.

**Hewitt, John P.** 2000. *Self and Society.* 8th ed. Needham Heights, MA: Allyn & Bacon.

**Jenkins, Richard.** 1996. *Social Identity.* London: Routledge.

**Johnson, Allan G.** 1997. *The Forest and the Trees: Sociology as Life, Practice, and Promise.* Philadelphia: Temple University Press.

**Keller, Suzanne.** 2003. *Community: Pursuing the Dream, Living the Reality.* Princeton, NJ: Princeton University Press.

**Kincaid, Harold.** 1996. *Philosophical Foundations of the Social Sciences.* Cambridge, England: Cambridge University Press.

**Lancaster, Jane Beckman.** 1975. *Primate Behavior and the Emergence of Human Culture.* New York: Holt, Rinehart & Winston.

**Lane, Harlan.** 1976. *The Wild Boy of Aveyron.* Cambridge, MA: Harvard University Press.

**Liebow, Elliot.** 1967. *Tally's Corner.* Boston: Little, Brown.

**Linden, Eugene.** 1993. "Can Animals Think?" *Time,* March 22, 54–61.

**Lindesmith, Alfred R., Anselm L. Strauss, and Norman K. Denzin**. 1999. *Social Psychology.* 8th ed. Thousand Oaks, CA: Sage.

**MacIver, Robert M.** 1931. *Society: Its Structure and Changes.* New York: Ray Long and Richard R. Smith.

**McCall, George J., and J. L. Simmons**. 1978. *Identities and Interactions.* New York: Free Press.

**Mead, George Herbert.** 1925. "The Genesis of the Self and Social Control." *International Journal of Ethics* 35: 251–277.

———. 1934. *Mind, Self, and Society.* Chicago: University of Chicago Press.

**Newman, Katherine.** 2004. *The Social Roots of School Shootings.* New York: Basic Books.

**Nisbet, Robert.** 1953. *The Quest for Community.* New York: Oxford University Press.

**Oliner, Pearl M., and Samuel P. Oliner**. 1995. *Toward a Caring Society.* Westport, CT: Praeger.

**Rosenberg, Morris.** 1979. *Conceiving the Self.* New York: Basic Books.

**Rousseau, Nathan.** 2002. *Self, Symbols, and Society: Classic Readings in Social Psychology.* New York: Rowman & Littlefield.

**Rubington, Earl, and Martin S. Weinberg**. 2003. *The Study of Social Problems: Seven Perspectives.* 6th ed. New York: Oxford University Press.

**Shibutani, Tamotsu.** 1961. *Society and Personality: An Interactionist Approach to Social Psychology.* Englewood Cliffs, NJ: Prentice Hall.

———. 1986. *Social Processes: An Introduction to Sociology.* Berkeley: University of California Press.

**Spitz, R. A.** 1945. "Hospitalism: An Inquiry into the Genesis of Psychiatric Conditions in Early Childhood," pp. 53–74 in *The Psychoanalytic Study of the Child,* edited by Anna Freud, et al. New York: International University Press.

**Sumner, William Graham.** [1906] 1940. *Folkways.* Boston: Ginn & Co.

**Turnbull, Colin.** 1972. *The Mountain People.* New York: Simon & Schuster.

**Warriner, Charles K.** 1970. *The Emergence of Society.* Homewood, IL: Dorsey Press.

**Weber, Max.** [1905] 1958. *The Protestant Ethic and the Spirit of Capitalism.* Translated and edited by Talcott Parsons. New York: Scribner's.

**Wheelan, Susan A.** 1994. *Group Processes: The Developmental Perspective.* Boston: Allyn & Bacon.

**White, Leslie A.** 1940. *The Science of Culture.* New York: Farrar, Straus & Giroux.

**Wolfe, Alan.** 2001. *Moral Freedom: The Impossible Idea That Defines the Way We Live Now.* New York: W. W. Norton.

# How Is Society Possible?

## The Basis for Social Order

### Concepts, Themes, and Key Individuals

- ❑ Society, social organization, and social order
- ❑ Society, nation, and nationalism
- ❑ Social interaction and social patterns
- ❑ Culture, social structure, and social institutions
- ❑ Socialization, commitment, integration
- ❑ Social conflict

Who are we? Who are you?

Among your many identities, you probably call yourself an American. Because of your birth in or your migration to the United States, this has become your society. Technically, you live in the United States, because "America" includes Canada, Mexico, Central America, and South America. However, we often forget some technicalities, and refer to ourselves as Americans, even though we mean citizens of the United States. Many of us will enter a neighborhood of poverty and wonder, "Is this America, too?" Or we enter a gated community, gape at houses selling for many millions of dollars, and ask again, "Is this America?"

We are born not knowing who we yet are, pushed by various needs and biological responses, able to survive only because of human adults around us. Slowly we learn who we are, and we come to realize that getting through life involves dealing with others around us in one way or another. Social interaction forms us, and over time we learn that we exist within a family, in various groups, in community, and in American society. Over time, we come to take for granted the fact that we are American, until something such as war, terrorism, oppression of minorities, critics, or a parade shakes us out of our slumber and causes us to wonder about what living in this society actually means. Our parents and schools teach us the origins of our society, its rules, its great qualities, its problems.

Some of us wonder about just how this society began. What started it? Why did it become what it is? Why is it that it continues over many generations? Can it end? What does it mean for our lives if it ends? Does it just disappear and something replaces it, or does it simply evolve into something different?

Why does any society come about? How is it possible for any bunch of individuals to put aside their differences and agree to work together, passing down that cooperative bond to the next generation? What is the role of force? Interdependence? Socialization? Human nature? It seems to many of us that it is a real wonder that society continues at all.

Sociologists have wondered a lot about these questions. "How is society possible?" wonders Georg Simmel, a contemporary of Max Weber and Émile Durkheim. Earlier in the history of philosophy, Thomas Hobbes asked, "How is order possible?" What factors go into the creation and perpetuation of society? Certainly Marx, Weber, and Durkheim were inspired by this question. As much as any other, this is the one question that created the discipline of sociology, and, in discussions among thoughtful sociologists, this is the one that still comes up time and time again. Hobbes tended to answer the question by the use of force; sociologists do not deny force, but their answer is more complex and leads to a more peaceful and willing acceptance of social order. All would probably agree that where society is able to exist over time, it almost seems miraculous. Why doesn't it simply collapse, given the many problems people seem to have as they try to get along?

## Society Is a Social Organization

Society is not the same as a nation. A nation is a political organization of people; a political organization includes government, law, army, and physical boundaries. We can usually date when a nation begins. The boundaries of nations are sometimes close to encompassing an entire society; usually, however, a nation is formed that includes several societies or simply a part of one society. We began as a nation when we ratified the United States Constitution.

*Society is a social organization of people.* Other smaller social organizations exist within society, including groups, formal organizations, and communities. Each organization has its own history, culture, structure, identity, and sometimes even its own language and institutions. Because other social organizations exist within society, they are influenced by that society and subject to the changes that go on in that society. *Society is the largest social organization that individuals identify with and are socialized into; they are constantly affected by its social patterns.*

Societies exist through historically developed social patterns that become taken for granted and are only sometimes formalized. In societies, people have come to know one another through ongoing social interaction; through sharing, communicating, and cooperating; through creating a common identity and social commitment; and through similar ways of thinking and acting.

Over time, nations can make themselves into societies. This is what happened to the United States after the Civil War. The societies of the North and South increasingly became one society through ongoing social interaction, interdependence, and shared ways. England created the United Kingdom—a nation—and over several years to some extent Ireland, Scotland, and Wales continue as societies. The nation of the Soviet Union failed to create one society. China constantly struggles with this issue; it will probably be a long time before the nation of China becomes one society.

People living in societies often want to make themselves into nations. Nationalism is a claim by people living in a society that they, too, have a right to their own nation, to have their own political order. If a society is already a nation, then nationalism becomes a commitment that people have toward the nation they are in. If no nation yet exists, as in the case of the Kurds, then nationalism is a claim that one should exist.

Societies and nations are never neat and perfect in their boundaries. Rarely are they exactly the same. Even in the United States some people will identify two or three or more societies living together in one nation. Some would argue that the division caused by segregation has created two Americas, not one. The Middle East, the former Soviet Union, and the former Yugoslavia are constantly determining what constitutes a nation. The conflicts between Sudan and Darfur, Syria and Lebanon, Israel and Palestine, China and Tibet, Pakistan and India, North and South Korea are examples of struggles over what determines a society and what is a nation. Nations work because the political system works. They sometimes work because over time the political order is able to increase social interaction, communication, and understanding throughout the nation. And sometimes over time the nations develop their own social patterns such as social structures, cultures, and social institutions that slowly take over people's lives. In short, sometimes nations become societies.

Societies are different from nations because ultimately societies are held together by the ongoing social interaction of people and by social patterns that develop as people act and work together. Societies are more than simply governments, laws, boundaries, and armies, although they may certainly include these entities. Societies should be understood as social organizations, and like all other social organizations (such as groups and communities), they are characterized by ongoing social interaction and social patterns.

It is important at this point to turn our attention to the qualities that build society. We will begin with social interaction, then examine the social patterns of culture, social structure, and social institutions. We then turn to the importance of loyalty to society as well as the contributions that social conflict and social change make to the continuation of society.

## Society Is Possible through Social Interaction

At the heart of the sociological approach to social unity is the importance of social interaction, people acting back and forth with one another in mind: cooperating, communicating, sharing, arguing, negotiating, compromising, influencing, competing, trading, and understanding one another. People must interact for society to begin and for it to continue. Where social interaction ends, society ends. Where it eventually divides a society, two or more societies are created.

Interaction is the building block of society. Consider for a moment what interaction means (and it is not an easy concept to grasp): Actors take account of one another when they act. I act with you in mind; you act with me in mind; I act with you in mind again. What I do at any one point depends on what you do, and vice versa. This is easy to see when we consider two people: I say hello to you; on hearing me, you say hello back; when you say hello back, I inform you that I'm depressed; when you hear that I am depressed, you ask me what's wrong. Back and forth we talk. Each of us reacts to the acts of the other; the other, in turn, acts back.

Interaction is also easy to see in a group; for example, consider a football team. If we concentrate only on the eleven players, we see the quarterback telling the others the play, we see players altering their acts as they see what other players on their own team are doing. A guard misses a block, so a back picks up the block; a receiver goes out for a pass, and the quarterback sees an opportunity and throws a pass to that receiver. In the huddle the receiver declares to the quarterback: "Good pass." This is social interaction.

Of course, we can also see that there is ongoing interaction between the teams on the field, and from a distance we can observe interaction among a number of teams. For example, because most of the teams play most of the other teams, we can declare that all the teams in the league interact. We can see coaches among the teams meeting and drawing up rules, and referees holding meetings to help ensure that the league has consistency.

It is harder to observe interaction in a larger area, such as a neighborhood, but it is there. Sidewalks, stores, street corners, playgrounds, and hundreds of other places provide occasions for people to interact with one another. Everyone does not interact with everyone else at the same time, but if we observe carefully, we see a pattern of crisscrossing interaction among people within the area, which is more intense and continuous than that between those people and people outside the area. That is one reason we declare, "That's a neighborhood." We can say the same about the larger community: There is crisscrossing interaction within the community that is far more intense and continuous than the interaction with those outside.

Society, too, is defined in part by this interaction. When people from several communities interact on a continuous basis and when that interaction is far more intense and continuous among them than with outsiders, we see the beginning of society. Look at the opposite position: When there is no interaction, there can be no society; when interaction is segregated into two or more distinct entities, we must say that there is more than one society among those people. This was the point of the Kerner Commission report on riots in the United States in 1968: America had become two societies, segregated, each with different problems, each with different interests. Whether we are one or two or three or more societies is debatable, but here I am only trying to make the point that to be a society there must be ongoing interaction.

Why is interaction important to society? In a large part, this is because human interaction is symbolic. Symbolic interaction means that people's actions are usually meant to communicate something to others, and that the others who are objects of the communication constantly try to understand the meaning of those actions. Interaction is not simply physical responses to stimuli. Because we intentionally communicate, individuals can share with others their interests, concerns, values, demands, ideas, intentions, and feelings. Because we try to understand what others communicate, we have an opportunity to learn something from others, leading either to disagreement or, more usually, to sharing. Ongoing interaction that involves intentional communication and understanding facilitates cooperation and the negotiation of disagreement, both essential for the development and continuation of society. The significance of symbolic communication cannot be understated:

1. Communication brings a means of knowing one another, making possible consideration of the other's needs and helping to ensure that one's own needs are expressed. It brings a process known as "taking the role of the other," understanding the world from the perspective of others in the situation.

2. Over time, communication makes possible "shared understanding" among people. This shared understanding includes a way of handling disagreements and compromising among people's various interests.

3. Communication brings a basis for continuing cooperation, a way of handling problems together as they arise.

4. Communication brings a means by which people new to the interaction can be socialized so they know how to act in the interaction.

5. Finally, communication lets people know when their acts are unacceptable. It is a means of telling others that they are breaking the rules, that they are not going by the established group procedures, or that their acts are wrong.

In each instance, symbolic communication contributes to the functioning of society. To be outside the communication channels of society (that is, to be separated from interaction with others) is to be outside of society itself. If large numbers are outside that interaction—if they interact among themselves and are isolated from everyone else—then the larger society's maintenance is made more difficult.

The United States exists as a society in part because people continuously interact (for example, through travel, mail, telephones, the Internet, television, radio, newspapers, and business deals). Through symbolic interaction I begin to understand the problems of the individual in the inner city, the lives of wealthy corporate executives, and the ideas of my political leaders. And through symbolic interaction with others, I let them know my ideas, my interests, my values. Although I rarely agree perfectly with any of these people, over time an underlying agreement usually arises among us: Poverty is a tragedy in American life; capitalism is a healthy American institution even though there are serious problems associated with it; a college education is a necessity. Sometimes continuous interaction will bring serious disagreements among us, but more often it brings understanding and some agreement. When there is continuous interaction over time, we come to "think like Americans," to adopt certain values (such as individualism), to believe certain core ideas (such as "time is money"), and to accept certain customs and morals (Sunday is a day away from work, drug abuse is harmful, incest is wrong). This is what is meant by a people's culture, one of the other reasons that human society is able to exist. Culture arises in symbolic interaction; it is learned from others in symbolic interaction; it disappears without symbolic interaction.

Knowledge of others, being understood by others, and sharing culture are basic to all cooperation in society. Consider any service in society—medical care, television, distributing and selling goods, education. These work because each actor understands his or her part in relation to relevant others in society. The storeowner understands what to do in relation to, among others, customers, potential customers, advertising agencies, distributors, wholesalers, producers, and federal, state, and local governments. As infants are born into society and are socialized through interaction, they come to learn what to do within the cooperative order and what not to do. And as they violate society's rules, they are told through interaction (with parents, teachers, police officers, members of the clergy, or other representatives of society) that their acts are unacceptable.

The tragedy of the hurricane that destroyed much of New Orleans underscores the importance of social interaction in society. People were isolated. Communication was impossible for many. They neither knew what was happening in other parts of the city nor were they able to send out communications to others that they were alive. Those who tried to help were unable to do much until they were organized through social interaction. Otherwise, cooperation was impossible. It was only through the social interaction of the Red Cross and Salvation Army, interaction among people in the local, state, and federal governments, thousands of volunteers from all over the world, neighbors, friends, families, radio, television, and newspapers that made the very first steps in rebuilding New Orleans. The city began to be built again only as conditions allowed citizens to interact, communicate, share their problems and hopes, feel part of the community once again, and start to develop the social patterns that make order possible once again.

## Society Depends on Social Patterns

Social interaction is the first of two qualities that are necessary for any form of organization, including society. A set of social patterns is the second.

Almost every sociologist believes that as people interact, social patterns will develop among them and become an important influence over their actions. Indeed, these patterns distinguish a "bunch of individuals" from some form of organization such as a society.

A *social pattern* means that social interaction is made regular, it is regulated, and a stability is established whereby individual actors know what they are to do in relation to one another. Social patterns are routines, common expectations, predictable behaviors, and ways of thinking and acting that have been established so that ongoing cooperation is made possible. People get used to the ideas, rules, and actions that they use over and over, and people depend on their continuation so that interaction runs smoothly. The longer the interaction and the more isolated it is, the more likely the patterns will take hold. As new people enter the interaction they learn the patterns. Patterns are anchored in the past. They are taken for granted.

Sociologists sometimes identify three important patterns: *Culture, social structure,* and *social institutions.* Any cooperative order demands a certain degree of self-control in line with these patterns that exist.

### *Culture*

Culture is one of the social patterns in society. It arises in social interaction. It is taught in social interaction. It controls individuals as they interact. Culture is made up of three smaller sets of patterns: (1) rules, (2) values, and (3) beliefs.

For there to be cooperation, there must be *rules,* and individuals must be willing to guide their actions according to these rules. Societies are guided by customs: for example, when to have sex, with whom to have it, how one should feel about it, how it should be done. Societies are guided by laws: how old one's sex partner may be, what gender one's sex partner must be, what relatives must be excluded from marriage, under what circumstances one must refrain from sex. Societies are guided by taboos (prohibitions with severe punishment): what relatives must be excluded as sexual partners. Societies are guided by morals: how many sex partners are right, whether it is right to have sex outside of marriage, and whether the individual has a moral obligation to respect the wishes of his or her partner. Societies are guided by procedures: the role of foreplay, the best ways to have intercourse, what to do after the sexual act is over. Societies also have informal expectations: who should be assertive in the relationship, who should take the responsibility for preventing unwanted pregnancies, and who should remain a virgin. All of these rules—customs, laws, taboos, morals, procedures, and informal expectations—matter to the individual and to society. They tell individuals how to act; they tell them how others expect them to act; they tell them how to expect others to act. They also work to control the individual and help to ensure cooperation in interaction. In short, they aid society's continuation through regulating individual action according to rules that most people understand.

Besides rules, culture also includes *values* (what people are committed to, what they consider to be important in their lives), and agreement over values allows for more cooperative interaction. For example, a society may value materialism, individualism, and family life. These values influence action: They encourage people to work hard in order to make money for themselves and their immediate family. They encourage people to go to school to get an education in order to make money for themselves and their future family. Because these are shared social values, many people will take this same direction in society, facilitating cooperation. Without some shared sense of what is important, organization would become more tentative and less united, with individuals going in whatever direction they decided to go, and cooperation would be made far more difficult. Shared values make it much easier for people to understand one another's actions, again facilitating cooperation, because others know what to expect from them. Values are standards we apply to specific situations. They guide what we choose to do. "They are unquestioned, self-justifying premises that account for much of the consistency in responses to recurrent situations among those who share a culture" (Shibutani 1986: 68).

Culture is also made up of a shared set of *beliefs*. People may believe that hard work leads to material success or that a college education leads to a good job. A common belief in American society today is that marriage leads to a fulfilling life. "The free market system is the most effective economic system" seems to be an important belief in American culture. We have also come to believe that "a good government is one that stays out of the affairs of the individual" and that "people can become anything they want in our society." "We have classes in society, but people are able to easily move up or down." These beliefs may or may not be true—that is not the point. They are beliefs that are important in this society, and thus they have become part of society's culture. We are all taught them, and most of us will accept them, unless, of course, others around us reject them and develop a culture that is contrary to the dominant one. Such shared beliefs influence people's actions, and order and cooperation are made easier.

Marx saw through these patterns of culture. He maintained that a people's rules, values, and beliefs are exaggerations of reality and that there are generally understandable reasons why particular exaggerations occur. Much of culture, he wrote, is ideology, or ideas that act to defend society as it exists, including its inequality of power and privilege. An ideology is not created by all people in interaction; it tends to be created and expounded by those who have power in society. To say that culture binds society together meant to Marx that certain ideas are created by and for the powerful and that these ideas are taught to most people. These ideas are given the name *culture*, but in fact they are ideology. They do work to keep order in society, but they work because they defend the inequality that exists. Most sociologists would agree with Marx to some extent. If we examine culture carefully, we can see that the rules, values, and norms tend to be exaggerations that operate to protect the powerful in society and, in that way, help to establish social order.

Culture, then, means that people in society agree on many important matters—rules, values, and beliefs—and this agreement fosters the continuation of society. Perfect agreement is far from being possible or even desirable, but general agreement is not only possible but also of central importance to society. Of course, individuals may disagree with the dominant culture of society, and they may even interact with others and develop a culture among themselves contrary to the dominant societal culture. If disagreement becomes widespread and critical of the dominant culture, then a serious challenge to the culture may arise, undermining one of the important bonds of society. Normally this brings conflict and change, with a new dominant culture emerging, different but with many ties to the old.

### *Social Structure*

Social structure is another important social pattern that makes society possible. As people interact over time, they establish *relationships,* they *position* and *rank* themselves in relation to one another, and they learn and enact *roles* in the interaction. Structure organizes people's actions in relation to one another. As in culture, people understand what others expect them to do, and they understand what others are supposed to do.

A social structure is *a set of positions* (or what we also call *statuses, social locations, locations,* or *status positions*) that arise in interaction. People fill these positions in relation to one another. They are students (in relation to teachers), members of the middle class (in relation to the working and upper classes), men (in relation to women), quarterbacks (in relation to the rest of the team and the coach), first-year employees (in relation to old-timers and employers), and presidents (in relation to vice presidents, secretaries, treasurers, and general membership). There are thousands of positions in society. People come to learn what is supposed to be done in each position they enter or may enter, and together the positions create order out of what would otherwise be chaos.

Positions are important because they bring with them roles, perspectives, identities, and inequality.

First, each position has a *role* attached. A role is a set of expectations that other people have for action in that position. The individual who assumes the position learns to enact that role. Children are taught roles in preschool classes: "Here is what a nurse does; here is what a father does; here is what a little boy does; here is what a firefighter does; here is what a doctor does." Through learning this, children are able to know what people in these positions are expected to do, and if they encounter one or become one of these they will know what they should expect or do. Every position brings a role: wealthy class, woman, retired, father, friend, student, and a private in the army, for example.

Second, a position brings a *perspective.* A perspective is an angle on reality, how one is supposed to see reality. Third, a position brings an *identity,* a name others call us, a name we call ourselves, and a name we present to others. When we become a teacher, an unemployed person, or a married person our identity changes.

Fourth, for good or for bad, social structure *ranks* positions by attributing power, prestige, and privilege to the position. Our rank is higher than some, lower than others. In the army's structure, the general has more power, privilege, and prestige than the sergeant or private; in our class structure, those in the upper class have more power, privilege, and prestige than those in the working class.

Structure sorts people. Structure distributes people throughout society, people learn appropriate behaviors and ways of thinking, and people learn who they are and learn to fit their actions into the whole complex system. People are organized, labor is divided, inequality established. People learn how to act cooperatively with those in the other positions, and they learn to think about themselves and others according to their positions. By controlling the individual in an organization, cooperation is made easier.

There is no claim here that people take on the position they deserve or earn. Positions are gained on the basis of birth, interaction, talent, or luck, and the relative importance of these depends on the degree to which the structure is open (individuals can move up or down on the basis of their achievement). No matter how open the structure is, much talent is wasted and people inherit positions they are not qualified for; therefore, the system works against solving society's problems and creates anger among those who feel that the distribution is unjust. In spite of this, however, social structure also contributes to the continuation of society through making control, socialization, and cooperation more systematic and complete.

Structure aids society in a second way. It builds an interdependence among the actors and through this interdependence creates a commitment to the whole. Durkheim best describes this process. As we each do what we are expected to do in our various positions, others become dependent on us. As we deal with others in their positions, we become dependent on them. When I was a professor at Moorhead, I was dependent on students, the president of my college, and the state legislature, for example. They also needed me to teach sociology. Of course, I was also dependent on people in the Fargo–Moorhead Symphony Orchestra, the Minnesota Vikings, and National Public Radio for my entertainment; on those working at Hornbacher's Grocery and Walgreen's Drugstore to provide many of my simple daily needs; and on those in the police department and courts to protect my family and me. This exchange of services—this mutual dependence—tied us all together, and each individual became more and more conscious of his or her place in the whole society. Indeed, out of such interdependence will grow a recognition of a higher social morality that must prevail if our mutual services are going to continue. Thus, a common morality results, a tie to a moral whole: society. Durkheim (1893) writes that the "division of labor"—what I am here calling *social structure*—balances individual self-interest with a higher system of rules:

> We may say that what is moral is . . . everything that forces man to take account of other people, to regulate his actions by something other than the promptings of his own egoism, and the more numerous and strong these ties [the interdependence of positions] are, the more solid is the morality. (P. 331)

Structure (relationships, positions) sits alongside culture (agreed upon rules, beliefs, values), and they reinforce one another, both contributing to control over the individual, cooperation and interdependence among individuals, and order in society. Both are social patterns; both developed over time through social interaction; both are necessary for the continuation of society.

Social institutions contribute a whole other set of social patterns that exist alongside social structure and culture.

## Social Institutions

Societies exist over time because people have worked out ways of dealing with ongoing situations. Such established ways are called *social institutions*. Indeed, every group develops its own ways, and that includes every formal organization. Grocery stores have computerized checkouts and numbered boxes for grocery pickup, and a family might have Friday evening meals or a Christmas Eve celebration. All of these are grooves people follow. They sometimes become rituals and seem to be almost sacred, and they are patterns that keep action working smoothly in organizations over time.

If society is going to work, it must have ways to produce and distribute goods, control disruptive behavior, socialize the young, regulate sex, defend itself, carry on business with other societies, encourage the performance of all the necessary roles, and develop adequate means of transportation and communication. It must minimally satisfy individual members' needs. Sociologists have long debated exactly what list of functions must be met for society to continue. Although no perfect agreement has emerged, all have recognized that each society develops its own ways that allow it to function, solving basic problems as they arise. These patterns are *social institutions,* the third set of social patterns that make society possible. We could illustrate many examples, but here we will simply describe institutions that socialize and integrate individuals who exist in society.

**Socializing Institutions** How do we create willing, hardworking individuals who accept society's ways? How do we create people who understand the right things and have the right skills to function in our society at this particular time in history? This cannot be a natural process, because every society and every period in its history is different. It cannot be a natural process because new problems arise demanding that new types of people capable of acting in new situations are necessary. This problem demands a social process. Nature sets us up: We are helpless, so we must learn that survival and success depend on learning and accepting society's social patterns. The world we are born into tries to be ready for us. It has social institutions set up to form us into productive members of society.

Over a long period of time, the United States has developed institutions to socialize its population, native citizens and immigrants, young and old, upper classes and disadvantaged classes. Educational institutions are developed specifically for this and include, for example, kindergarten, the elementary school, the comprehensive high school, the two-year college, the state university, night school, continuing education, vocational education, public and private schools, and civic education in the high school. Religious institutions include Sunday school and the theological seminary; and media institutions such as television, daily local and national newspapers, and weekly magazines cooperate in the socialization effort. None of these institutions is inevitable—over time they are what developed in our society, and these help society to continue and make our society different from others. Social change makes some of these less and less important, and new institutions are continually being created to socialize people into society. Computerization has rapidly become a central institution, changing the way we learn things in this society. The Internet, Google, and online colleges are becoming dominant socializing institutions.

George Herbert Mead described the process of socialization in two stages. First, we interact with and learn from individuals he calls *significant others.* As we develop into adulthood, we take our significant others and see them as a group, a generalized whole, what he calls a *generalized other.* What individuals want of us and teach us becomes shaped into a consistent whole, and, as this is done, we become members of society rather than just individuals being shaped by other individuals. Through this process we are increasingly able to cooperate with others; society's rules become our own, and we internalize those rules. Socialization ultimately means that we develop the ability to control ourselves according to the society's rules and thus take part in cooperative actions. All societies develop institutions to accomplish this, beginning with family institutions and then spreading to educational, religious, media, and even economic and political institutions.

**Integrating Institutions** Another example of a problem that societies must consider is *integration.* How can we hold individuals together into one whole people? How can we keep individuals committed to one another and to the society at large? How can we form lasting social relationships that matter to individuals? What institutions develop to meet this need? Or, to ask this in another way, what institutions function to effect the integration of society?

To some extent, the public school and the family contribute to integration, but so do the law, the courts, and the prison system. All of these together encourage conformity and punish nonconformity. Political leaders help bring us together, as do mass transportation and modern communication. Voluntary organizations from churches to the Democratic Party are important for integration, because they bring the individual into society, control the individual, and help attach the individual to the whole.

**Other Institutions** Societies have many needs, many problems that must be handled; thus they must develop a wide variety of institutions. According to one sociologist, Talcott Parsons, besides socialization and integration, a society must develop institutions that allow it to adapt successfully to its physical and social environment, to develop and work cooperatively toward goals, and to keep its population relatively satisfied with their lives. All societies develop political, economic, religious, legal, military, familial, educational, health, and recreational institutions to help ensure that such needs are met successfully. Government has to work; that is, it must efficiently achieve societal goals. It must arbitrate disputes and enforce rules to ensure social control. It must develop relations with other societies. It must provide aid to those who are unable to care for themselves. The society's economic institutions must effectively produce and distribute goods; legal institutions must regulate people's activities and settle disputes; and religious, educational, and familial institutions must help maintain individuals' satisfaction with their lives. Society works in part because it has developed such institutions, and, overall, they work.

### *A Brief Summary*

Recall we have described two qualities that are necessary for the creation and continuation of society: (1) symbolic social interaction and (2) social patterns that arise from that interaction. Three social patterns emerge: culture, social structure, and social institutions. Together these patterns make choices for the individual, order and coordinate action, and work out the ongoing problems that confront society. There are, however, two more qualities that make society possible: loyalty to society and the existence of positive social conflict and social change.

## Society Is Made Possible through Feelings of Loyalty

Almost every definition of society ends with a recognition of the importance of feeling and commitment. Society exists in part because people feel something positive about it. The whole takes on significance to them, and they feel good about being part of that whole. Individuals are willing to cooperate with others in spite of their own individual interests or wishes. Volunteerism supplements force in guaranteeing the ongoing cooperation. Of course, no society exists in which everyone feels that commitment; but without some widespread feeling of loyalty and belonging, leaders of society are required to rely on force for conformity—and inefficiency, anger in the population, and instability often result. Ferdinand Toennies, an important European sociologist writing in the nineteenth century, describes two types of societies, each based on a different kind of loyalty. In more traditional societies, commitment is based on a "feeling of community," an emotional bond in which the individual feels that he or she is part of something larger. A sense of "we" prevails, and a belief that my efforts are important not for me, but for *us*. In German, this is called *Gemeinschaft,* and Toennies's description is almost identical to the feeling of "we" that Charles Cooley identifies when he describes the primary group (a small, relatively permanent, intimate, and unspecialized group). Many gang members have this strong sense of loyalty, as do small religious and political groups (for example, al-Qaeda), and some societies such as Nazi Germany, England in the nineteenth century, and Japan prior to World War II. Strong nationalism comes close to this feeling: "My society is very important to me. My life is part of it. I will defend it against all enemies. I get my importance as a human being in part because I belong to it." The September 11 terrorist attacks tapped into this, and we created throughout society a strong sense of loyalty, a feeling that even with the disagreements and conflicts within the United States, we are in the end one society and nation, and it is a good one that we have created.

In every society there will be people who feel a real sense of community and see themselves as something great. Many Americans do; on occasion, almost *all* Americans do. Toennies—and most sociologists—recognize that commitment to modern society is more often characterized by a "conditional loyalty." More thought is involved in the commitment: "I am loyal to certain principles. I give my commitment so long as society meets my needs. My society is important, but so am I as an individual." Thus, what enters into our feeling of commitment is a belief that society does, in fact, work in our interests. Instead of *Gemeinschaft* (a sense of "community"), we have *Gesellschaft* (an "association" of people) in which contract and reason are more prevalent. Instead of feeling part of a primary group in which a sense of "we" prevails, people feel part of a secondary group, in which a sense of "I" prevails, with loyalty depending on whether the society meets their needs.

All societies are a mixture of both kinds of loyalty, some able to get more emotional commitment, others relying more on conditional loyalty. In modern society, conditional loyalty seems more prevalent. But in Nazi Germany we saw a modern example of strong emotional ties to society. As individualism increases in society, real emotional commitment to the whole becomes more difficult. And, as Erich Fromm reminds us, as people develop fewer and fewer ties with one another and with society as a whole, many will try to turn away from individualism, seeking stronger commitments to larger groups and society in order to rediscover something from the past that has somehow been taken from them.

Modern industrialized societies such as the United States face a dilemma that is impossible to ever resolve but that is an important key to many of the problems that face the individual: How much loyalty to the whole? How much individualism and freedom? Nothing has been more important to my own life than seeking freedom. Yet can freedom exist among many people in a society where people are unwilling to give loyalty to the community and refuse to follow the social patterns even minimally? Can freedom exist without commitment to the whole? And if we give commitment to the whole, can we pursue our own dreams and develop our own ideas and morals? Durkheim asks over and over: Can freedom exist without a shared agreement as to what is right and wrong?

Institutions are meant to create and maintain a strong commitment to society. Public schools, families, religion, and political leaders try to socialize us so that we *feel* good about being part of our society. In times of tragedy, such as the attack on the World Trade Center in New York, the political leaders, schools, and media work to bring a people together in commitment to society, and national symbols of all kinds—the flag, the president, the national anthem, "God Bless America"—accomplish this purpose. Defining some people as outsiders, terrorists, evil, or anti-American serves to bring loyalty to society, as does punishment or war to those defined as threats. Rituals of all kinds help bring people together into a society, integrating them and causing them to feel a sense of belonging. Ritual is an action whose purpose is not purely instrumental (goal-directed) but that communicates something among people that is symbolic of "the whole." Ritual is social action, and its purpose is to bind people together and to bind them with the past. All the various institutions include rituals, which reaffirm, dramatize, and encourage loyalty to society (Wuthnow 1987: 140). Annual rituals commemorating Pearl Harbor, September 11, the American Revolution, and birthdays of important heroes in American history serve this function.

Institutions help maintain conditional loyalty by delivering services to people in the society. Government must prioritize goals and convince us that these goals are being achieved. Schools must teach; economic institutions must produce prosperity, employment, and a bright future; and the courts must justly punish. Religion and family must bring some meaning and security to individuals. For conditional loyalty to exist in society, people must perceive that their society works, that the institutions do an adequate job in dealing with problems. This is especially important in modern society.

Without loyalty, if a society were to work, it would have to rely to a great extent on force. Max Weber, in his brilliant analysis of authority, shows how relying on force alone brings serious disadvantages to society. Too much effort has to go into surveillance of the population; fear is costly, and constantly punishing the population is a waste of talent. Weber emphasizes that voluntary obedience is the basis for stable systems of power in society and is thus dependent on loyalty. He calls such systems *authority* and defines it as "legitimate power." In short, loyalty to society brings a willingness to obey legitimate representatives of that society, so long as they, too, conform to the rules. Without a legitimate system of authority, there would be a continuous refusal by the population to follow rules, which would cause leaders to turn to force. It is impossible to determine how many people in a given society recognize the system of power as legitimate, and it is also impossible to determine how many people must do so for society to continue, but any society is clearly at a great disadvantage if it is without a power structure that is considered legitimate in the eyes of a large portion of the population.

Emotional commitment, conditional loyalty, institutions that encourage loyalty, institutions that are successful in meeting the needs of people in society, and legitimate authority all combine to create voluntary acceptance of society and its ways and thus help society continue.

## Conflict and Change Help Preserve Society

Loyalty and order can be exaggerated. All societies also depend on change; change, in turn, depends on social conflict. Georg Simmel emphasizes in his analysis that social conflict is inevitable in every social organization from the small group to the society. Marvin Olsen describes social organization as a "process" rather than an entity, a changing rather than a stable thing. Instead of conflict being made into a negative quality, it is described as a necessary and positive one. In any organization, change will bring conflict; conflict then will bring change; change, in turn, will bring more conflict. Conflict highlights problems and often points to necessary change. Conflict causes people to reassess their society. If problems are addressed successfully, social stability is more assured, and the society is able to more effectively achieve its various goals.

Conflict needs to be recognized as an opportunity, an opportunity to identify and deal with social problems and change society's patterns so that they meet the needs of more people and, over time, a society can become even more stable and effective.

Early nineteenth-century slavery was an important problem to a growing democratic society. Before the Civil War, there was conflict over this issue throughout society, a conflict that became increasingly serious. Slave owners and political leaders from slave states exerted their agendas; abolitionists and free states exerted theirs. Increasingly, the problem of slavery was impossible to ignore, and through attempted compromise the federal government decided to put off the problem by admitting states to the union in a balanced way so that neither side had an advantage in the legislative branch. The conflict became more obvious and serious, the question of states' rights and national power came to the fore, a presidential election led to the victory of Abraham Lincoln, and the conflict that became the most destructive war in our history broke out. No longer was the problem itself addressed, but the desire to destroy the other side became central. Through victory in that destructive conflict, changes were made that freed the slaves. After the Civil War, however, that conflict continued. No side was able to get exactly what it wanted, but change did take place. Poverty among the African-American population and an increasingly segregated society changed the society and its problems, conflict continued, and through the twentieth century and into the twenty-first, conflict—peaceful and sometimes violent—created an ongoing attempt by many to address the problem of racial oppression in the United States.

Without conflict, there would have been little change. With conflict, there was change. There is every reason to believe that this process will continue for a long time. Indeed, the conflict between a multitude of people in our society continuously highlights and causes us to address problems: poor and nonpoor, wealthy and nonwealthy, city people and rural people, men and women, employers and employees, the elderly and the nonelderly, professors and students, tenured professors and nontenured instructors, Microsoft and non-Microsoft computer companies, pro-choice and pro-life groups, religious sects, religious cults, and established churches, Republicans and Democrats, major political parties and third parties, environmentalists and industrialists.

A democratic society in many ways has an important advantage: Conflict is encouraged more than in other societies; problems can be identified and dealt with more effectively, and what is constructive and peaceful conflict does not have to become destructive and violent. However, it is essential to recognize that conflict and change are both necessary and inevitable parts of every society, and without them, the stability of society is threatened and its ability to achieve goals is lessened.

## Summary and Conclusion

Societies exist because of *social interaction.* Without interaction, there is no society; with segregated interaction, there are several separate societies. Social interaction simply means that people act with one another in mind. Social interaction is symbolic. People communicate. They understand one another. They share various aspects of the world in which they live.

Societies exist because of what people share in symbolic interaction. Over time, people create *social patterns* in that interaction: culture, social structure, and institutions. *Culture* binds people together because they come to agree on several important matters: beliefs, values, and rules. *Social structure* distributes people in society, locates them, controls them, teaches each how to act in relation to the others, develops interdependence, and facilitates cooperation.

*Institutions* develop to solve society's problems. People are socialized through institutions, society is integrated, people are rewarded and punished, goods are produced and distributed, goals are developed and worked for, and people are protected.

Societies are also able to exist because people *feel loyalty.* They feel that they belong. They believe the institutions work. They are willing to obey those in positions of power because they regard them as legitimate representatives.

Finally, societies exist because they are able to *change* and respond to *conflict.* Their members solve problems rather than ignore them; they devise creative solutions rather than try old solutions over and over again. A complex mixture of social change and social patterns works successfully.

We live in a complex world. It is sometimes difficult to know what has gone wrong in our society or in others. Examine the world carefully: You will be able to identify symbolic interaction, culture, social structure, social institutions, and feelings of loyalty scattered here and there. You will see that social conflict and societies evolve new patterns that work to solve new problems. This can be a good beginning for you to unravel the mystery introduced in this chapter: How is society possible? How is it able to exist? This is also one way for you to appreciate the approach that sociologists take to understanding important issues that exist in our world.

## Questions to Consider

1. Is the United States a single society or is it really several societies?

2. How is society possible if diversity is one of its values? Can it be possible if immigration is encouraged? Can it be possible if cultures are significantly different within its population?

3. What is segregation in society? How does segregation undermine society?

4. If it is not culture and agreement that hold society together, then what does?

5. What, after all, is the role of conflict in society? Is conflict necessary for society to exist? Is conflict destructive of society's existence?

6. Is society a cooperative order or is it an arena of continuous conflict?

7. How would an experienced police officer answer the question: How is society possible?

## References

The following works examine the meaning of society and the general problem of social order. Each either makes a general philosophical examination of society or looks at the more specific ways in which the continuation of society is made possible.

**Aberle, D. F., A. K. Cohen, A. K. Davis, M. J. Levy, Jr., and F. X. Sutton**. 1950. "The Functional Prerequisites of a Society." *Ethics* 60: 100–111.

**Altbach, Philip G., Robert O. Berdahl, and Patricia J. Gumport.** 1999. *American Higher Education in the Twenty-First Century.* Baltimore: Johns Hopkins University Press.

**Altheide, David.** 2006. *Terrorism and the Politics of Fear.* Landham, MD: Rowman & Littlefield.

**Apraku, Kofi.** 1996. *Outside Looking In: An African Perspective on American Pluralistic Society.* Westport, CT: Praeger.

**Ballantine, Jeanne H.** 1997. *The Sociology of Education.* 4th ed. Englewood Cliffs, NJ: Prentice Hall.

**Barlett, Donald L., and James B. Steele**. 1996. *America: Who Stole the Dream?* Kansas City, MO: Andrews & McMeel.

**Bellah, Robert N., Richard Madsen, William M. Sullivan, Ann Swidler, and Steven M. Tipton**. 1985. *Habits of the Heart: Individualism and Commitment in American Life.* Berkeley: University of California Press.

**Berger, Bennett M.** 1995. *An Essay on Culture: Symbolic Structure and Social Structure.* Berkeley: University of California Press.

**Berger, Peter L.** 1963. *Invitation to Sociology.* Garden City, NY: Doubleday.

———. ed. 1998. *The Limits of Social Cohesion.* Boulder, CO: Westview Press.

**Berger, Peter L. and Thomas Luckmann**. 1966. *The Social Construction of Reality.* Garden City, NY: Doubleday.

**Berry, Jeffrey.** 1997. *The Interest-Group Society.* 3rd ed. New York: Longman.

**Bluestone, Barry.** 1995. *The Polarization of American Society.* New York: Twentieth Century Fund.

**Blumer, Herbert.** 1962. "Society as Symbolic Interaction," pp. 00–00 in *Human Behavior and Social Processes,* edited by Arnold Rose. Boston: Houghton Mifflin.

———. 1969. *Symbolic Interactionism: Perspective and Method.* Englewood Cliffs, NJ: Prentice Hall.

**Bok, Derek.** 1996. *The State of the Nation.* Cambridge, MA: Harvard University Press.

**Bowles, Samuel, Herbert Gintis, Melissa Osborne-Groves**. 2005. *Unequal Chances: Family Background and Economic Success.* New Brunswick, NJ: Princeton University Press.

**Brown, Michael K., Martin Carnoy, Ellioitt Currie, Troy Duster, David B. Oppenheimer, Marjory M. Schultz, and David Wellman**. 2003. *Whitewashing Race: The Myth of a Color-Blind Society.* Berkeley: University of California Press.

**Charon, Joel M.** 1997. *Symbolic Interactionism: An Introduction, an Interpretation, an Integration.* 6th ed. Englewood Cliffs, NJ: Prentice Hall.

———. 2002. *The Meaning of Sociology.* 7th ed. Upper Saddle River, NJ: Prentice Hall.

———. 2009. "An Introduction to the Study of Social Problems," pp. 1–12, in *Social Problems: Readings with Four Questions,* edited by Joel Charon and Lee Vigilant. 3rd ed. Belmont, CA: Wadsworth Cengage Learning.

**Charon, Joel M., and Lee Garth Vigilant**. 2009. *Social Problems: Readings with Four Questions.* 3rd ed. Belmont, CA: Wadsworth Cengage Learning.

**Cooley, Charles Horton.** 1902 [1964]. *Human Nature and the Social Order.* New York: Schocken Books.

———. [1909] 1962. *Social Organization.* New York: Schocken Books.

**Coontz, Stephanie.** 1997. *The Way We Really Are: Coming to Terms with America's Changing Families.* New York: Basic Books.

**Coser, Lewis.** 1956. *The Functions of Social Conflict.* New York: Free Press.

**Currie, Elliott.** 1998. *Crime and Punishment in America.* New York: Metropolitan Books.

———. 2005. *The Road to Whatever: Middle-Class Culture and the Crisis of Adolescence.* New York: Henry Holt and Company.

**Dahrendorf, Ralf.** 1958. "Toward a Theory of Social Conflict." *Journal of Conflict Resolution* 2: 170–183.

———. 1959. *Class and Class Conflict in Industrial Society.* Stanford, CA: Stanford University Press.

**Davey, Joseph Dillon.** 1995. *The New Social Contract: America's Journey from Welfare State to Police State.* Westport, CT: Praeger.

**Davis, Kingsley.** 1949. *Human Society.* New York: Macmillan.

**Derber, Charles.** 1996. *The Wilding of America: How Greed and Violence Are Eroding Our Nation's Character.* New York: St. Martin's.

**Deutsch, Morton.** 1973. *The Resolution of Conflict.* New Haven, CT: Yale University Press.

**Durkheim, Émile** 1893 [1964]. *The Division of Labor in Society.* Translated by George Simpson. New York: Free Press.

———. 1895 [1964]. *The Rules of the Sociological Method.* Translatedby Sarah A. Solovay and John H. Mueller. New York: Free Press.

———. 1915 [1954]. *The Elementary Forms of Religious Life.* Translated by Joseph Swain. New York: Free Press.

**Dyer, Joel.** 1997. *Harvest of Rage: Why Oklahoma City Is Only the Beginning.* Boulder, CO: Westview Press.

**Ehrensal, Kenneth.** 2001. "Training Capitalism's Soldiers: The Hidden Curriculum of Undergraduate Business Education," pp. 97–113, in *The Hidden Curriculum,* edited by Eric Margolis. New York: Routledge/ Taylor & Francis.

**Elkin, Frederick, and Gerald Handel.** 1995. *The Child and Society.* 5th ed. New York: Random House.

**Erikson, Kai T.** 1976. *Everything in Its Path.* New York: Simon & Schuster.

**Eshleman, J. Ross.** 1999. *The Family: An Introduction.* 9th ed. Needham Heights, MA: Allyn & Bacon.

**Etzioni, Amitai.** 2001. *The Monochrome Society.* Princeton, NJ: Princeton University Press.

**Ewen, Stuart.** 1976. *Captains of Consciousness.* New York: McGraw-Hill.

**Fenn, Richard K.** 2001. *Beyond Idols: The Shape of a Secular Society.* Oxford, England: Oxford University Press.

**Festinger, Leon.** 1956. *When Prophecy Fails.* Minneapolis: University of Minnesota Press.

**Forschi, Martha, and Edward J. Lawler,** eds. 1994. *Group Processes: Sociological Analysis.* New York: Nelson-Hall.

**Fromm, Erich.** 1941. *Escape from Freedom.* New York: Holt, Rinehart & Winston.

**Gamson, William A.** 1968. *Power and Discontent.* Homewood, IL: Dorsey.

**Geertz, Clifford.** 1965. "The Impact of the Concept of Culture on the Concept of Man," pp. 00–00 in *New Views of the Nature of Man,* edited by John R. Platt. Chicago: University of Chicago Press.

**Goldman, Alvin I.** 2002. *Pathways to Knowledge: Private and Public.* New York: Oxford University Press.

**Goode, Erich.** 2002. *Deviance in Everyday Life: Personal Accounts of Unconventional Lives.* Prospect Heights, IL: Waveland.

**Hall, John A., and Charles Lindholm.** 2000. *Is America Breaking Apart?* Princeton, NJ: Princeton University Press.

**Hall, John R., and Mary Jo Neitz.** 1993. *Culture: Sociological Perspectives.* Englewood Cliffs, NJ: Prentice Hall.

**Hallinan, Maureen T.** 2005. *The Socialization of Schooling.* New York: Russell Sage Foundation.

**Hardin, Russell.** 1995. *One for All: The Logic of Group Conflict.* Princeton, NJ: Princeton University Press.

**Hedges, Chris.** 2002. *War Is a Force That Gives Us Meaning.* New York: Public Affairs.

**Hertzler, Joyce O.** 1965. *A Sociology of Language.* New York: Random House.

**Ignatieff, Michael**, ed. 2004. *The Lesser Evil: Political Ethics in an Age of Terror.* Princeton, NJ: Princeton University Press.

**Keller, Suzanne.** 2003. *Community: Pursuing the Dream, Living the Reality.* Princeton, NJ: Princeton University Press.

**Kerner, Otto (Chair), United States Commission on Civil Disorders.** 1968. *Report.* Washington, DC: U.S. Government Printing Office.

**Kinkead, Gwen.** 1992. *Chinatown: A Portrait of a Closed Society.* New York: HarperCollins.

**Koonigs, Kees, and Dirk Druijt**, eds. 1999. *The Legacy of Civil War, Violence and Terror in Latin America.* London: Zed Books.

**Kosmin, Barry A., and Seymour P. Lachman**. 1993. *One Nation under God: Religion in Contemporary American Society.* New York: Harmony Books.

**Lancaster, Jane Beckman.** 1975. *Primate Behavior and the Emergence of Human Culture.* New York: Holt, Rinehart & Winston.

**Lenski, Gerhard.** 1987. *Human Societies: An Introduction to Macrosociology.* 5th ed. New York: McGraw-Hill.

**Lichtenstein, Nelson.** 2006. *Wal-Mart: The Face of Twenty-First Century Capitalism.* New York: New Press.

**Liebow, Elliot.** 1967. *Tally's Corner.* Boston: Little, Brown.

**McCall, George J., and J. L. Simmons**. 1978. *Identities and Interactions.* New York: Free Press.

**McCord, Joan**, ed. 1997. *Violence and Childhood in the Inner City.* New York: Cambridge University Press.

**Mead, George Herbert.** 1925. "The Genesis of the Self and Social Control." *International Journal of Ethics* 35: 251–277.

———. 1934. *Mind, Self and Society.* Chicago: University of Chicago Press.

**Miller, W. Watts.** 1996. *Durkheim, Morals and Modernity.* Montreal and London: McGill-Queen's University Press/UCL Press.

**Mills, Nicolaus.** 1997. *The Triumph of Meanness: America's War against Its Better Self.* Boston: Houghton Mifflin/Sage Foundation.

**Neuman, W. Lawrence.** 2004. *Basics of Social Research: Qualitative and Quantitative Approaches.* Boston: Pearson.

**Newman, Katherine.** 2004. *The Social Roots of School Shootings.* New York: Basic Books.

**Olsen, Marvin E.** 1978. *The Process of Social Organization.* 2nd ed. New York: Holt, Rinehart & Winston.

**Parsons, Talcott.** 1961. *Theories of Society.* New York: Free Press.

**Ravitch, Diane, and Joseph P. Viteritti**. 2001. *Making Good Citizens: Education and Civil Society.* New Haven, CT: Yale University Press.

**Rousseau, Nathan.** 2002. *Self, Symbols, and Society: Classic Readings in Social Psychology.* New York: Rowman & Littlefield.

**Rubington, Earl, and Martin S. Weinberg**. 2003. *The Study of Social Problems: Seven Perspectives.* 6th ed. New York: Oxford University Press.

**Scaff, Lawrence A.** 2000. "Weber on the Cultural Situation of the Modern Age." Chapter 5 in *The Cambridge Companion to Weber,* edited by Stephen Turner. New York: Cambridge University Press.

**Schwab, William A.** 2005. *Deciphering the City.* Upper Saddle River, NJ: Pearson Education, Inc.

**Scott, W. Richard.** 2001. *Institutions and Organizations.* Thousand Oaks, CA: Sage.

**Seligman, Adam.** 1995. *The Idea of Civil Society.* Princeton, NJ: Princeton University Press.

**Shibutani, Tamotsu.** 1955. "Reference Groups as Perspectives." *American Journal of Sociology* 60: 562–569.

———. 1961. *Society and Personality: An Interactionist Approach to Social Psychology.* Englewood Cliffs, NJ: Prentice Hall.

———. 1986. *Social Processes: An Introduction to Sociology.* Berkeley: University of California Press.

**Shils, Edward S., and Morris Janowitz**. 1948. "Cohesion and Disintegration in the Wehrmacht in World War II." *Public Opinion Quarterly* 12: 280–294.

**Simmel, Georg.** 1950. *The Sociology of Georg Simmel,* edited by Kurt W. Wolff. New York: Free Press.

**Skolnick, Arlene, and Jerome Skolnick**. 2000. *Family in Transition.* 11th ed. Needham Heights, MA: Allyn & Bacon.

**Skolnick, Arlene S., and Jerome Skolnick**. 2007. *Family in Transition,* 14th ed. Boston: Allyn & Bacon.

**Stacey, Judith.** 1996. *In the Name of the Family: Rethinking the Family Values in the Post Modern Age.* Boston: Beacon.

**Strauss, Anselm L.** 1978. *Negotiations: Contexts, Processes and Social Order.* San Francisco: Jossey-Bass.

**Sumner, William Graham.** 1906 [1940]. *Folkways.* Boston: Ginn & Co.

**Toennies, Ferdinand.** 1887 [1957]. *Community and Society.* Translated and edited by Charles A. Loomis. East Lansing: Michigan State University Press.

**Tonry, Michael.** 2004. *Thinking about Crime: Sense and Sensibility in American Penal Culture.* New York: Oxford University Press.

**Travis, Jeremy, and Michelle Waul.** 2003. *Prisoners Once Removed: The Impact of Incarceration and Reentry on Children, Families, and Communities.* Washington, DC: Urban Institute Press.

**Turow, Scott.** 2002. *Ultimate Punishment.* New York: Farrar, Straus & Giroux.

**Volti, Rudi.** 1995. *Society and Technological Change.* 3rd ed. New York: St. Martin's.

**Waddock, Sandra A.** 1995. *Not by Schools Alone: Sharing Responsibility for America's Reform.* Westport, CT: Praeger.

**Walker, Henry A., Phyllis Moen, and Donna Dempster-McClain,** eds. 1999. *A Nation Divided: Diversity, Inequality, and Community in American Society.* Ithaca, NY: Cornell University Press.

**Warriner, Charles K.** 1970. *The Emergence of Society.* Homewood, IL: Dorsey.

**Weber, Max.** 1905 [1958]. *The Protestant Ethic and the Spirit of Capitalism.* Translated and edited by Talcott Parsons. New York: Scribner's.

———. 1969. "The Social Psychology of the World Religions," pp. 00–00 *in Max Weber: Essays in Sociology,* translated and edited by H. H. Gerth and C. Wright Mills. New York: Oxford University Press.

**Wheelan, Susan A.** 1994. *Group Processes: The Developmental Perspective.* Boston: Allyn & Bacon.

**White, Leslie A.** 1940. *The Science of Culture.* New York: Farrar, Straus & Giroux.

**Whyte, William Foote.** 1949. "The Social Structure of the Restaurant." *American Sociological Review* 54: 302–310.

**Wolfe, Alan.** 1998. *One Nation, After All.* New York: Penguin Books.

———. 2001. *Moral Freedom: The Search for Virtue in a World of Choice.* New York: Norton.

**Wolff, Edward N.** 1996. *Top Heavy: A Study of the Increasing Inequality of Wealth in America.* New York: Twentieth Century Fund Press.

**Wrong, Dennis H.** 1994. *The Problem of Order: What Unites and Divides Society.* New York: Free Press.

**Wuthnow, Robert.** 1987. *Meaning and Moral Order: Explorations in Cultural Analysis.* Berkeley: University of California Press.

# 4 Why Are People Unequal in Society?

## The Origin and Perpetuation of Social Inequality

**Concepts, Themes, and Key Individuals**

- ❑ Human inequality
- ❑ Source and perpetuation of inequality
- ❑ Marx, Weber, and Michels
- ❑ Division of labor, social power, and social conflict
- ❑ Privilege, prestige, and social power
- ❑ Culture, socialization, and force

The French Revolution in the late eighteenth century brought cries for liberty, equality, and fraternity. It was a revolution that aimed to destroy great gaps that existed between the royalty, high clergy, and nobility on the one hand and everyone else on the other hand. The thinking that inspired the revolution was a product of the cultural movement known as the *Enlightenment.* It was the philosophers of the Enlightenment who recognized that injustice originated from social inequality. They wondered why such inequality existed throughout Europe, and they developed theories to explain its seeming inevitability. Whatever else their answers were, the Enlightenment philosophers seemed to agree that the *nature of society itself created inequality* and that one purpose of government was to limit such inequality. The strong tendency for societies to develop inequalities is also recognized by sociologists, most of whom believe that ignoring them has serious consequences. An important goal in sociology has always been to somehow come to terms with social inequality and to seek an understanding as to why it occurs.

Every time we interact with one another, inequality emerges in some form or another. Individual qualities, for example, not only will differentiate us from one another but also often will become the basis for inequality between us. We will be unequally handsome, intelligent, outgoing, talented in athletics, and even cool. Where such qualities matter, inequality will exist. When we compare ourselves on more social qualities, we will see many others as richer, more successful, or friendlier. It is hard to escape inequality and the perception of inequality in our lives.

The United States is a society that prides itself on a democratic ideal that includes equal opportunity for all. Yet if we are honest about it, we cannot ignore the great inequalities that persist and cause us to fall short of this ideal. In 2005, for example, *the top one-fifth of the population received 50.34 percent of the income (47.28 percent after taxes), whereas the bottom one-fifth received 3.42 percent of all income.* The top 40 percent of the population received 73.37 percent of the total income (60 percent of the bottom population received 26.63 percent). (U.S. Census Bureau, Current Population Survey, 2006 Annual Social and Economic Supplement, p. 7) Because of various forces at work in American society today, the income of the wealthy is rising at a much higher rate than the incomes of the rest of the population.

Total wealth is much more difficult to measure, but there is general agreement that it is even more unequal than income.

–In 2004 the *wealthiest 1 percent* of the U.S. households owned 33.4 percent of the total wealth (27.1 percent in 1989).

–In 2004 the *wealthiest 5 percent* of the U.S. households owned 57.5 percent of the total wealth (49.4 percent in 1989).

–In 2004 the *wealthiest 10 percent* of the U.S. households owned 69.5 percent of the total wealth (62 percent in 1989).

–In 2004 the *bottom 90 percent* of the U.S. households owned 30.4 percent of the total wealth (32.9 percent in 1989).

–In 2004 the *bottom 50 percent* of the U.S. population owned 2.5% of the total wealth (3.0 percent in 1989)

(Arthur B. Kennickell, 2006, p. 11)

Much of what sociology has done is to understand and document the various types of inequality. It is not an easy task. Weber writes that we are unequal in three orders, or arenas, in society: the economic order, the social order, and the political order. We might translate this into class (economic order); race, occupation, education, gender, and ethnic group membership (social order); and political position (political order). Furthermore, sociologists point out, within every organization from university to place of employment we also enter positions that are unequal. Even when we interact in small groups or even with one other person, we develop a system in which our informal positions are almost always unequal. In groups, there are leaders, followers, and even scapegoats. Similarly, we find differences between organizations of people. Management in a corporation is more or less powerful than a labor union. Harvard and Stanford have more prestige—sometimes even more power—than most other universities. The National Rifle Association has great power in certain matters, as do the American Medical Association and the National Association of Manufacturers. Political parties are rich or poor. Some churches survive on a few funds, whereas others are quite wealthy.

Why does inequality exist? Whereas most of us are tempted to answer this in terms of human nature, biological superiority or inferiority, supernatural forces, or a just free-market system, sociologists will, of course, look at the nature of our social life. To understand inequality among people in society, it is necessary to understand social organization itself, to see social inequality as a central quality in society, to see social forces at work that make it almost inevitable in human life. The sociologist will generally examine two aspects of the problem:

1. the reasons why inequality *arises in the first place*, and
2. how inequality is *perpetuated over time*—why, once established, it tends to continue in society, and thus why it is so difficult to limit its development or alter its importance.

Sociology emphasizes the significance of social structure. By doing so it is predictable that the sociologist will go back time and time again to structure as the reason that inequality exists. Social structure is a social pattern that involves unequal ranks; thus, wherever it develops, a permanent system of inequality is established, and individuals become ranked throughout society and throughout every social organization. It is, therefore, neither the nature of human beings that creates inequality nor natural superiority or inferiority or supernatural forces, so much as it is social interaction and the creation of social patterns. In a sense, sociologists take a view that maintains that so long as people have to live in society—and that seems to be our essence—inequality will be central part of life. Even Karl Marx, who dreamed of and predicted a world of equality someday, saw its possibility only with the disappearance of society and its permanent social patterns.

## Why Does Inequality Emerge in the First Place?

Social interaction develops social patterns. Once created, these patterns hang on. One of these patterns is social structure, a pattern that is almost always a system of inequality. That is what we need to focus our attention on.

What processes create a permanent social structure? The division of labor, social conflict, and the institution of private property.

### *The Economic Division of Labor and the Rise of Social Inequality*

It is important to note that almost every sociologist who tries to explain inequality brings in the process called "the division of labor." To Karl Marx this is central and it is economic: Economic activities eventually lead to a division of labor in society, where people do increasingly different things from one another. Division of labor inevitably brings advantages to some people over others. Some activities are more valued, some allow for control over others, some give people advantages in the way they live their lives, some lead to increasing centralization of control of the private property necessary for investment used for further accumulation of wealth. If one person farms and others do not, then the others become increasingly dependent on that one. And if one person farms successfully (through skill, luck, exploitation, or cheating), then that farmer is able to accumulate economic resources—for example, land, laborers, and capital (money to be used for investment)—that give him or her advantages over those with fewer of these resources. If the farmer is able to employ workers (a critically important division of labor), then a system of inequality begins in earnest. Often, the division of labor simply arises because some societies conquer other societies and establish a division of labor that favors those who are the victors.

To Marx, the employer controls employees. The boss exploits the workers, gains at their expense, and becomes increasingly wealthy and powerful in relation to them. The employer over time is able to consolidate a favored position in society and advantages accumulate. Once this process begins in society, it cannot be easily stopped. Division of labor itself encourages social inequality, and the division between employers and employees is especially important.

### *The Organizational Division of Labor and the Rise of Social Inequality*

Marx focuses almost entirely on the economic division of labor, but division of labor can be broadened to include other types of activities. Division of labor and therefore inequality can be found in families, in friendship groups, in schools, in politics, in churches—indeed, wherever there is a social organization. We might notice division of labor will exist in such places because a division will arise between those who lead and those who follow. As an organization becomes large or complex (differentiated in functions), someone arises to make sure that events go smoothly, takes care of day-to-day decision making, and guarantees that the organization works to achieve its goals. Indeed, such individuals are often necessary to represent the interests of the organization in relation to other organizations. Coordination of activities and successful achievement of goals normally mean having a leader or set of leaders. *Once leadership positions are created, a division of labor has been established.*

Why will this necessarily bring inequality? Robert Michels (1876–1936) explains this process and believes it is inevitable. Once a leader is chosen in any organization (and it does not seem to matter how), certain forces are set in motion that give the leader advantages over everyone else. In fact, leadership usually accrues to a small number of individuals whom Michels calls the *elite*. Positions of leadership give the elite great advantages: more information about the organization, the right to make decisions on a day-to-day basis, and control over what others in the organization know. Over time, these members of the elite separate themselves from the rest in the organization and purposely create ways to hold onto their positions. Those who are not leaders eventually become less and less capable of criticizing the leadership—and less and less willing. Whoever enters the higher positions in the future will enter positions that are inherently more powerful. Michels's pessimism concerning the possibility of equality in organization has been called the "iron law of oligarchy," which means that wherever organization exists, so will a few people have power in relation to everyone else. To say "organization" is to say the rule by a few, whether we call it a democracy or a dictatorship. In the end, this division of labor itself means inequality in power.

### *The Intentional Division of Labor and the Rise of Social Inequality*

Sometimes a division of labor is created on purpose, and a system of inequality exists from the beginning of organization. When one society conquers another and a new social structure is set up, almost always a structure purposely places the losing group into positions different from and lower than those who have won. Slavery and colonialism in the nineteenth century are two examples.

In fact, when people come together to create their own organization—a business, a club, a league of sports teams—a division of labor is purposefully created, and along with it comes an explicit and clear system of inequality. Friends, relatives, and loyal workers are rewarded, and privileges and power will be given to favored individuals—perhaps men rather than women, whites rather than nonwhites, and people we like, trust, or simply believe will help the organization. A new business will set up a board of directors with a president, a vice president, and so on. A new restaurant will open with a manager, an assistant manager, a headwaiter, and a chief dishwasher.

Max Weber described a tendency in modern society to create highly "bureaucratic" structures that were increasingly efficient, clearly ranked, and well managed. The positions created within these structures were formally laid out and listed both qualifications and responsibilities. Informal, traditional, emotional, and less rational inequality in an organization would be replaced as much as possible so that the organization would become effective and efficient. Although Weber knew that such bureaucratic organization would often fall short of what was intended, he believed people would find the goals of efficiency and a rational system of inequality important, even though the complexity of an organization increases considerably with its size. To Weber, this trend would exist in every modern society as well as in almost every organization within a modern society. A bureaucratic authority structure will exist whenever we enter an organization, and we will all be required to fill formal positions within that structure and know our responsibilities and power. In this sense, *social inequality will be created on purpose by people who want to organize and control other people in order to get work done in a highly rational way.*

Weber saw bureaucracy as a system of organization that would come to dominate the future, and, if we look around, we can see how right he was. Almost every large organization we enter is another example of bureaucracy. Weber's brilliant analysis of bureaucracy reminds us of some central points: (1) Bureaucracy is a system of organization set up on purpose to get things done as efficiently as possible. (2) Bureaucracy is a system of inequality created so that responsibilities and lines of authority are clearly distributed. Obedience to authority is a central value. (3) Bureaucracy is a form of organization whose inequality is regarded by actors as necessary and legitimate. It is a system of control carefully created to ensure that the commands of the few are carried out by the many. (4) Once formed, bureaucracy is almost impossible to dismantle because the division of labor becomes a necessary tool to achieve the goals of the organization, and because those at the top have the means to control other people in ways unimagined in other types of organization.

Almost every classical sociologist underlines the importance of the division of labor for the creation of inequality. An organization in which there are no leaders or a society in which everyone does the same tasks seems like an impossible dream (or nightmare). The trick is to create a division of labor that has equal positions, and this seems to be impossible.

### *Social Conflict, the Emergence of Winners and Losers, and the Rise of Social Inequality*

Thus far, we have focused on the division of labor as an explanation for the rise of social inequality. We discussed how the economic division of labor, the organizational division of leaders and followers, and the intentionally created division of labor all lead to inequality. In addition to the division of labor—economic, organizational, and intentionally created—there is another source: social conflict.

Conflict means the struggle by actors over something of value. Where there is struggle, some actors win and some lose, or, in most cases, some simply get more of what they want than others. Conflict occurs when there is scarcity: Not everyone can obtain what he or she wants because there is simply not enough to go around. Conflict also occurs when some people monopolize what is valued in society, and, as a result, others are denied. In either case, scarcity or unequal distribution is the result as some are able to increase their possessions and others cannot.

Victory in conflict is explained best by understanding the role of power: Those who win have more power than their opposition and are therefore able to win in the interaction. This might be personal power (based on intelligence, strength, attractiveness, guns, or wealth, for example). Often it is an organization, group, or society that enters into conflict and wins through superior power (better efficiency, more people, greater loyalty among members, better technology, better leaders, or more weapons or wealth, for example).

When there is conflict and some people begin to win, they are able to achieve their goals better than the opposition. They are normally able to build on their victory and increase their advantage over others, which in turn allows them to build up even greater power. Eventually, those who win are able to create a *system of inequality*, a social structure where they are at the top, and a culture and set of institutions that work to protect them. This system of inequality helps to ensure that they (as well as their group and their descendants) will continue in this advantaged position. Victory becomes institutionalized; that is, it becomes established in the way society operates. Those who win create a system that helps to guarantee their continued success. Thus, the Europeans come to America, conquer the Native American population, and establish a treaty and reservation system that guarantees continued ownership of the land and subservience by the Native American people. Two businesses compete for a market. One eventually moves ahead in that competition, and over time the one that is ahead attempts to protect that favored position through instituting a distribution system, a pricing system, and an advertising program that will continue its domination of the market. Normally, the result is a fairly permanent system of inequality between those businesses.

Building a permanent system of inequality is advantageous to the dominant group in several ways. First, it protects the advantages the group already has. Thus, a system of law and government, if it is heavily influenced by the rich and powerful, will help protect their property and their privileges against others in society. Second, it places the dominant group in a favorable competitive position for jobs, education, and housing. Third, it allows members of the elite to use those in lower positions as laborers, renters, and consumers, thus making life easier and increasing their wealth. We might truthfully maintain that the system of inequality *protects the favored position* of the powerful, *increases their competitive edge,* and allows them to *exploit* those who are in less powerful positions.

Imagine the world as a place of continuous social conflict: All individuals, groups, organizations, and societies struggle for whatever is valued. As some win, others lose, and over time a fairly permanent system of inequality emerges. Some will be rich, some poor; some will be powerful, some powerless. An upper class emerges, and a poor class develops. Men rather than women come to control the economic and political order, and all kinds of laws, ideas, customs, and institutions arise to continue that control. Whites dominate non-whites; Protestants dominate Catholics. In each case, a stable system of inequality favors the powerful. This happens in every organization and group. Individuals engage in conflict in newly formed groups, from juries to clubs to families. In each case, as people win, they establish ways to protect their interests. *In short, the victors in social conflict are normally able to ensure their favored position.*

The clearest examples are instances of extreme domination. Until the early 1990s, South Africa had been characterized by laws, customs, and ideas that openly declared the separation of the races and excluded blacks from equal participation in society and from equal protection by the law. Such was also the case with American slavery. To different degrees, Christian, Jewish, and Muslim societies have always had a strict code governing the actions of men and women, a code in which women have been systematically excluded from full participation in the political, educational, religious, and economic orders. When the Nazis came to power in Germany, they made it illegal for Jewish people to hold decent jobs, passed laws stripping Jews of citizenship, and took over Jewish property. They eventually established an organized system of resettling European Jews in concentration camps, where the vast majority was systematically murdered.

Donald Noel (1968) highlights the role of conflict in his theory of the origin of ethnic stratification. Three conditions are necessary. First, groups that have separate cultures and identities come together. Second, there is competition for a scarce resource, or there is an opportunity for the exploitation of one group by the other (both are examples of conflict). Third, one group has more power than the other and is able to exert itself successfully in the conflict.

As parties win in conflict, a system evolves that essentially perpetuates the resulting inequality:

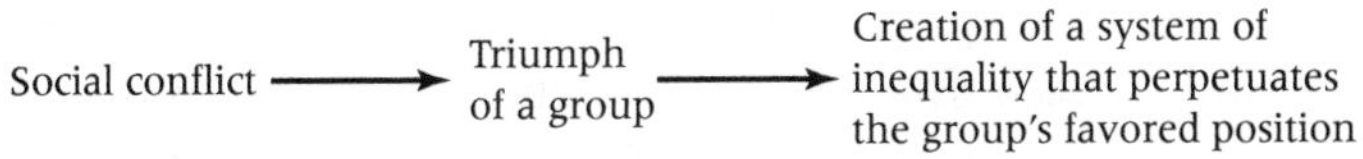

Social conflict → Triumph of a group → Creation of a system of inequality that perpetuates the group's favored position

Two explanations of social inequality have been presented here: the division of labor and social conflict. As an organization or society divides tasks among its people, some will become more powerful than others. As people engage in social conflict, some will win and some will lose, and over time a system of unequal power will be created. There is no suggestion that those who win are evil and selfish (sometimes they are and sometimes they are not), but most of us who succeed will be motivated to preserve the kind of world within which we were successful and will do things to protect that world.

### *The Social Institution of Private Property, the Unequal Distribution of Privilege, and the Rise of Social Inequality*

Power, luck, and ability bring accumulated wealth and other privileges. If it were possible to distribute valued items—privileges—equally, then conflict would be lessened considerably, and inequality would not become so institutionalized and permanent. However, as Marx reminds us, equal distribution of material goods is impossible when the institution of private property exists. Then people are able to own whatever they can get. People are no longer governed by satisfying one another's basic needs, but by increasing their privileges and their power in relation to others. Jean-Jacques Rousseau (1755) describes this most dramatically:

> The first man who, having enclosed a piece of ground, bethought himself of saying, "THIS IS MINE" and found people simple enough to believe him, was the real founder of civil society. From how many crimes, wars and murders, from how many horrors and misfortunes might not anyone have saved mankind, by pulling up the stakes, filling up the ditch, and crying to his fellows, "Beware that the fruits of the earth belong to us all, and the earth itself to nobody." (P. 207)

If everyone owned everything, if no one had more of a right to material things than anyone else, then inequality would be insignificant. In most societies, however, private property exists and is cherished. Some people, in conflict over material things, win and accumulate slaves or land or money. It becomes *theirs.* The more of such things some people acquire relative to others, the greater is the inequality between them. Weber calls such things *life chances.* They include all the benefits the actor receives because of his or her position in society or a social organization—for example, income, housing, office space, health care, and opportunities for education. *We become unequal, therefore, in both power (our ability to achieve our will) and privileges (the benefits we receive), and both inequalities arise from social conflict and private property.*

An inequality of privilege results also from the division of labor. Employers make more money than employees; doctors make more than nurses; rock stars make more than teachers. Why should this be? Obviously, those who gain powerful positions in the division of labor are in the best position to increase their privilege. Owners of factories have more power than others—and therefore more opportunity—to increase what they receive in society. Those who have less powerful positions can increase their privileges only through organizing and taking a bigger share from those above them.

Another reason the division of labor leads to the unequal distribution of privilege has something to do with market conditions: Positions are given different amounts of privilege in an organization because of a combination of (1) importance to that organization (the most important positions tend to get the most privileges); (2) the amount of training and sacrifice one must go through to prepare for those positions (the more the training and sacrifice, the greater the privileges); and (3) the scarcity of people for those positions (the fewer the people qualified and seeking the position, the greater the privileges). Conversely, any position that is not important, takes little training, and has many people competing for it will be paid much less. After all, positions that are essential must be filled; to fill them, people must be attracted; to attract them, privileges must be used. *Market conditions and social power combine to ensure that the division of labor creates a system of unequal privilege.*

### *The Interplay of Power, Privilege, and Prestige*

As noted, social conflict ultimately creates both unequal power and unequal privileges. Because of private property, individuals are able to perpetuate and expand on what they have accumulated. Thus, the division of labor, social conflict over what is valued, and the right to private property create a relatively permanent system of inequality. Over time, surgeons or rock stars or corporate executives become advantaged within the division of labor as well as in the social conflict and are able to earn more than laborers, nurses, and third violinists in the symphony orchestra. Over time, those in these positions are able to increase their advantage and hand it down to their children—and those who enter these positions in the future inherit the advantages. Those who become owners, employers, financiers, presidents, and corporate lawyers also come to be winners in the social conflict that takes place in society. Of course, those who are less fortunate in the conflict become the exploited, the unemployed, the working poor, the migrant workers, the unskilled laborers, and the blue-collar workers, assigned positions with little power and privilege.

The division of labor, the right to private property, and the resulting social conflict together create positions that have various amounts of both power and privilege. Power and privilege are linked: More power tends to bring even more privilege; more privilege brings even more power. Out of this a relatively permanent stratification system with high and low positions is created.

One additional point: Another advantage that arises from positions within social structure is *prestige,* the honor that comes to be associated with positions. Those positions that give people power and privilege usually give them prestige also. *Individuals are judged by others according to the positions they occupy.* Are they men or women? White or nonwhite? Professional or blue collar? Secretary to the president or secretary to the vice president? People accord high prestige to the top executive and dishonor to the homeless in society. Officers in the army have more prestige than instructors, and the rich more prestige than the poor. Differences in power, privilege, and prestige, work together to create an established system of inequality. All three arise from the division of labor, social conflict, and right of private property. Although sometimes power, privilege, and prestige are not linked, usually they are, so advantages in any one tend to bring advantages in the others.

### *Summary*

As people interact, an unequal structure tends to be created. This seems to arise from three sources: the division of labor, social conflict, and the institution of private property. Over time, advantaged positions increase their power, privilege, and prestige. Individuals who enter these positions are given the power, privilege, and prestige associated with these positions, and the children of those in these positions are advantaged.

Of course, most people do not come out of the division of labor, social conflict, and private property advantaged. Many are greatly disadvantaged; most are in between. But the process is the same. Those who lose in social conflict and in the division of labor are disadvantaged and have little accumulated wealth, power, and prestige. They hand down these disadvantages to those who take these positions, and their children will also be disadvantaged.

Social inequality arises from the development of a social structure. It is not human nature, supernatural forces, survival of the fittest, or nature that creates social structure, but the fact that in all social interaction certain important processes encourage a system of inequality we are calling social structure. The processes that create this inequality are the division of labor, social conflict, and the institution of private property. Once established, individuals are born into or enter into unequal positions. And if for some reasons we organize our own group that does not yet have positions, as a division of labor and social conflict will arise as well as privileges associated with victors, and then a new social structure will arise—and voilà!—a relatively permanent system of inequality.

## Why Does Inequality Continue?

Once a system of inequality has been established, it is difficult to alter. Of course, it changes slightly over time, but it tends to perpetuate itself. Five mechanisms seem to work to cause this stability:

1. Efforts of the powerful
2. Social institutions
3. Culture
4. Socialization
5. Instruments of force

Let us examine each in turn.

### *Efforts of the Powerful*

Marx and Michels (the "iron law of oligarchy") explain how the powerful protect the system of inequality. Those who gain high position have the resources to protect themselves. Those who are not favored within the system have fewer resources to protect themselves and little ability to change a system that keeps them low. In short, inequality is perpetuated through social power, and those who have power are those who benefit from the system of inequality.

Marx highlights this in his theory of society. When some people own the means of production, he argues, they will have great power. They will use this power to protect their positions and increase their wealth. Thus, once economic inequality is created, there will be a strong tendency for the rich to get richer and effectively protect the whole social structure as well as their positions in that structure.

Those who own the means of production are appropriately called the "ruling class" by Marx. Control over large businesses gives them control over people's jobs, the communities people live in, the products that are made, the economic decisions that affect the society, even the world. Control means that any decisions made will probably help the rich and powerful.

Marx goes much further, however. Control over the means of production—economic power—is translated into other types of power. Economic power influences government: the rules government goes by, the people who fill its positions, and the laws it makes. The ruling class influences media, the schools, the courts, and almost every other sector of society.

Why does this happen? Simply put, it is in everyone's interests to influence successfully the direction of society. I want to—so do you. But I will try to influence it differently from you. For example, you will try to lower tuition, but I will try to raise faculty salaries. Feminists, African Americans, lawyers, unions, ministers—to name but a few—all have their own agendas, and all would like to see their needs met. The rich and powerful have interests, too; but the difference between them and everyone else is that they have greater resources to use in ensuring that their needs are, in fact, met. Thus, although the society never works completely in their interests, the tendency is for it to work in ways that are more consistent with what the powerful want.

Remember: The question we are considering is, how is inequality perpetuated over time? How is the system able to continue? We have one answer here: Inequality is perpetuated because those who are wealthy and powerful are in the best position to ensure that their interests—the protection of their wealth and power—are met throughout society.

Michels agrees with Marx in that he, too, argues that those who have power will protect the system of inequality. Michels emphasizes the political side of the coin rather than the economic. He is not interested in ownership of the means of production but in people who lead society. He simply makes the point that leaders in any organization will over time become increasingly separated from everyone else in the organization. They will have a stake in keeping the positions and in continuing their policies. They will try to ensure that their positions remain theirs, and that the system of inequality is maintained in their favor. Although they might be elected democratically, once they are in their positions, there is a strong tendency for them to regard their positions as "theirs," and they will tend to institute policies and pursue goals that are consistent with the belief. Leaders eventually unite and form a self-supporting elite that distinguishes them from everyone else. Others in the organization tend to trust them, and they, in turn, become more and more interested in perpetuating the system that favors their positions.

To some extent, Marx and Michels agree: Those who have power (economic or political) develop interests different from everyone else's (to maintain the inequality that exists), and they are in the best position to influence society to work in these interests. We must therefore begin to understand the perpetuation of inequality by recognizing that it is in the interests of the most powerful to do what they can to maintain the system that favors them, and their power gives them the ability to do so.

### *Prevailing Social Institutions*

The result of this control by the ruling class is the creation of *institutions,* the ongoing and legitimate ways of doing things in society. Institutions, as we saw in Chapter 3, are the established procedures that help to ensure the continuation of society and thus the established system of inequality.

Over time, for example, political institutions are created in society to make laws, carry out those laws, and interpret those laws. The United States has separate legislative, executive, and judicial branches of government to do this. Other societies, including England (which is at least as democratic as the United States), do not separate these powers. The political system in the United States is characterized by a two-party system, an electoral college, federalism, separation of powers, and civilian control over the military. These are our political institutions, our ways of dealing with political matters. We also have economic institutions such as multinational corporations, a federal reserve system, a stock market, private property, and private enterprise. We also have certain educational, religious, health care, military, kinship, and entertainment institutions. The United States as a society works—or does not work—because of its institutions. One primary reason the communist system of the former Soviet Union failed was because its institutions could not solve the problems it faced in the early 1990s. It did not work.

It is important to realize that institutions generally work for the society as it is. If they seem to work, they continue; if they seem to work in the interests of the powerful, they are especially encouraged. *Once a society develops a system of inequality, the prevailing institutions tend to work in such a way that the inequality is maintained or even increased.* It is easier to see this process in other societies than in our own. Saudi Arabia is a society where almost everything that exists works to maintain the wealth and power of a few families and the dominance of men over women. That is the way the government, the economy, the religion, the military, and the family work in that society. Apartheid in South Africa, the forced separation of the races for purposes of domination by the whites, was maintained through a complex set of institutions. In China, government, military, education, and media combine to help ensure the continued dictatorship of a small party.

Unless equality is a value that a society truly pursues, the institutions will normally protect and expand inequality. Poverty continues in the United States because institutions are not truly set up to deal with this problem. Our tax system does not substantially redistribute wealth, it does little to effectively limit the wealth that one can achieve, and in the past two decades it has actually contributed to greater inequality. Our schools, government, welfare system, and economic system may be wonderful in some ways, but they tend to protect the system of inequality that prevails and keep people in the class positions of their birth. Institutions maintain our segregated society, they generally support the inequality between men and women, and they protect the power and privileges of a political and economic elite. This is why sociologists tend to see the perpetuation of inequality built into society itself. It is rare to see a society whose real purpose is to maintain a system of equality.

Recall the warning of the philosophers of the Enlightenment: Unless a society really makes efforts to create and maintain equality, the tendency will be toward a state of inequality. It is easy to see why this is true when we recognize the tendency for social institutions to protect and expand social inequality.

*Thus, the second reason why inequality becomes perpetuated in society is that institutions generally work in that direction, partly because the powerful have the greatest impact on the nature of societal institutions. Those who are advantaged will also tend to believe in institutions that helped their own success, so it is easy to understand that they will defend institutions that have favored their own success.*

### *Culture: The Acceptance of Inequality*

Almost all institutions teach and reinforce the culture of society. *Culture becomes an important conservative force in society.* Over time, culture influences people to accept a system of inequality as natural.

Many will come to accept the inequality that exists, not because they necessarily like their situation but because this is what they were born into and have become influenced to believe it is the way the world must be. Always there is rebellion by some, but socialization by parents, political leaders, religious leaders, media leaders, and teachers is a powerful tool: we are influenced to take on *their* language, *their* rules, *their* values, and *their* expectations. It is always difficult to know how effective socialization really is because what looks to the outsider to be acceptance may well be hidden anger and rejection of what culture teaches. And there will always be some critics, social movements, and revolutionaries who act out against inequality, even where socialization may be successful for the vast majority. Many of the most emotional discussions I have had with others have been over the success of socialization: Have women been socialized to accept a subordinate position in a given society or are they simply fearful and quietly angry? How about African Americans in our own society? How about students? workers? How about China? Cuba? Mexico? The answer is always a complex one, and it probably depends on the time period we are looking at, but it is a mistake to ignore the effectiveness of culture in getting most people in a society to accept the system of inequality.

Culture provides justifications for inequality. In the United States we are taught that people will be justly rewarded for hard work:

"If you work hard, you can rise to the top." We tend to believe that the system of inequality is somehow just and democratic, rewarding those who ought to be rewarded. We are taught and we come to believe that some people have a moral right to keep all the marbles they can get their hands on, even though others must survive without any marbles at all. We are taught that if one does not make it, it is his or her own fault, and we are taught that all people in society can be anything they want to be. Peter Berger points out that most societies actually develop two ideologies that serve to protect inequality. One legitimates the position of the upper classes, usually arguing that these people are somehow superior or more deserving. (For example, they are more talented, more hardworking, more educated, more naturally superior.) The other ideology tries to explain and justify poverty. (For example, poverty is a consequence of sin, laziness, or irresponsibility, and good behavior by the poor will eventually be rewarded in the afterlife or next life.) The people living around me believe that they deserve the affluence they enjoy; I quietly believe that we are fortunate. I wonder, who really "deserves" what? I believe that culture is largely an attempt to assure all of us that what we get out of life we actually deserve.

In European societies, inequality was justified for centuries through arguing that it was God's plan. God chose the rulers and also favored an upper class whose purpose was to lead the masses. Although revolution eventually destroyed this idea, it held on for a long time. Indeed, in much of the world today, inequality is still seen to be God's will; in most cases, people are taught to spend their time and energy doing other things besides trying to change society to make it more equal.

For a long time, people in the United States denied the existence of either a rich upper class or a class of poor people: "We are the land of equal opportunity for all." To believe this is to deny the effects of inequality, and such an idea works well to protect the inequality that actually exists. It is difficult to identify exactly what American culture consists of, but list the most basic ideas, values, and morals that we believe in and you will see that they work to uphold the system of inequality that prevails: "The poor do not really want to work [and thus deserve their fate]." "Capitalism with little government regulation and taxes is the most just and efficient economic system." "People have a right to make what they can, keep all that they make, and pass down to their children all that they keep." "We may be a class society, but everyone has the chance to strike it rich." "People are naturally competitive, selfish, and lazy. We all work to beat the other guy, we all take whatever we can get, and we have to be paid well if we're going to work."

Or consider the ideas that accompany other kinds of inequality: "Women are not naturally capable of competing in the political and economic world." "God did not mean for women to work outside the home." "Women have a moral obligation to obey their husbands." "Blacks are naturally inferior to whites." "God meant for blacks to obey whites." All of these ideas—and many more—have existed in our own society and have worked to retain systems of inequality. Similar ideas exist in other societies. For centuries, in India people were thought to be born into castes in which they are expected to stay their entire lives. How can such a system be justified? The answer is that people come to believe that their position in the next life depends on how well they accept their position in this life. Caste is seen as a test; acceptance of position becomes a moral virtue. And when people become too sophisticated to believe such ideas, new ideas arise to justify inequality: "Women are naturally different from men and will do better than men in some things, such as rearing children, and not in others, such as mathematics." "I believe that blacks should be equal to whites, but the fact that they are not is their own fault." "It is morally right for us to dominate nonwhites in society because of their cultural inferiority—they just do not accept the right values and ideas, and thus deserve their positions."

In *Oppression,* Turner, Singleton, and Musick carefully show how the dominant beliefs in the United States concerning African Americans have changed as the relationship between blacks and whites has changed (1984: 170–176). We always work out a new ideology that justifies inequality. Before 1820, whites described blacks as uncivilized heathens, the curse of God, and ill-suited for freedom. After 1820 and before the Civil War, slavery was justified as good for both blacks and whites, an institution that civilized and protected African Americans. After the Civil War and before World War I, while all areas of life became segregated, blacks were described as inherently inferior, and thus segregation became necessary for the protection of whites. From World War I to 1941, as African Americans moved north, black inferiority became a "scientific fact," and segregation was described as natural, distinctive, and desired by both races. After World War II, as discrimination and segregation were increasingly attacked, ideas favoring inequality were fewer. After 1968, however, inequality again became justified: "There is black inferiority, and it is due to the African Americans themselves, especially their lack of motivation." Many Americans have recently come to believe that "we have moved too fast toward equal rights and that society has done enough."

Culture is also important in what it does not teach. The early and mid-1990s saw the end of Soviet domination of Eastern Europe and the rapid dismantling of communist institutions within the Soviet Union. The U.S. media responded by paying tribute to certain American institutions and values, especially private property, a free market, and freedom to own one's own business. Somewhere along the line these values have come to dominate our culture more than equal opportunity, social justice, respect for the individual, and cultural pluralism. It is not that we no longer believe in the latter; it is simply that the former values have come to dominate our thinking. By ignoring equality and social justice, culture tends to support inequality and lack of social justice. It does not encourage people to question the extent to which these inequalities exist in society.

Even in organizations we see ideas that serve to justify inequality there: "The leader knows what is going on; the rest of us just don't have enough information to make intelligent decisions." "Democracy, where everyone votes, is inefficient." "The real world isn't equal. Why should this organization practice equality? We have to run this place like a business, don't we?" "You want justice? Life is simply not just."

It should come as no surprise that those at the top try to ensure that their ideas, values, and rules prevail in society. Think of the society giving rise to many ideas. Which ones are believed, and which ones are rejected? This question is not answered simply, but be aware that ideas, values, and rules must have sponsors, groups that push for their acceptance. Those groups with the most power will have the best chance for their ideas to be accepted. To some extent Marx is correct: The rich not only produce the goods of society, but also to a great extent produce (and spread) the society's culture. And, of course, when ideas, values, and rules are put forth that are not consistent with those of the elite, they will be opposed and have less chance for acceptance.

*Inequality prevails in society, then, because it is supported by culture; and culture, in turn, most reflects the ideas, values, and norms of the most powerful in society.*

### *Socialization: The Acceptance of Place*

Besides the fact that people are socialized to accept the system of inequality itself, they are also socialized to accept their own position. This is a complex process. We learn who we are early in life. Our neighbors, parents, and teachers tell us in overt and covert ways our ranks in society and what we have a right to expect from life: "People like us don't do those things." "Marry your own kind." "Go to Harvard." "Be satisfied with the college in your own town." We are taught what to expect from life if we work hard (generally slightly above our present rank, whatever that may be), but rarely do we expect great things without a realistic model to go by. Wealthy business executives socialize their children to expect wealth, fame, and power. Lawyers socialize their children to expect a professional position. Of course, we are not simply the result of what our parents expect, but they, together with teachers (who teach in class-based schools) and friends (who tend to come from our class-based neighborhoods), show us where we are in society and teach us to expect approximately that level.

In *Schooling and Capitalist America,* Bowles and Gintis (1976) highlight how legitimating inequality and teaching people their positions is an integral part of our educational system. Schools sort students into academic tracks, which then distribute students into the occupational system and ultimately into the economic system. Schools generally teach discipline, hierarchy, and obedience, and they teach students to expect little independent control over their work. Working-class students learn obedience; upper-middle-class students learn leadership and innovation.

Or witness, too, women and minorities, who are socialized to accept subordinate positions in most societies. Such socialization is usually successful, but not always. Inevitably, some refuse their position, some figure out ways to make great leaps to overcome their subordinate position, and some succeed. Most who try do not succeed, not because of lack of effort or intelligence alone, but because real opportunity is denied by factors related to class and minority positions. And for those who try and do not succeed, it becomes increasingly difficult to hold onto aspirations that seem to be out of reach. Here is one of the most important ways we are all socialized: We are taught to change our sights, and this means accepting the position that most people like us are expected to stay in.

Legitimate authority is one of the concepts that Weber introduces to sociology. Why is it that we willingly obey others over us? he asked. His answer is that most of us *believe* that they have a *right to command* us. Socialization induces people to feel a part of a community, and to *feel an obligation* to obey the people who represent that community. We come to believe that we must accept the prevailing system of inequality as right if we are to exist as a community. A stable system of inequality is built into the community's tradition or law that loyal people feel an obligation to follow. Our system of inequality appears to most citizens to be *legitimate,* and most feel a moral obligation to obey those above them.

*Thus, the system prevails through the collusion of the individuals who are socialized into society. Socialization brings the acceptance of a culture that justifies inequality, and it normally brings an acceptance of one's relative position in the system of inequality.*

### *Society's Instruments of Force*

Of course, some individuals refuse to accept their place in society and try to improve their position in any way they can, including going outside the law. Often, they realize that the system works against them and that to make it they cannot try the normal channels. Through their acts, they threaten the legitimacy of the prevailing order. Procedures are instituted to discover, control, and punish such individuals. Police, courts, and prisons work to protect more than people; they also protect the system of inequality.

Some groups refuse to accept the system itself and organize to overthrow it. In our society, such groups have some leeway: They usually have the right to say what they want and to write what they want. When they act outside the law to alter the system, we call them "revolutionaries," and we use force to stop them. All societies draw lines, and all try to control groups with force if they threaten the established order, including the prevailing system of inequality.

Although crime and revolution upset order and create hardship for all people in society, those at the top of the system of inequality have the most to lose. It is their favored status that is most at risk.

They have the most property to lose. The advantages they enjoy are threatened. It is vital to them that what they worked for or inherited be protected. They therefore take an active interest in politics, law, and law enforcement. Marx goes so far as to say that the state's purpose is to protect the ruling class. At the very least, we can see that legitimate instruments of force are important ways in which those at the top are able to protect themselves and maintain the system of inequality.

Every organization establishes instruments of social control that protect the structure. Colleges use grades, threats of suspension, and, sometimes, refusals to cooperate with threatening students. A business can fire, demote, or refuse to promote. A family can send the offender to his or her room or "ground" the individual for two weeks. An informal group can simply let the member know that he or she is not liked and may not invite the individual to future activities. Of course, there is some individuality and some freedom, but threats to the prevailing system of inequality and one's position in it are almost always reacted to.

*Violence and threats may be the last resort, but they always exist in society to protect the system and keep individuals in place. This is the fifth way inequality is protected and perpetuated.*

## Summary and Conclusion

To understand why inequality exists, it is important to consider how inequality arises in the first place and then consider how it is protected and perpetuated.

Inequality arises from the division of labor, social conflict, and the institution of private property. The *division of labor* allows some to create favored positions in society and leads to a ranking system. *Social conflict* over valued things leads to winners and losers, allowing the winners to establish and protect the positions they are able to win in society. The *institution of private property creates and encourages unequal privilege* and allows those who capture high positions through the division of labor or social conflict (or both) to accumulate wealth and be able to protect their positions, hand down what they accumulate to their children, and achieve power, privilege, and prestige in their positions. Those who eventually leave their positions through rising up, losing what they have, or dying are replaced by others who then inherit whatever power, privilege, and prestige exist in those positions. Of course, these processes are relevant not only to the high positions in organized life but also to every position from the lowest to the highest—except for those who have little to protect because their positions are so low.

Inequality continues over time for many reasons. *The efforts of those at the top of the society*—in the economy, in government, in education, in the criminal justice system, in religion, in education, and in the media—help protect and perpetuate it. All organizations besides society will also have some at the top, and they also will do what they can to protect and perpetuate their positions. The *social institutions,* the basic ways in which things are done, operate to uphold the existing inequality. The *socialization of people* into a culture that justifies inequality is an important factor, as is the successful socialization of people to learn and accept their positions. Finally, *instruments of force* are sometimes used to ensure that the social structure continues intact.

Some inequality is probably inevitable. People must work hard to prevent its emergence; once it emerges, they must work even harder to control it. Robert Michels argues that organizations with leaders will inevitably develop a system of inequality and ultimately that system will be quite difficult to eradicate. Although Marx believed that with the destruction of capitalism equality would prevail in society, the twentieth century did not prove him right. Indeed, the situation seems far more complicated. In the Soviet Union, China, or Cuba, nations where private property was abolished, there still arose a stable system of inequality, perhaps not based on ownership of property as much as on political leadership, occupation, and control over (rather than ownership of) property.

To claim that inequality is inevitable does not mean that people should also claim that poverty and hardship must be accepted or that tyranny must be tolerated. The question for all human beings should be, *How much inequality* is to be tolerated in society or in an organization? *How much* inequality is necessary? beneficial? democratic? humane? moral?

To realize that inequality is inevitable also means that those people who are dedicated to principles of equality have a difficult task ahead, because so much in society seems to encourage and protect a system of inequality. In this respect, equality is like freedom: Far from being automatic, it is possible only with eternal vigilance.

## Questions to Consider

1. Why is social equality so difficult to establish and maintain?
2. Exactly why does the division of labor encourage social inequality?

3. What are the advantages that leaders in any organization or society have that allow them to keep their positions?

4. What ideas in American culture can be used to defend the systems of inequality that exist in American society?

5. How would a wealthy individual answer the question: Why are people unequal in society? How would a person in poverty answer the same question?

## References

The following works examine various forms of social inequality, focusing especially on class, race, and gender. All are good introductions to the general questions of why inequality arises and how it is perpetuated. Some look at inequality in all societies; some concentrate on the United States.

**Adam, Barry B.** 1978. *The Survival of Domination: Inferiorization and Everyday Life*. New York: Elsevier.

**Ancheta, Angelo N.** 1998. *Race, Rights, and the Asian American Experience*. New Brunswick, NJ: Rutgers University Press.

**Anderson, Eric.** 2005. *In the Game: Athletes and the Cult of Masculinity*. New York: State University of New York Press.

**Anderson, Elijah.** 1990. *Streetwise: Race, Class, and Change in an Urban Community*. Chicago: University of Chicago Press.

———. 1999. *Code of the Street: Decency, Violence, and the Moral Life of the Inner City*. New York: W.W. Norton.

**Aronowitz, Stanley.** 2001. *The Last Good Job in America: Work and Education in the New Global Technoculture*. Lanham, MD: Rowman & Littlefield.

———. 2003. *How Class Works: Power and Social Movement*. New Haven, CT: Yale University Press.

———. 2005. *Just around the Corner: The Paradox of the Jobless Recovery*. Philadelphia: Temple University Press.

**Barnes, Sandra L.** 2005. *The Cost of Being Poor: A Comparative Study of Life in Poor Urban Neighborhoods in Gary Indiana*. New York: State University of New York Press.

**Borchard, Kurt.** 2005. *The Word on the Street: Homeless Men in Las Vegas*. Reno, NV: University of Nevada Press.

**Baldwin, James.** 1963. *The Fire Next Time*. New York: Dial.

**Ballantine, Jeanne H.** 1997. *The Sociology of Education*. 4th ed. Englewood Cliffs, NJ: Prentice Hall.

**Baltzell, E. Digby.** 1964. *The Protestant Establishment: Aristocracy and Caste in America.* New York: Vintage.

**Banes, Colin, and Geof Mercer**. 2003. *Disability.* Malden, MA: Blackwell.

**Beeghley, Leonard.** 1983. *Living Poorly in America.* New York: Praeger.

———. 1995. *The Structure of Social Stratification in the United States.* 2nd ed. Boston: Allyn & Bacon.

———. 1996. *What Does Your Wife Do? Gender and the Transformation of Family Life.* Boulder, CO: Westview.

**Bellah, Robert N.** 1999. "The Ethics of Polarization in the United States and the World." *The Good Citizen,* edited by David Batstone and Eduardo Mendieta. New York: Routledge/Taylor & Francis.

**Berger, Peter L., and Thomas Luckmann**. 1966. *The Social Construction of Reality.* Garden City, NY: Doubleday.

**Bernard, Jessie.** 1987. *The Female World from a Global Perspective.* Bloomington: Indiana University Press.

**Blau, Peter M., and Marshall W. Meyer**. 1987. *Bureaucracy in Modern Society.* 3rd ed. New York: Random House.

**Blauner, Robert.** 1972. *Racial Oppression in America.* New York: Harper & Row.

**Blumberg, Rae Lesser.** 1991. *Gender, Family, and Economy: The Triple Overlap.* Thousand Oaks, CA: Sage.

**Boggs, Carl.** 2000. *The End of Politics: Corporate Power and the Decline of the Public Sphere.* New York: Guilford.

**Bottomore, T. B.** 1966. *Classes in Modern Society.* New York: Pantheon.

**Bowles, Samuel, and Herbert Gintis**. 1976. *Schooling and Capitalist America.* New York: Basic Books.

**Bowles, Samuel, Herbert Gintis, and Melissa Osborne-Groves**. 2005. *Unequal Chances: Family Background and Economic Success.* Brunswick, NJ: Princeton University Press.

**Bowser, Benjamin P.** 2006. *The Rise and Fall of Class in Britain.* New York: Columbia University Press.

**Brecher, Jeremy, and Tim Costello**. 1994. *Global Village or Global Pillage: Economic Reconstruction from the Bottom Up.* Cambridge, MA: South End.

**Carmichael, Stokely, and Charles V. Hamilton**. 1967. *Black Power.* New York: Random House.

**Cashmore, E. Ellis.** 1987. *The Logic of Racism.* London: Allen & Unwin.

**Chafetz, Janet Saltzman.** 1990. *Gender Equity: An Integrated Theory of Stability and Change.* Newbury Park, CA: Sage.

**Chambliss, William J.** 1973. "The Saints and the Roughnecks." *Society* 11: 24–31.

**Charon, Joel M.** 2009. "An Introduction to the Study of Social Problems," pp. 1–12 in *Social Problems: Readings with Four Questions,* edited by Joel Charon and Lee Vigilant. 3rd ed. Belmont, CA: Wadsworth Cengage Learning.

**Charon, Joel M. and Lee Garth Vigilant**. 2009. *Social Problems: Readings with Four Questions,* 3rd ed. Belmont, CA: Wadsworth Cengage Learning.

**Chin, Margaret M.** 2005. *Sewing Women: Immigrants and the New York Government Industry.* New York: Columbia University Press.

**Cohen, Mark Nathan.** 1998. *Culture of Intolerance: Chauvinism, Class, and Racism in the United States.* New Haven, CT: Yale University Press.

**Collins, Randall.** 1979. *The Credential Society: An Historical Sociology of Education and Stratification.* New York: Academic.

**Cookson, P. W., and C. H. Persell**. 1985. *Preparing for Power: America's Elite Boarding Schools.* New York: Basic Books.

**Coontz, Stephanie.** 2005. *Marriage, a History: From Obedience to Intimacy or How Love Conquered Marriage.* New York, NY: Viking.

**Crompton, Rosemary.** 1993. *Class and Stratification: An Introduction to Current Debates.* Cambridge, MA: Polity.

**Currie, Elliott.** 1998. *Crime and Punishment in America.* New York: Metropolitan.

———. 2005. *The Road to Whatever: Middle-Class Culture and the Crisis of Adolescence.* New York: Holt.

**Dahrendorf, Ralf.** 1959. *Class and Class Conflict in Industrial Society.* Stanford, CA: Stanford University Press.

**Dalphin, John.** 1987. *The Persistence of Social Inequality in America.* 2nd ed. Cambridge, MA: Schenkman.

**Danzinger, Sheldon, and Ann Chih Lin**, eds. 2000. *Coping with Poverty: The Social Contexts of Neighborhood, Work, and Family in the African American Community.* Ann Arbor: University of Michigan Press.

**Davey, Joseph Dillon.** 1995. *The New Social Contract: America's Journey from Welfare State to Police State.* Westport, CT: Praeger.

**Della Fave, L. Richard.** 1980. "The Meek Shall Not Inherit the Earth: Self-Evaluation and the Legitimacy of Stratification." *American Sociological Review* 45: 955–971.

**DePrie, Jason.** 2004. *American Dream: Three Women, Ten Kids, and a Nation's Drive to End Welfare.* New York: Penguin.

**Dobratz, Betty A., Lisa K. Waldner, and Timothy Buzzell**, eds. 2001. *The Politics of Social Inequality.* New York: JAI.

**Dohan, Daniel.** 2003. *The Price of Poverty: Money, Work, and Culture in the Next Mexican American Barrio.* Berkeley, CA: University of California Press.

**Domhoff, G. William.** 1998. *Who Rules America?* 3rd ed. Mountain View, CA: Mayfield.

**Dougherty, Charles J.** 1996. *Back to Reform: Values, Markets and the Healthcare System.* New York: Oxford University Press.

**Durkheim, Émile.** 1893 [1964]. *The Division of Labor in Society.* Translated by George Simpson. New York: Free Press.

**Dye, Thomas R.** 1994. *Who's Running America?* 6th ed. Englewood Cliffs, NJ: Prentice Hall.

**Edin, Kathryn, and Maria Kefalas**. 2007. *Promises I Can Keep: Why Poor Women Put Motherhood before Marriage.* Berkeley: University of California Press.

**Ehrenreich, Barbara.** 2006. *Bait and Switch: The (Futile) Pursuit of the American Dream.* New York: Metropolitan Books

**Eliade, Mircea.** 1954. *Cosmos and History.* New York: Harper & Row.

**Engels, Friedrich**. [1884] 1972. "The Origin of the Family, Private Property, and the State," in *Karl Marx: On Society and Social Change,* edited by Neil J. Smelser. Chicago: University of Chicago Press.

**Ewen, Stuart.** 1976. *Captains of Consciousness.* New York: McGraw-Hill.

**Ezekiel, Raphael S.** 1995. *The Racist Mind.* New York: Viking.

**Faludi, Susan.** 1999. *Stiffed: The Betrayal of the American Man.* New York: Perennial.

**Farley, John E.** 1999. *Majority-Minority Relations.* 4th ed. Upper Saddle River, NJ: Prentice Hall.

**Farley, John E., and Gregory D. Squires**. 2005. "Fences and Neighbors: Segregation in 21st-Century America." *Contexts* 4(1): 33–39.

**Feagin, Joe R., and Melvin P. Sikes**. 1994. *Living with Racism: The Black Middle Class Experience.* Boston: Beacon.

**Feagin, Joe R., and Hernan Vera**. 1995. *White Racism: The Basics.* New York: Routledge.

**Feiner, Susan F.**, ed. 1994. *Race and Gender in the American Economy: Views from across the Spectrum.* Englewood Cliffs, NJ: Prentice Hall.

**Freeman, Jo.** 1995. *Women: A Feminist Perspective.* 5th ed. Palo Alto, CA: Mayfield.

**Friedan, Betty.** 1963. *The Feminine Mystique.* New York: Norton.

**Friedman, Thomas L.** 2005. *The World Is Flat: A Brief History of the Twenty-First Century.* New York: Farrar, Straus & Giroux.

**Fuller, Robert W.** 2003. *Somebodies and Nobodies: Overcoming the Abuse of Rank.* Gabriola Island, BC: New Society.

**Galbraith, James K.** 1998. *Created Unequal: The Crisis in American Pay.* New York: Free Press.

**Galbraith, John Kenneth.** 1979. *The New Industrial State.* 3rd ed. New York: New American Library.

**Gans, Herbert J.** 1995. *The War against the Poor: The Underclass and Antipoverty Policy.* New York: Basic Books.

**Gilbert, Dennis, and Joseph A. Kahl.** 1997. *The American Class Structure in an Age of Growing Inequality: A New Synthesis.* 5th ed. Belmont, CA: Wadsworth.

**Glassner, Barry, and Rosanna Hertz.** 2003. *Our Studies, Ourselves: Sociologists' Lives and Work.* New York: Oxford University Press.

**Goode, Erich.** 2002. *Deviance in Everyday Life: Personal Accounts of Unconventional Lives.* Prospect Heights, IL: Waveland.

**Guillaumin, Colette.** 1995. *Racism, Sexism, Power and Ideology.* New York: Routledge.

**Hacker, Andrew.** 1997. *Money: Who Has How Much and Why.* New York: Scribner.

**Hare, Bruce R.** (Ed.). 2002. *2001 Race Odyssey: African Americans and Sociology.* Syracuse, NY: Syracuse University Press.

**Harris, Scott R.** 2006. *The Meaning of Marital Equality.* New York: University of New York Press.

**Hays, Sharon.** 2003. *Flat Broke with Children: Women in the Age of Welfare Reform.* New York: Oxford University Press.

**Healey, Joseph F.** 1997. *Race, Ethnicity, and Gender in the United States: Inequality, Group Conflict, and Power.* Thousand Oaks, CA: Pine Forge.

**Henslin, James M.** 2000. *Social Problems.* 5th ed. Upper SaddleRiver, NJ: Prentice Hall.

**Higley, Richard.** 1995. *Privilege, Power, and Place.* Lanham, MD: Rowman & Littlefield.

**Hill, Herbert, and James E. Jones, Jr.,** eds. 1993. *Race in America: The Struggle for Equality.* Madison: University of Wisconsin Press.

**Hochschild, Arlie Russell, and Anne Machung.** 2003. *The Second Shift.* New York: Viking Penguin.

**Hochschild, Jennifer L.** 1995. *Facing Up to the American Dream: Race, Class, and the Soul of the Nation.* Princeton, NJ: Princeton University Press.

**Howard, Judith A., and Jocelyn A. Hollander.** 1996. *Gendered Situations, Gendered Selves: A Gender Lens on Social Psychology.* Thousand Oaks, CA: Sage.

**Iceland, John.** 2003. *Poverty in America: A Handbook.* Berkeley: University of California Press.

**Johnson, Allan G.** 1997. *The Forest and the Trees: Sociology as Life, Practice, and Promise.* Philadelphia: Temple University Press.

**Johnson, Heather B.** 2006. *The American Dream and the Power of Wealth: Choosing Schools and Inheriting Inequality in the Land of Opportunity.* New York: Routledge.

**Kanter, Rosabeth.** 1977. *Men and Women of the Corporation.* New York: Basic Books.

**Keller, Suzanne.** 1963. *Beyond the Ruling Class: Strategic Elites in Modern Society.* New York: Random House.

———. 2001. *Community: Pursuing the Dream, Living the Reality*. Princeton, NJ: Princeton University Press.

**Keister, Lisa A.** 2005. *Getting Rich: America's New Rich and How They Get That Way*. New York: Cambridge University Press.

**Kennickell, Arthur B.** 2006. "Currents and Undercurrents: Changes in the Distribution of Wealth, 1989–2004," Washingon, DC: Federal Reserve Board, August 2, http://www.federalreserve.gov/pubs/oss2/scfindex.html

**Kerbo, Harold R.** 1999. *Social Stratification and Inequality*. 4th ed. New York: McGraw-Hill.

**Kerbo, Harold R., and John A. McKinstry**. 1995. *Who Rules Japan: The Inner Circles of Economic and Political Power*. Westport, CT: Praeger.

**Kinkead, Gwen.** 1992. *Chinatown: A Portrait of a Closed Society*. New York: HarperCollins.

**Kitano, Harry H. L.** 1991. *Race Relations*. 4th ed. Englewood Cliffs, NJ: Prentice Hall.

**Koonigs, Kees, and Dirk Druijt,** eds. 1999. *The Legacy of Civil War, Violence, and Terror in Latin America*. London: Zed.

**Lareau, Annette.** 2003. *Unequal Childhoods: Class, Race, and Family Life*. Berkeley, CA: University of California Press.

**Lenski, Gerhard E.** 1966. *Power and Privilege: A Theory of Social Stratification*. New York: McGraw-Hill.

**Lerner, Robert, Althea K. Nagal, and Stanley Rothman**. 1996. *American Elites*. New Haven, CT: Yale University Press.

**Lewellen, Ted C.** 1995. *Dependency and Development: An Introduction to the Third World*. Westport, CT: Bergin & Garvey.

**Lewis, Bernard.** 2002. *What Went Wrong? Western Impact and Middle Eastern Response*. New York: Oxford University Press.

**Liebow, Elliot.** 1967. *Tally's Corner*. Boston: Little, Brown.

**Lipset, Seymour Martin, Martin Trow, and James Coleman.** 1956. *Union Democracy: The Inside Politics of the International Typographical Union*. New York: Free Press.

**Lorber, Judith.** 2005. *Breaking the Bowls: Degendering and Feminist Change*. New York: Norton.

**Lukes, Steven.** 2005. *Power: A Radical View*. 2nd ed. Oxford, UK: Palgrave Macmillan.

**Malcolm X and Alex Haley**. 1965. *The Autobiography of Malcolm X*. New York: Grove Press.

**Mann, Coramae Richey**. 1994. *Unequal Justice: A Question of Color*. Bloomington: Indiana University Press.

**Marcuse, Herbert.** 1964. *One-Dimensional Man*. Boston: Beacon.

**Marger, Martin N.** 1987. *Elites and Masses*. 2nd ed. New York: VanNostrand Reinhold.

———. 1999. *Race and Ethnic Relations: American and Global Perspectives.* 5th ed. Belmont, CA: Wadsworth.

**Marx, Karl, and Friedrich Engels**. [1848] 1955. *The Communist Manifesto.* New York: Appleton-Century-Crofts.

**McCall, Nathan.** 1994. *Makes Me Wanna Holler.* New York: Random House.

**McCord, Joan**, ed. 1997. *Violence and Childhood in the Inner City.* New York: Cambridge University Press.

**McNamee, Stephen J., and Robert K. Miller, Jr.** 2004. *The Meritocracy Myth.* Landham, MD: Rowman & Littlefield.

**Merry, Sally E.** 2005. *Human Rights and Gender Violence: Translating International Law into Local Justice.* Chicago: University of Chicago Press.

**Michels, Robert.** 1915 [1962]. *Political Parties.* Edited and translated by Eden Paul and Cedar Paul. New York: Free Press.

**Mills, C. Wright.** 1956. *The Power Elite.* New York: Oxford University Press.

**Mintz, Beth, and Michael Schwartz**. 1985. *The Power Structure of American Business.* Chicago: University of Chicago Press.

**Myrdal, Gunnar.** 1944. *An American Dilemma.* New York: Harper & Row.

**Nash, Kate.** 2000. *Contemporary Political Sociology: Globalization, Politics, and Power.* Malden, MA: Blackwell.

***New York Times.*** 2005. *Class Matters.* New York: Times Books.

**Newman, Katherine S.** 1999. *No Shame in My Game: The Working Poor in the Inner City.* New York: Knopf.

———. 2004. *The Social Roots of School Shootings.* New York: Basic Books.

**Noel, Donald.** 1968. "A Theory of the Origin of Ethnic Stratification." *Social Problems,* 16: 157–172.

**Olsen, Marvin E.**, ed. 1970. *Power in Societies.* New York: Macmillan.

———. 1978. *The Process of Social Organization.* 2nd ed. New York: Holt, Rinehart & Winston.

**Ouchi, William G., with Lydia G. Segal**. 2003. *Making Schools Work.* New York: Simon & Schuster.

**Paap, Kris.** 2006. *Working Construction: Why White Working-Class Men Put Themselves—and the Labor Movement—in Harm's Way.* Ithaca, NY: ILR Press/Cornell University Press.

**Parenti, Michael.** 1998. *America Besieged.* San Francisco, CA: City Lights.

**Parrillo, Vincent N.** 2002. *Contemporary Social Problems.* 5th ed. Boston: Allyn & Bacon.

**Perrow, Charles.** 2002. *Organizing America.* Princeton, NJ: Princeton University Press.

**Perrucci, Robert, and Earl Wysong**. 2003. *The New Class Society: Goodbye American Dream?* 2nd ed. New York: Rowman & Littlefield.

**Phillips, Kevin P.** 1990. *The Politics of Rich and Poor.* New York: Random House.

———. 2002. *Wealth and Democracy: A Political History of the American Rich.* New York: Broadway Books.

**Piven, Frances Fox, and Richard A. Cloward**. 1993. *Regulating of the Poor: The Functions of Public Welfare.* 2nd ed. New York: Vintage.

**Reich, Charles A.** 1995. *Opposing the System.* New York: Crown.

**Risman, Barbara J.** 1998. *Gender Vertigo: American Families in Transition.* New Haven, CT: Yale University Press.

**Rosen, Bernard Carl.** 1998. *Winners and Losers of the Information Revolution.* Westport, CT: Praeger.

**Rotenberg, Paula S.** 2002. *White Privilege: Essential Readings on the Other Side of Racism.* New York: Worth.

**Roth, Louise Marie**. 2006. *Selling Women Short: Gender and Money on Wall Street.* Brunswick, NJ: Princeton University Press.

**Rousseau, Jean-Jacques**. [1755] 1913. "A Discourse on the Origin of Inequality," pp. 00–00 in *The Social Contract and Discourses,* translated by G. D. H. Cole. New York: Dutton.

**Rousseau, Nathan.** 2002. *Self, Symbols, and Society: Classic Readings in Social Psychology.* New York: Rowman & Littlefield.

**Roy, William G.** 1997. *Socializing Capital: The Rise of the Large Industrial Corporation in America.* Princeton, NJ: Princeton University Press.

**Ryan, William.** 1976. *Blaming the Victim.* Rev. ed. New York: Vintage.

**Schlosser, Eric.** 2001. *Fast Food Nation: The Dark Side of the All-American Meal.* Boston: Houghton Mifflin.

**Schwartz, Barry.** 1994. *The Costs of Living: How Market Freedom Erodes the Best Things in Life.* New York: Norton.

**Scott, W. Richard.** 2001. *Institutions and Organizations.* Thousand Oaks, CA: Sage.

**Segrave, Kerry.** 1993. *The Sexual Harassment of Women in the Workplace, 1600 to 1993.* Jefferson, NC: McFarland.

**Shapiro, Thomas M.** 2003. *The Hidden Costs of Being African American.* New York: Oxford University Press.

**Skolnick, Arlene, and Jerome Skolnick**. 2000. *Family in Transition.* 11th ed. Needham Heights, MA: Allyn & Bacon.

**Staub, Ervin.** 1989. *The Roots of Evil: The Origins of Genocide and Other Group Violence.* New York: Cambridge University Press.

**Steinberg, Stephen.** 1995. *Turning Back: The Retreat from Racial Justice in American Thought and Policy.* Boston: Beacon.

**Suro, Roberto.** 1999. *Strangers among Us: Latino Lives in a Changing America.* New York: Knopf.

**Thrasher, Frederic.** 1927. *The Gang.* Chicago: University of Chicago Press.

**Tichenor, Veronica J.** 2005. *Earning More and Getting Less: Why Successful Wives Can't Buy Equality.* New Brunswick, NJ: Rutgers University Press.

**Tilly, Charles.** 1998. *Durable Inequality.* Berkeley: University of California Press.

**Titchkosky, Tanya.** 2003. *Disability, Self, and Society.* Toronto: University of Toronto Press.

**Travis, Jeremy, and Michelle Waul**. 2003. *Prisoners Once Removed: The Impact of Incarceration and Reentry on Children, Families, and Communities.* Washington, DC: Urban Institute Press.

**Trotter, Joe W., with Earl Lewish and Tera W. Hunter**, eds. 2004. *African American Urban Experience: Perspectives from the Colonial Period to the Present.* New York: Palgrave Macmillan.

**Tucker, Robert C.**, ed. 1972. *The Marx-Engels Reader.* New York: Norton.

**Turner, Jonathan H., Royce Singleton, and David Musick.** 1984. *Oppression: A Socio-History of Black–White Relations in America.* Chicago: Nelson-Hall.

**Turow, Scott.** 2002. *Ultimate Punishment.* New York: Farrar, Straus, & Giroux.

**U.S. Bureau of the Census.** 2006. *Current Population Survey, Annual Social and Economic Supplement.* Washington, DC: Government Printing Office.

**U.S. Congress, Joint Economic Committee.** 1986. *The Concentration of Wealth in the U.S.* Washington, DC: U.S. Congress, Joint Economic Committee.

**Useem, Michael.** 1984. *The Inner Circle.* New York: Oxford University Press.

**Van den Berghe, Pierre.** 1978. *Race and Racism: A Comparative Perspective.* New York: Wiley.

**Venatesh, Sudhir Alladi.** 2006. *Off the Books: The Undergound Economy of the Urban Poor.* Cambridge, MA: Harvard University Press.

**Waldinger, Roger David, and Michael I. Lichter**. 2003. *How the Other Half Works: Immigration and the Social Organization of Labor.* Berkeley: University of California Press.

**Walker, Henry A., Phyllis Moen, and Donna Dempster-McClain**, eds. 1999. *A Nation Divided: Diversity, Inequality, and Community in American Society.* Ithaca, NY: Cornell University Press.

**Wallerstein, Immanuel.** 1974. *The Modern World-System.* New York: Academic.

**Weitzman, Lenore J.** 1985. *The Divorce Revolution: The Unexpected Consequences for Women and Children in America.* New York: Free Press.

**Western, Bruce.** 2006. *Punishment and Inequality.* New York: Russell Sage Foundation.

**Whyte, William Foote.** 1949. "The Social Structure of the Restaurant." *American Sociological Review* 54: 302–310.

**Williams, Christine L.** 2006. *Inside Toyland: Working, Shopping, and Social Inequality.* Berkeley: University of California Press.

**Wilson, William Julius.** 1987. *The Truly Disadvantaged: The Inner City, the Underclass, and Public Policy.* Chicago: University of Chicago Press.

———. 1997. *When Work Disappears: The World of the New Urban Poor.* New York: Knopf.

———. 1999. *The Bridge over the Racial Divide.* Berkeley: University of California Press.

**Winant, Howard.** 1994. *Racial Conditions: Politics, Theory, Comparisons.* Minneapolis: University of Minnesota Press.

**Witt, Griff.** 2004. "As Income Gap Widens, Uncertainty Spreads." *Washington Post,* September 20.

**Wolff, Edward N.** 2007. "Recent Trends in Household Wealth in the United States: Rising Debt and the Middle-Class Squeeze." Working paper No. 502. The Levy Economics Institute of Bard College (www.levy.org).

**Wright, Erik Olin.** 1978. *Class, Crisis and the State.* London: NLB.

———. 1994. *Interrogating Inequality: Essays on Class Analysis, Socialism, and Marxism.* New York: Verso.

———. 1997. *Class Counts: Comparative Studies in Class Analysis.* Cambridge: Cambridge University Press.

**Young, Alfred A.** 2004. *The Minds of Marginalized Black Men.* Princeton, NJ: Princeton University Press.

**Zimbardo, Philip.** 1972. "Pathology of Imprisonment." *Society,* 9: 4–8.

**Zweigenhaft, Richard L., and G. William Domhoff.** 1998. *Diversity in the Power Elite: Have Women and Minorities Reached the Top?* New Haven, CT: Yale University Press.

# Are Human Beings Free?

## The Power of Society over Human Thinking and Action

### Concepts, Themes, and Key Individuals

- Society and freedom
- Social forces
- Social construction of reality
- Deviance, ideology, and language
- Society, socialization, and the control of self
- Society as the origin of freedom

For much of my life, I have pondered the question of freedom. As I studied world history, I became aware of the human struggle for freedom, yet certain questions arose that were bothersome to me. Is freedom real? How can there be freedom when so much of what we do is caused by forces we hardly even understand? Perhaps, I thought, freedom is an illusion that people are taught so that powerful people can manipulate them. As I studied psychology and sociology I discovered new and more subtle ways freedom seemed to be limited. Sigmund Freud, Karl Marx, and Erich Fromm showed me new dimensions to the problem of freedom, challenging my understanding of what the concept of freedom means, why freedom is so difficult to achieve, and why so many of us fear freedom in our lives.

As I studied sociology seriously, I found myself questioning the whole idea of freedom. The more I understood my own life sociologically, the more I learned about all the social forces that formed me. Eventually, I discovered that the sociological perspective was not as simple as I thought. I still believed that freedom was an important goal for society, and I sought to find ways to live what I considered a freer life. Freedom continued to be an important value in my teaching and in my family. As I studied the work of George Herbert Mead and Herbert Blumer, both important, impressive social psychologists, I began to understand that freedom is possible, even within the perspectives of psychology and sociology. Mead and Blumer taught that society is so central to the human being, that it not only socializes us into its ways, but it also gives us the tools to act back on it and make it possible for us to control our own destiny. The wonder of society is that it creates us, and, at the same time, it develops our language system, our selfhood, and our mind, all necessary for freedom.

Freedom is quite difficult to study. I constantly alter my views as I learn more. Yet I have learned a few ideas about freedom that I am relatively sure about. Perhaps it will help if I briefly list these so you can understand better the way I handle freedom in this chapter. These ideas are:

1. *It is impossible to determine whether human beings can be free.* I do believe that some freedom is possible, but any attempt to prove freedom through reason or empirical evidence is impossible. All we can do is uncover many of the ways human beings are not free and then hope or have faith that something is left that we might call freedom.

2. *To understand freedom, it is critical to recognize that freedom needs to be analyzed in terms of degree.* There is no absolute freedom; there are instead degrees of freedom. Some societies are freer than others. Some individuals are freer than others. Some acts are freer than others.

3. *Most people have a highly exaggerated view of how much freedom they actually have.* We think we are free because representatives of society continually remind us we are free; we build a society that assumes people are free and therefore responsible for their own actions. We tend to take freedom for granted without critically examining what that actually means.

4. *There are two aspects of freedom to understand: freedom of thought and freedom of action.* It is especially difficult to

achieve any freedom of thought, but without such freedom there can be no freedom of action. And even with freedom of thought there still may not be much freedom of action.

5. *Humans are almost always influenced by society, yet they also have an ability to control their own lives to some extent.* Always there is social influence: almost always we have some freedom.

## The Meaning of Freedom

### *Freedom: Control, Understanding, and Choice*

What happens when someone does something we do not like? We get angry; we tend to blame the individual for what he or she has chosen to do. This tendency assumes that they somehow had control over their acts and acted deliberately: "He knew what he was doing!" "She knew that others would be hurt. She does not care." "It was his fault that he got her pregnant." "No, it was her fault." In a sense, we justify our own anger toward what others do by assuming that they had control over their actions. After World War II, the judges at the Nuremberg Trials ruled that human beings had acted immorally and that the excuse that they had only been following orders was unacceptable. Instead, they were found guilty of choosing to do evil things against humanity. "They knew what they were doing, and they could have said no."

To a great extent, American culture especially emphasizes this view of human action. Individual responsibility is big. After all, life is the consequences of individual choice; we choose what we think and do. Much of Western religion assumes that punishment and reward await us after death primarily because of the free choices we make during life. Even those who come out of horrible conditions such as abuse, poverty, or war are assumed to be free actors.

For me, it makes most sense to define freedom as one's ability to *control* his or her life. The issue is one of "who is in charge?" Does the individual actually choose what he or she believes? Does the individual really choose what he or she does? To control one's own thinking and acting is a matter of *cause*: what causes, what directs, what shapes, what influences the human being? Is it the individual actor or is it something else? If we are *conditioned,* if we *respond* to stimuli, if we are *trained* or *oppressed,* if we are *unconscious,* if we are *brainwashed,* then we are not in control, something else is. To be free means that the individual is *active,* not passive, is *self-directed* rather than directed, *chooses direction* rather than influenced.

If something or someone controls the actor, then the actor does not control himself or herself. If it is family, society, unconsciousness, emotions, impulse, habit, social class, culture, neighborhood, government, the world marketplace, heredity, or a myriad of other factors that cause what we do or think, or cause us to not adequately understand our situation, or severely limit our choices, then freedom is not real.

One more point. To be active, to direct oneself, to make choices, to cause what one does *assumes* a thought process, a consciousness involved, an understanding of one's self and one's world. The ability to take charge of one's own life does not make sense to me if thinking is absent.

### *Freedom: The Sociological View*

Sociologists are caught in a great dilemma: They want to believe that human beings can be free, yet they understand too well how all-powerful society seems to be. Sociologists like to claim that "society shapes the individual, but the individual also shapes society." A nice thought, but almost all sociological thought and work shows the many ways in which society shapes the individual actor. Marx believed that humans could someday be truly free, yet he spent his life showing how powerful society is. Durkheim was a champion of individual freedom within society, yet his work emphasizes the power of social forces on every aspect of life. Weber, too, believed in free action, yet almost all of his work showed how we live in an iron cage of bureaucracy, within a culture within which we are socialized into a system of inequality, and within a society whose leaders are protected by an embedded authority structure—all powerful causes of what we think and do. Peter Berger shows us that the individual lives within a society, which he calls "a prison in history," yet he tells us that sociology can help "liberate us."

Sociology is largely a scientific perspective, so its purpose is really to uncover and show us how important social forces, social interaction, socialization, social structure, culture, and social institutions shape our lives. C. Wright Mills in *The Sociological Imagination* (1959) argues that to think sociologically is to see oneself located in both society and history, to understand that one exists within a social and historical context, within forces that act down upon us and forces that have arisen from a distant past. One may experience personal problems—a bad marriage, overwhelming debt, unemployment, a midlife crisis—but such problems need to be perceived in the larger context of society if one is to understand why they occur. So for example, to live in a society in which child abuse or spouse abuse is common and has historically been legitimated means that one lives within social forces that encourage many of us to be abusers and many of us to end up as victims.

Mills is important to sociologists because he turns our attention to understanding social problems. He links social problems to the nature of society itself. People do not simply get divorces or commit suicide or commit crime in a random manner. Societies encourage or discourage such acts, and the results are relatively stable rates of suicide, rates of divorce, rates of violent crime—strongly suggesting that there are encouraging and discouraging forces at work in a given society. We have stable birthrates, death rates, rates of migration, unemployment, and school-dropout rates—also suggesting a patterning of actions in society that lead a certain number of people to act in a certain way. Indeed, when rates change, political leaders and the public demand to know why. To ask the question "why" is to assume that there are reasons behind the change, that there are causes and forces other than simply free will.

Poverty is an example of how society makes a big difference in what people become. Although the general public often argues that poverty exists because people freely choose that direction, sociologists rarely make this claim. Not all children have an equal chance of being poor: If one is born into poverty, it is much more likely that one will continue to be poor in adulthood. If one is Hispanic, African American, or a woman, one is more likely to be poor in America than if one is white, Anglo, or a man. Is this a matter of free choice or social forces that control many of our directions in life?

The public, however, tends to see poverty as resulting from free will. For example, Joe R. Feagin (1975) studied American beliefs about the causes of poverty and found, not surprisingly, an emphasis on individual will rather than on social cause. By far, people regarded reasons such as "poor money management," "lack of effort, talent, or ability," and "loose morals and alcoholism" as good explanations of poverty (more than 80 percent in each case). Bad luck, being taken advantage of, failure of private industry, discrimination, and poor schooling were significantly less important (35 to 60 percent in each case). Clearly, the more personal the reason, the higher it was rated.

Social patterns and human problems that are socially produced are difficult to deny. Do people who are brought up in a community where public education fails the vast majority freely choose to drop out of school? Do women who are brought up in a community where many other teenage women get pregnant freely choose to get pregnant? Do people freely choose occupations or jobs that eventually lead them to be laid off, or do they freely choose to live in communities knowing that their major businesses eventually have to move out? Do people freely choose to live in a society where their racial group has minority status? If one becomes a farmer and loses the farm that has been in the family for more than 100 years to large corporate farms within a worldwide market system, do we conclude that the individual's outcomes arose by free decisions made?

Sociology, then, always examines social forces acting on the individual and is highly skeptical of human freedom in any absolute sense. Sociology remains suspicious of claims of freedom without examining action in the context of society.

### *Freedom: Free Thought and Free Action*

To repeat: Freedom has to do with controlling one's life. It means that control belongs to the actor rather than to some other force.

It is critical to recognize that a great deal of what we call freedom involves *thinking.* Freedom assumes a thinking organism: someone who exercises control through thinking, someone who makes choices through thinking through situations, someone who takes charge through thinking of his or her actions and their consequences. To act without thinking is to act without freedom.

However, if our thinking is a result of other influences, our freedom becomes limited. Without control over our thoughts freedom is empty. If society shapes our ideas, values, and norms, our actions are not our own. If, for example, I decide to live life for the purpose of accumulating wealth, is this a free choice if everyone around me is involved in pursuit of material gain? If I demand the right to speak or write, but what I communicate is simply what I have been told, are my freedom very meaningful?

It is important to realize that free action is a second aspect of freedom. Even if our thoughts are freely derived, it is possible that we are not able to act according to these thoughts. Something prevents our actions.

We might define *action* as movement toward our environment—that is, doing things in situations we encounter. I may oppose my government inside my head, but unless I can act on that and protest its policies, my freedom is highly limited, and makes little difference in my actions. I may decide that clothes are not right, but unless I am free to go naked, my free thinking has limited consequences for action. I may choose to marry, but I cannot act on that if I have no one to marry, or if I have personality qualities that deny me an opportunity to make myself attractive to someone. We all have expectations of ourselves and, if these expectations arise from free thought, that does not mean we are free to direct ourselves in our actions according to these expectations. I may believe that a university education is important, but unless I have the abilities and the finances necessary, I am not able to enter a university, and, if I do enter, my freedom to get a degree is severely limited.

I may decide on my own that I shall be a doctor, a lawyer, a teacher, or a musician, but my actually becoming one of these depends on much more than thinking. It depends on whether my abilities match my thinking, on the competition in the university, on the opportunity to survive in the marketplace, on whether or not I have met the formal and informal qualifications, and, in many societies, on my gender, my race, my ethnic group identity, my social class, and even my religion. It may even depend on a professor who is unkind or unjust or simply jealous or exploitative of my talents.

Society shapes much of what we believe and what we do. Few sociologists claim that there is no freedom, but almost all point out the many ways in which freedom is limited by socialization, social patterns, and various means of social control. In the next two sections we will examine the two aspects of freedom discussed here: freedom of thought and freedom of action.

## Freedom and the Control of Thought

### *Reality Is Socially Constructed*

Why do I believe in God? Is it something that I freely chose to believe?

Is it a belief that has been proved to me? Is it something that human beings believe naturally? Is it something that I have accepted from parents? Is it something that people around me believe? Is it something I need to believe? Is it a truth that has been revealed to me from a supernatural power? If my life had been different, if I had been born at a different time or place, would I still believe what I do about God?

We believe a lot of things. For a moment consider what you believe. What do you believe about the death penalty? About human nature? About capitalism? Individualism? Freedom? A meaningful life? Good music? High school education? About the frequent displays of violence on television? About men and women, romance, sex, and marriage? About the Middle East? About the future of the United States in the world?

It is one thing to believe something because we have carefully investigated its truth or falsehood, but if you really examine what you have come to believe, you will notice that much of it is simply embedded in a social context. The society, community, neighborhood, family, peers, media, and school have a lot to do with what you believe. You are influenced. The year you were born and the years you went to high school and college are important contexts within which your beliefs were formed. The occupations of your parents, your gender, your ethnic group, and your social class are also important. If you lived in New York, Iowa, Texas, Wyoming, or California your beliefs have been formed by a different culture. If you have been surrounded by wealthy businesspeople, farmers, lawyers, electricians, or artists, then your ideas have been influenced.

The sociological approach to the question of truth is sometimes called the "social construction of reality," a phrase that arose from a wonderful book by that name written by Peter Berger and Thomas Luckmann (1966). The basic argument is the following:

1. *Reality*—the real world apart from us—may, in fact, exist, but it is extremely difficult, probably impossible, to know exactly what is "true" about that reality.

2. Instead of humans simply responding to a reality that imposes itself on them, *humans work to understand* that reality. The truth about reality is examined, interpreted, debated, and hopefully understood in a way that is useful for our actions.

3. Capturing the truth of reality is a *social task.* It is through social interaction that we come to learn about reality. For humans, reality is built socially—hence the phrase "the social construction of reality." Experience is understood,

objects are understood, people are understood, situations are understood, nature is understood by ideas developed in social interaction.

4. This is not to deny the fact that some groups do a much better job than others in capturing the truth about reality. Humanity has come to understand a great deal about the universe through its social life. It is never a perfect understanding, yet what we know can and probably has in fact brought us closer to capturing the reality that actually exists.

5. It also does not deny the possibility that some of us are able to go beyond social reality and develop our own unique realities, but this usually builds on some social reality.

To understand how beliefs are formed and how our thinking is socially controlled to a great extent, it is important to examine (1) culture, (2) language, (3) social structure, (4) social power, (5) knowledge, and (6) clear thinking. Beliefs are not built in a vacuum; they arise in social interaction, are perpetuated in social interaction, and are changed through social interaction.

### *Culture Becomes Our Reality*

All organizations—groups, formal organizations, communities, and societies—develop a unique set of ideas, values, and rules that come to be useful for achieving organizational goals and for solving the problems that must be dealt with. We might call this a "shared perspective" or, as we have seen, "culture." Berger and Luckmann (1966: 66–67) point out that a hunting society will know facts about hunting, will develop a set of rituals around hunting, and will teach ideas concerning how to be a successful hunter. On the other hand, a community that is morally opposed to eating meat will know a different set of facts about hunting and will teach ideas about how to survive without eating meat.

Culture contains our taken-for-granted truths, a set of assumptions that we generally accept without serious question. A given culture may make religious assumptions or scientific ones about the universe. It may emphasize progress or place great importance on tradition. It may be committed to the individual or to the collective. Culture will tell us to value freedom, materialism, family, or art. It will teach us to work hard, take it easy, compete, cooperate, exploit others, or love others. Culture is a broadly general guide to how people are supposed to believe in a given organization, and usually it seems to people in the organization to be "common sense" or "natural" or "truth with a capital T" rather than a socially constructed approach to reality.

Deviance is a result of our designation of others who are different; deviance results from our own commitment to certain cultural standards and "truths." Society will teach us about homosexuality, filling us with "facts," values, and morals that we use to react to a gay couple we might encounter. Society has given us words—such as *gay, queer, faggot, lesbian,* and so on—that we use to apply to people who are homosexual. It teaches us reasons why people are gay (choice, weakness, illness, biology, upbringing, and so on), and it shows us why such actions are moral or immoral. Most of us are influenced by society's perspective—its culture—so when we see a gay couple, this perspective becomes our guide for selecting what we see and believe about them.

Of course, not all of us think the same way. If we are part of the gay community, we will learn a different perspective than if we are a member of a fundamentalist church. If we become part of a university community, chances are we will become more tolerant of human differences and thus see homosexuality simply as the sexual orientation of some people. If we are a part of the community of psychiatrists, then we will see another reality; and if we are sociologists, we will see yet another. But that is exactly the point! We all interact, we take on the cultures of the social organizations in which we exist, and those cultures influence how we think about the world in which we live.

There are many reasons we tend to take on the culture of the organization within which we exist. First, its culture is functional for the group—it works for the group—so if we are part of the group, we will be attracted to its culture. Second, the culture of our own organization is what we are most likely to know—it is here that we interact on an ongoing basis, and that interaction will isolate us from those outside it. Our cultural ideas, values, and rules become our taken-for-granted truths, and it is easy to slip into the belief that our culture is true and that others are false. Third, we take on our culture because we almost always seek some affirmation that our ideas are true. Like it or not, other people with whom we interact usually become our measuring stick for what we believe.

A culture is *not* something that a society or group accidentally believes. Culture is functional; its ideas, values, and rules fit in relation to that particular group. If we are a society of inequality, then we develop ideas to justify that inequality or help us ignore its existence. If we are a capitalist society, then we develop a culture that values competition and profit. If we have enemies, then it is important for us to dehumanize them by calling them "terrorists." If we are active in the right-to-life movement, then it is important for us to believe that the fetus is a human being; if we are pro-choice, then it is important to believe that the fetus is not yet a human being. Karl Mannheim wrote that "even the categories in which our experiences are... collected and ordered" depend on the position in society of the group whose ideas we use (1929: 130). There are reasons why a particular group believes in its core cultural elements, apart from their truth or falsehood. It is both amazing and frightening to recognize that we all have probably rejected or forgotten a large number of "truths" simply because we have changed our social lives.

In the 1920s and 1930s fascism—a set of beliefs, norms, and values—was developed in Germany and Italy. Fascism is a *political ideology* (a culture that exaggerates certain elements in the world in order to justify a certain political program). The culture of fascism taught a view of war and power as bringing out the best in humankind, a belief that various categories of people are naturally unequal, and a belief that democracy and freedom indicate weakness. Fascism became a central part of German culture because it worked: It explained the failure of Germany in World War II and the causes of the worldwide depression of the 1930s, and it appealed to people's general discontent. It was also consistent with several themes contained in traditional German culture: strong nationalism, militarism, and authoritarianism. It worked for certain categories of people such as German industrialists and various political opportunists who were able to use it to their advantage. Fascism worked because it seemed to explain the present, was consistent with the past, and gave hope for the future. We might even examine carefully the ideas that have dominated our own history and recognize that they too worked well for what we did as a society. They both influenced and justified our actions in our history.

Cultures are tricky. Once people believe in one, they will find it difficult to accept evidence that challenges it. An internal logic is at work, a selective interpretation of evidence, a tendency to interpret experience in line with what people already believe. If we are able to step back from what we believe for a moment and see culture for what it really is, we will recognize that what we all believe is in large part the product of our social life; therefore, those who disagree with our basic beliefs are neither fools nor free actors. They are usually a part of a social world that sees reality differently from us. What they believe may or may not be truer than what we believe; but all of us should not be as tempted to be certain that we have the truth.

The culture of an organization—be it a group, formal organization, community, or society—is a powerful reason why humans do not freely arrive at their own "truths." It is one reason why human beings are not as free as they sometimes think (by the way, it is actually our society's culture that we are taught that we live in a society of freedom, and most of us tend to accept what we have been told).

### *Language and the Control of Thought*

Human beings use language to teach others how and what to think. Language is what the individual ultimately uses to think with. Ultimately, language sets the parameters within which individuals think.

Clearly, individuals at birth have potential to learn language and begin to imitate sound and learn words early on. But the language that individuals are filled with depends on their social interactions, which continue through their lives. The words they learn—the number, type, and use as well as the importance of using words in general—depend on the society and community within which they grow up. These will become the words that they use to divide up reality, to make sense out of their world. We see through words; we understand through words; we think with words. Although some people maintain that thinking might exist without words, it is critical to understand that words are a central part of most human thinking.

There is a long history of the use of words for thinking within social science. Benjamin Lee Whorf (1941) put forth this position many years ago:

> The real world is to a large extent unconsciously built up on the language habits of the group. . . . We see and hear and otherwise experience very largely as we do because the language habits of our community predispose certain choices of interpretation. (p. 250)

Awareness, consciousness, problem solving, understanding, discussion with oneself, interpretation, taking control of our actions, and creatively examining and assessing situations are all aspects of thinking that involve language. Language then becomes an important guide to what we think; the language we learn and use limits what we think.

A person in a religious community learns words pertaining to religious concerns, and then he or she is in the position to see people in that context. Some families divide the world into *Christian* and *non-Christian;* others divide it into *believers* and *nonbelievers;* others into *Jewish* and *non-Jewish;* still others into *Muslim* and *infidel.* These words become the basis for thinking about people. Some leaders remind us that the world is made up of *white* and *nonwhite* people; others use more judgmental words. To enter school is to learn vocabularies that open worlds previously unknown to us, and the university introduces us to sets of words that certain academic communities (such as sociology, physics, or psychology) continually use. If we enter the world of a motorcycle community, the world of a gay community, or a society in Asia or Africa, then we will find a new language, a new emphasis on what the community considers important in dividing reality.

It is through language that all organizations form and teach their dominant ideas, values, and norms. Ideas are made from the words within the language learned. The emphases in the language create the emphases in the ideas. Thus, in a capitalist society we are likely to use certain words over and over: *competition, free enterprise, profit, individual effort, private property,* and *marketplace.* Around these words will grow a set of ideas that are reinforced over and over. We may have ideas concerning socialism, but such ideas are reinforced less—unless, of course, they are used negatively in relation to our commitment to capitalism. Political leaders know that the use of words is an important way to influence people's thinking: Words are carefully chosen in order to gain support. "War on terrorism," "evil societies," and "preemptive war" are all phrases used by leaders to help us "understand" and support the policies toward those we wish to battle. "Death taxes," "secularists," "fascists," "liberal professors," and "evolutionists" are phrases used to influence us to swallow simple explanations about highly complex issues.

The emphases we use in our language help create the values we hold, and they reinforce the rules that we are supposed to follow.

How different our language would be if we lived in a monastery, a prison, the army, or on a farm, and how different would be the way we think about the world around us.

Language and the culture that it teaches and creates are important controls over the way human beings think—and if freedom depends on thinking, then both language and culture set important limits to our freedom.

### *Social Structure and the Control of Thought*

Social structure, like culture and language, controls much of what we think. This means that our view of reality arises from where we are positioned in organized life. People have perspectives that are attached to their positions. Perspectives are points of view, angles from which one sees reality. One angle is from the top, another from the bottom. The upper class and the poor look at reality from different places or positions. Leaders and followers in groups have different perspectives, and so do professors and students. Whereas culture controls our thinking by causing all of us in an organization to believe alike, structure controls our thinking by positioning us in relation to others and causing us to see the world through our positions.

We believe, for example, in ideas that arise from the fact that we are men or women. Part of how one becomes a woman is to learn to "think like a woman," in many societies to be taught to think one is subservient to men or created for men, in other societies to think that falling in love, marrying, and having children are necessary actions for a fulfilling life. In most societies, to become a man is to learn to "think like a man"—to believe, for example, that making money and gaining prestige, power, and privilege in the economic world are necessary achievements for a fulfilling life. Femininity and masculinity are ways of thinking that include expectations for certain categories of people. In our more open society today, of course, these distinctions have increasingly become more complex, but the increasing complexity does not erase their existence. It may be easier to see social structure as an influence on thinking in a religious fundamentalist culture. It is probably inconceivable to most of us that traditional Arab women can allow themselves to be subservient to men, to reject doing the kinds of things that men do, to be satisfied with being solely wives and mothers, and to cover their faces and bodies when they go out in public. If we were to ask them, however, we would find a belief system that explains it. And once we understood that belief system, we would understand the logic of such actions. Women are not simply forced to do such things; they are taught a perspective that justifies it. And their angle of vision becomes part of their very being; they come to think differently from the men with whom they interact. Let us not forget that in any society women and men have different positions and therefore will have different perspectives on life. The perspective of women will influence individual women's ideas concerning pregnancy, abortion, marriage, birth control, menstruation, occupational opportunities, American history, and professional athletics.

But go beyond gender. A factory worker thinks differently from a manager, a boss thinks differently from an employee, an owner from a manager, a bookkeeper from a secretary. Each has a position in the social structure, and each has a different way of looking at reality. Imagine society as having thousands of positions, occupational and otherwise. We interact and find ourselves in these positions. And what happens to us? We come to think of the world according to these positions: high school graduate, dentist, artist, general in the army, ex-prisoner, member of the upper class, African American, rock star. Each position we hold molds what we believe. It is extremely difficult to be in a position and play the role that is expected and yet escape the tyranny of the perspective that goes with that position. The suddenly successful rock star matter-of-factly declares, "I'm just the same person I always was; I think the way I always did." Oh, yeah? Fame and fortune bring the need for security; they bring isolation from the larger society; they bring new tastes in clothes, cars, and homes; they bring a view of oneself as a star; they bring a belief that one deserves recognition and respect that is different from that given to other people. Over time, the pressures are extreme to change the way one thinks to fit the position.

Why do we do it? Why must we necessarily take on the perspective of the position that we occupy and allow our thinking to be controlled? (Of course, I hear all the silent readers of this view of the human being crying, "No—this does not happen.") Perhaps you are able to overcome these pressures. However, these pressures do exist, and they tend to limit our freedom of thought.

*The first reason we take on the perspective of our position is that acting appropriately in a position demands appropriate thinking.* We learn over time that if we are going to succeed in a position, we had better change our thinking, at least temporarily; and the longer we desire to do well in that position, the more likely we will think as the position requires.

*The second reason is, like it or not, other people who have a relationship with us within the social structure and who interact with us teach us how we should think in that position.* If we join a firm, we are slowly taught what people in our positions are supposed to think about those above us, below us, and equal to us.

*Probably the most important reason why positions so powerfully influence our thinking is that each one is, in fact, a different angle in the organization.* A position points us to look at reality; it is an eyeglass through which we see. If I am a man, I do not normally think like a woman. A student does not think like an instructor. A worker does not think like a boss. I cannot know what it really means to be a nonwhite in American society, a general in the army, or the president of the university if I am not in these positions.

Human beings change their thinking when they occupy new positions, when they go from student to graduate, from single to married, from working class to middle class. If I am careful, I can sometimes come close to understanding someone else's position's perspective, but I still must *use* my own angle on reality. In the end, most of us do not even recognize how significant our positions are for how we think about the world. It occurs without our understanding the process; before we realize that our perspective changes, we are already taking for granted a new way of thinking.

Sociologists study social structure along with culture to understand the control that society and other organizations have on the way we think. Class, race, and gender are positions that we obviously examine. Occupations; political and corporate positions; positions in gangs, families, and committees; and dominant and minority positions are studied extensively. Let us for a moment, however, expand the importance of position to include *age cohort,* the individual's generational position in society. The baby boom generation is that category of people who were born right after World War II and who began to reach their fifties in the mid-1990s. This generation is in a different societal position from those born in another decade, and that forms much of the way people think. When someone is born matters because it places the individual within a historical period when a generation begins to form its perspective, and it ties the individual to that generation's perspective throughout one's life. Those who experience war in their generation have a different view of war and peace than those who do not. Those who had job opportunities when they graduated from college have a different view of work and the future than those who did not. We learn our views of family and sex at a certain point in our society's history. "I can't help it. When I grew up, that's how people thought." Some of our most basic beliefs—such as what a successful life entails—are influenced by our generational position in society. Immigrants to the United States typically change their thinking according to the generation they are: The first generation settles but tends to keep its traditional perspective; the second generation works hard to "become American"; the third generation is well assimilated into American society; and then the next generation looks back and tries to recapture to some extent the tradition that has been forgotten.

### *Social Power and the Control of Thought*

Social structure has a lot to do with power. Positions give people not only a way to think about the world but also *social power,* or the ability to achieve something in relation to others. In relation to thinking, it is important to understand that high positions give individuals the opportunity and ability to influence how other people think in society. Marx and Mannheim note how the powerful create ideologies—exaggerated and even outmoded perspectives on reality—that are used to defend the status quo, that is, their own positions in society. Dictators teach people to obey, arguing that order is necessary now and that obedience and sacrifice will someday bring prosperity for all. The upper class teaches people that what it has it deserves and worked hard for. Slave owners teach slaves and non–slave owners various justifications for slavery, and racists inevitably create a whole set of ideas they try to teach others in order to justify racial inequality. The ideas are often sophisticated and not always seen for what they actually are—a defense of racial inequality. Even those who are slaves or victims of racism may actually come to believe them.

The perspective of the poor is usually a complex mixture of beliefs developed in their position as they interact with other poor people and the ideas developed by wealthier individuals with whom they come into contact directly (in interaction) or indirectly (through media, work, or renting property). The poor often become conservative in society, even though they have little to gain from the way society is. Because it is not the poor who control advertising and educational or political institutions, it is not their ideas that are taught in society, so they, like everyone else, become influenced by the ideas of those who do control these avenues of influence. People in powerful positions have the best opportunity to create the ideas that others come to believe. Brainwashing is an exaggeration, but the role of social power in the creation of what people come to believe is a critical aspect of all organizations.

### *The Control of Thought: Limited Understanding*

What we know makes a difference to our ability to think freely. We accumulate knowledge through a mixture of experiences, formal learning, informal learning, reading, discussion, imitation, and trial and error. No one understands everything, and no one can understand everything in a given situation. No one has every perspective that can be used in a situation; no one is able to understand all the possible choices in the situation. What we learn is little and it is a very limited sample of knowledge. We sometimes think we understand when we do not; usually our cultural bias stands in the way of understanding. Sometimes what we learn from others is not accurate; sometimes there is no opportunity in our community to understand on our own but we are expected to simply accept what others tell us. Sometimes our understanding is so firm and unchangeable that we are not willing to change it when new evidence is put forward.

Knowledge about the universe as well as an understanding of that knowledge is important for freedom. It is necessary for choice, working out situations we encounter, and rationally controlling our actions appropriately. That is why a free society that encourages debate, criticism, exploration of truth, and a plurality of perspectives is so critical for individual freedom. As all of us are limited in our knowledge and understanding, it is important to recognize that as knowledge and understanding increase, more freedom has a better chance. Those who lack knowledge and understanding of the situation in which they act are especially limited in their freedom of thought, and this will have implications for the choices they make for their action.

### *The Control of Thought: Our Inability to Think Carefully*

Thinking is not simple. It is something that people must learn. If one is free, there needs to be a critical, logical, analytical, open examination of what others teach us and what we come to believe. This is extremely important. Without this, one is likely to accept and simply memorize ideas. One is likely to become lazy in his or her pursuit of truth. One does not have tools with which he or she can evaluate the truth or falsehood of any idea one might encounter.

Social psychologists study the many ways people are influenced to accept ideas that they hear or read or even work out in their own thinking. It is important to question what we are taught and what we come to believe. It is important to know what constitutes good evidence and to recognize the ways people are influencing us not by evidence, but by tricks, by false logic, by making themselves attractive. It is important to understand our emotional commitments, our values, our biases, our culture, our positions in structure in order to evaluate how we have arrived at our views. In some ways learning is buying something from someone who is selling; freedom of thought must include a knowledgeable thinking process. One must habitually question, evaluate, and think about the knowledge and thinking he or she uses.

This is one of the most important skills that formal education should emphasize. Understanding philosophy, natural and social science, humanities, history, mathematics, speech, good literature, language, learning to write, and many other liberal arts classes should all be aimed, in part, to learning how to think. Actually, the term "liberal arts" suggests "liberation." Liberal arts is not a mass of knowledge we need to learn and memorize. Instead, for many teachers good thinking is the real essence of liberation.

### *The Control of Thought: A Summary*

Thinking is an important aspect of human behavior. If the human being is in control of what he or she does, then his or her thinking is central to that control. We think through the culture, language, knowledge, and understandings that we learn. We think according to the positions we fill in the social structure and according to what we learn from those in powerful positions in the structure. Our thinking is controlled by outside social factors, and because of that control it is difficult to argue for the existence of free thinking. However, this is only part of the story. Freedom is more than free thinking. Even if our thought might have a degree of freedom, then we must go further and ask what, if anything, limits our *actions?*

## Freedom and the Control of Action

A *free actor* is one who is able to think freely. A free actor is also one who *acts* freely. A free society encourages both free thinking and free acting.

Free thinking and acting are linked in several ways. People might be able to think freely but be severely limited in acting freely in their environments. Clearly, some slaves arose above their situation and were able to think more freely than other slaves, yet their actions were still probably controlled as much as the others'. In the extermination camps in World War II, there were undoubtedly many people who understood that they were being marched to their death, and in that sense their understanding was freer than others', but the situation was such that it was impossible for them to move freely out of that situation. Henry David Thoreau was imprisoned for civil disobedience, yet he wrote, "I saw that, if there was a wall of stone between me and my townsmen, there was a still more difficult one to climb or break through before they could get to be free as I was" (1849: 295). Thoreau thought he was free because he was not influenced by his society's impulse for war. He did not listen to what politicians, military leaders, newspaper reporters, and ordinary citizens were saying about the virtues of war. He was critical of people clamoring for violence without really understanding the implications of war. He may have acted freely by disobeying the authorities, but he was sent to prison. He declared to himself and others that this was a free act because he was in control of what he thought and how he acted. However, he came up against a society that interfered with his actions. In his cell he could be freer in his thinking than others, but he could no longer act as freely as he would if he were outside the cell. Freedom of thought certainly characterized those who opposed the Soviet state; but for most it was almost impossible to act on that freedom because they lived in a society whose leaders tolerated little criticism.

Free action has to do with movement. One is free if one can move without being controlled externally or internally—only the actor understands and controls what he or she does. When movement is interfered with, one is not free. Perhaps this is what people mean when they say "free as a bird" because flying seems to be action that is not interfered with (although the bird is not really free if it is, in fact, controlled by instinct, natural environment, or conditioning by experience or by humans).

We therefore come to the second way in which society controls us. It not only controls what we think but it also restrains us, directs us, and controls much of what we do, how we act. Even if society allows us to think freely (an assumption that is usually, in fact, highly exaggerated), action is always constrained, and what we actually do is directed by forces other than free choice by the actor. Some people maintain that such forces are minor, especially in our "free society"; others, such as most sociologists, regard such forces to be significant for all actors. *We act in a world where what we do as well as what we think is a product of much more than our free choice.*

### *Control of Thinking and the Control of Action*

The first step in understanding the control of action is to recall the control of thinking. To the extent that culture, language, social structure, and the powerful in society control what we think, to that extent our action is also controlled. I go to school, and I try hard to succeed. I memorize, take tests, write and rewrite papers, and discuss the material with others and with myself. Yes, it may seem that I am in control of my own life. But look at society: Its emphasis on achieving in formal education, its way of evaluating learning, its emphasis on education as a means for achieving material success, its individuals who are held up as model students, its demands for grading on a normal curve, its definition of learning, its division of subjects in the university, its definition of intelligence, as well as the power of its own professors—all of these affect what I actually think and then how I act as a student. Culture is intertwined with action: Culture not only is important to what we think but also actually controls what most of us do.

In *One-Dimensional Man* (1964), Herbert Marcuse paints a picture of our modern industrial society as a place where the media dominate thinking and action and where protest becomes almost unthinkable. Materialism and affluence, the dominant message of the media, tell us what seems important in our lives and directs our attention to pursuing more wealth. Questions about quality of life, liberty, equality, and general human welfare are left behind. It is not that a group of conspirators decides that this must be the message. It is much more subtle: The total message, the whole atmosphere, the implied, taken-for-granted values send one dominant message, and everything else is muted. The message is affluence; the message is accumulating wealth. The result is actions that aim toward material success and consumerism rather than critical thinking and protest against social injustice.

Our thinking includes our thinking about ourselves. Thinking about who we are and judging our worth is thinking that is highly influenced by others, and this influences how we act. To abuse a child may influence that child to see himself or herself in a negative light, to see an individual who is without worth. This has all kinds of implications for action. To participate in schools where one is unable to achieve is to come to perceive oneself, at the very least, as a poor student and, in some cases, as a person without much intelligence. To see oneself as a doctor, lawyer, or teacher; to judge oneself as ugly or beautiful; and to regard oneself as worthy or unworthy have been influenced in our social interaction with others who are close to us, and this continues throughout our lives. This thinking of ourselves has tremendous importance for how we act in our environment. If I think I am a man, a janitor, or an academic, then I tend to act like one. If I doubt my worth—or have no confidence in myself—then my actions will be influenced. James Baldwin (1963: 18), an African-American author, writes in a famous letter to his nephew: "Remember, James, ... you can only be destroyed by accepting what the white world calls a nigger." Baldwin understood that defeat comes to those who think of themselves as defeated. All of our thinking about ourselves is important for what we do in the world. It is misleading to hold someone simply responsible for his or her acts when the acts arise directly from a negative self-image fostered by interaction with others. It is also misleading to reward someone who does well in life without recognizing the importance of how he or she has learned to like, trust, and value himself or herself through interaction with others.

### *Social Institutions and the Control of Action*

Actions are guided by more than thinking. We also learn how to act in our homes, neighborhoods, and society. We learn to follow *institutions.* We marry because society has created that groove for us. We go out on dates, go steady, cohabitate, get engaged, get married, have children, and get divorced because these are the various kinship institutions—or grooves—that society has developed over many years for people like us to follow. For many of us, society's institution of remarriage and sometimes serial marriage brings us back to the institution of marriage. We vote among candidates, we attend party caucuses, we vote in primaries, we go to $1,000-a-plate dinners (or, more usually, $25 bean feeds or barbecues), and we send in campaign contributions because these are the various political institutions that society has developed. We pray, take communion, get baptized, attend Christmas Mass or synagogue on Saturday because these are the institutions set up in our community. There are economic, judicial, educational, health, and recreational institutions as well. Do we freely choose to watch television and use computers, or do we do these because they are dominant American recreational and educational institutions today?

Recognize that there are many ways of doing anything: being educated, going to war, ensuring peace, governing society, selling goods, being entertained, transporting ourselves, worshiping God, being treated for illness, forming relationships, clothing or cleaning our bodies. Over time, every society develops legitimate accepted ways that people get used to and normally follow without even thinking about options. These are our institutions: although not all of us allow ourselves to follow them, and sometimes a few of us will work to change them, they exist as forces that direct much of what we do. Institutions take away choice; instead of our having to make choices among so many options, institutions are there for us to follow; they are roads that are difficult to walk away from. And when we do, there are consequences, and often the consequences make life hard on us. For most of us, turning away—by free choice or by choices influenced by other factors—will end up sacrificing a much easier way of doing things.

### *Socialization and the Control of Action*

*Socialization* is the process by which we are taught to think and act the way we do, and it is the process that teaches us the institutions we are expected to follow. It is accomplished through many agents: parents, siblings, friends, teachers, peers, books, movies, neighbors, clubs, gangs, the police, and employers, to name some of the most important. Realize that such agents direct and reinforce what they want us to do, and they disapprove of acts that they do not like. Our actions are directed through an obstacle course of smiles and frowns, approval and scorn, praise and anger, fines and payments, A's and F's, promotion and demotion, getting rich and going broke, gaining elective office, and going to prison. Through it all we learn the directions that others approve of. We may turn our backs on them occasionally, but for most of us most of the time, the *rewards and punishments* matter.

However, socialization goes much further than simple reinforcement. It also arises from *opportunities that are made available to us.* Through the acts of others we are exposed to some things in the world and excluded from others. Parents may not expose children to reading. Friends may expose them to alcohol or illegal drugs. A community may not have a respected dance school that encourages ballet, but, if it does, it may encourage women but not men to participate. A neighborhood may have violent gangs that tempt youngsters to break the law. Socialization is not only reinforcement but also the subtlest influence of the opportunities offered by our socializers. If the opportunity is offered to us, it becomes a possibility for choice in our lives. If it is not offered, it is far more difficult to choose. If the opportunity is not there, if it does not exist among the people around us, then how do we suddenly decide to go in that direction? And if somehow we do decide, what happens to us when others around us continue to discourage us?

*Role modeling* also plays a large part in our socialization. Identification with others and imitating how they act are important ways in which we are socialized. How is a woman supposed to decide to become a lawyer or doctor if only men follow this path? How is a woman supposed to decide that having children is not the only way to live a full life if all the women she sees and knows pursue this direction? In the 1950s when I was still in high school, how were my female friends to know that they, too, could become baseball and basketball stars, Ph.D.s and political leaders, college presidents, and Supreme Court justices? Few female models existed in these positions.

Rewards and punishments, opportunities open to us, and role models are important ways socialization directs our actions. The poorest children in the United States grow up with few role models who have steady jobs, stable families, and successful and fulfilling lives. Parents are often unemployed and feel defeated. How can children be taught that school is important in a community where few people succeed in the educational system? How can children be socialized to succeed when people around them seem to fail? It is easy for those of us who are not involved to blame the adults; however, they, too, are victims of a society without sufficient role models, opportunities, and encouragement; they, too, are victims of poverty and often racism that they find almost impossible to escape.

### *Social Positions and the Control of Action*

Social structure and the positions that make it up also control individual action. We are all assigned a class position at birth and learn what that means and how to act in that class. The actions we see and take on as our own are those appropriate to our class position. Interest and activity in politics are influenced by class, as is the probability of divorce. Gender-role expectations differ according to class, and so does choice of religion. Educational achievement, health care, child-rearing practices, and likelihood of criminal behavior depend on class, at least in part. Sexual behavior, dating, family life, eating, drinking, dress habits, and language—all are influenced by class.

Class, like socialization, has to do with the options that the individual has in life. Our opportunities are formed by our class position. For example, class will influence the schools we go to and the jobs we can realistically consider in life. The choice of neighborhood, lawyer, and doctor is influenced heavily by class, and thus so will the safety of person and property, the likelihood of being convicted and sent to prison, and the likelihood of certain illnesses and early death.

For many people, poverty is a trap that is difficult to escape. It focuses people's attention on bare survival, on getting enough to eat and a place to live. The focus is taken away from working toward long-range dreams, getting a high school diploma, training for a decent job, saving for a rainy day. Poverty for most means dependence on others for one's own survival. One does not generally control one's own existence if one is poor. Instead, shelter, protection, food, clothing, and medical care are all in the hands of others. However, the wealthy, less dependent on finding bare essentials, still have much of their lives laid out too: values, education, aspirations, marriage, occupation, neighborhoods, and so on. Wealth brings the ability to choose from more options in life, but real freedom over thought and action is still quite limited.

And look at what gender does to us. To be a woman or a man in society is to learn a host of appropriate behaviors. Do we hold our books up against our chest or down at our sides? Do we play an assertive or a passive role in sexual encounters? Do we work to make it in the occupational world, in the world of the family, or both? Stereotypical actions related to gender characterize almost all societies; we do not have to act according to these, but there are consequences if we do and consequences if we do not. In every decision we make, position in society's gender structure shapes our actions. We do not yet live in a world where gender does not matter; it may change in each generation, but it does not disappear.

Positions in social structure rank us in relation to other people in power, privilege, and prestige. *Our rank matters in our actions.* Power influences the extent to which others can direct us in an organization; privilege influences the extent to which we can achieve and the extent to which we can act on real choices; prestige influences the extent to which we are honored by others in an organization; and honor is important for how people treat us, and how we, in turn, treat ourselves. Positions also have roles attached to them. Roles are the expectations others have of us in our positions. *Roles matter in our actions.* They are scripts we are expected to know and follow. Positions also give us identities: names that are applied to us by others, names we call ourselves, names we announce in our actions. *Identities matter in our actions,* And as we pointed out in the section on thinking, positions also give us perspectives to think with, and *perspectives matter in our actions.*

### *Social Controls and the Control of Action*

Finally, in considering the possibility for freedom, we must recall Thoreau's situation. Society punishes people who break the law. It puts people in jail so that they are unable to act as they choose. So it is with all of our actions: We are rewarded and we are punished for our actions. These are called *social controls.*

We have prisons and fines to punish those who act outside the rules of society; we have promotions and honors for those who are good citizens. Parents scream or snicker or spank or make children feel guilty; they also talk kindly, praise, hug, kiss, and make them feel good. Businesses fire, demote, threaten, and censure their employees; they also promote, praise, and give raises and bonuses. Friends, families, groups, and communities all exercise social sanctions—social controls—that encourage conformity.

Erving Goffman (1959) reminds us that in all social interaction there are rules we are expected to follow, and that there are social controls operating to ensure that actors follow them. Goffman believes we all act a part on a stage when we perform. Although most people realize that action is a performance, if it appears too phony, others will judge the actor harshly. Each actor knows this and is constrained to present a believable performance. On the other hand, the others are judging the performance, usually accepting the self that the actor is presenting, not embarrassing him or her by revealing qualities that might uncover the "real" person. Most actors recognize that phony performance as well as negative reactions undermine social interaction, so if we wish to continue the interaction we are constrained by these unstated norms. Respectfully listening to what others say, expressing anger in a positive manner, arguing without threatening the other are among the rules I have learned that facilitate ongoing social interaction. As long as I value the interaction, I must constrain myself accordingly.

There are many norms that make social life possible. That is why we are usually polite, communicate clearly, prevent embarrassment for ourselves or others, show respect, hide our weaknesses, show our strengths. Others are our prison guards to some extent, and life works only if we are willing to follow general rules. If we wish to belong there is much to obey; if we disregard the multitude of norms that social interaction demands, we face expulsion or the destruction of the organization itself.

## Is Any Freedom Possible?

Freedom is highly limited. The thrust of sociology is to show us that what we think and do are not simply something we determine, control, or choose. What we believe is created by our social life; what we do is influenced by our beliefs and by further social causes.

It is not just sociology that emphasizes the prison we are part of. Psychology has its controls, economics does, biology, anthropology, political science, and social psychology do as well. In fact all science assumes natural cause, and thus control rather than freedom is emphasized.

*Cause* is an extremely important word for all of us. Whenever we seek the cause of human actions we must either look for freedom or causal reasons outside of freedom. Why did Joel marry Susan? Why do Mexicans come to live and work in the United States legally or illegally? Why does someone succeed in the business world while someone else goes broke? Why did the United States enter war in Iraq? Why do we spend so much of our time watching television? Why did Art lose his farm? Why did George become homosexual? Why did John Schneider become an excellent teacher? Even religious people who normally believe that ultimately each of us has free will, usually want to know why people are turning away from organized in religion in Europe. Why are children no longer less polite or much more critical? Why are schools having trouble? The question of "why" is a search for reasons, and, like it or not, usually assuming free will becomes an empty understanding.

Yet, for many of us (including most sociologists, including myself), we still believe that it is possible for the human being to be active rather than simply passive, self-controlling rather than controlled, and making choices rather than simply responding to stimuli. Although freedom cannot be proved, we might at least examine the qualities that human beings encounter that might overcome the controls over their beliefs and actions.

For the sociologist, freedom is possible *only because of society.* The necessary prerequisites are *social.* Freedom is possible only through the creation of language, self, and mind, all created through social interaction.

Socialization occurs to a great extent through a language system, a highly complex set of symbols that are created and understood by people—not created by nature—and used *intentionally* to communicate to others and to our self. An intentional and complex language system allows us to teach others purposefully whatever we are able to understand ourselves (what we have been taught both through socialization and experience). In the process of learning through the language of others, the human being takes on that language from them, begins to understand as they do, and begins to discuss internally (think) with himself or herself about the environment he or she encounters. The thinking made possible with language that is understood and intentionally used allows us to arise from a simple stimulus–response association with our environment. Thinking allows us to take at least some control from our environment and control ourselves in that environment. Without thinking we are doomed to respond; with thinking, there is some chance of self-control and freedom—at least to some extent. With thinking, humans are able to define the situation they exist in, think about options for action, and control their own actions apart from simply the environment working on them.

This does not deny the existence of all the controls described earlier. Yes, our thinking and action are still controlled by many forces in society. On the other hand, as we are socialized by that society, we learn a special kind of language system, and that language system allows us to steal a part of control from society and to think and act to some extent on our own. Thinking with language allows us to understand, interpret, analyze situations, consider options, appraise morality and effectiveness of action, apply knowledge and past situations to the present, understanding consequences of actions, and appreciate the thinking and feelings of others with whom we are interacting. Steven Lukes describes the autonomous individual as one who "subjects the pressures and norms with which he [or she] is confronted to conscious and critical evaluation, and forms intentions and reaches practical decisions as the result of independent and rational reflection" (1973: 52). In spite of all the controls, we are able to discuss our own action with ourselves and to tell ourselves what must be done. Conforming to the expectations of society becomes to some extent a choice we can make. Determining to go outside what we have been taught by the socializers involves both free thinking and an ability to direct ourselves as we choose.

Two social psychologists, Herbert Blumer and George Herbert Mead, introduced the importance of society and the development of language. They created the image of the active, choosing, self-directing human being. Besides language they also focused on the development of *self* and *mind,* two qualities that developed with language and contribute to any freedom we have. *Mind* is the ability to talk to oneself, already described when we examined language. Mind is simply the activity in each of us that allows us to hold back, make choices, evaluate situations, and generally control our actions. *Self* is the actor's internal object that he or she acts toward. Self allows us to see ourselves in situations, to be able to act back on ourselves, become aware of ourselves, evaluate, direct, control, and judge ourselves. It is what one thinks and does through the use of language, self, and mind that allows for individual *as cause,* and it is society that is the creation of these qualities.

To end organized life is to end human life. Without social patterns, there would be chaos; and with chaos, there are no guides to action and cooperation is impossible. As romantic as complete freedom might sound to many of us, without social patterns it is hard to conceive of anything but destructive social conflict, power based on individual force, and a total disregard for the freedom of anyone but oneself.

Indeed, social organization actually thrives on the freedom of the individual. Our active rather than passive nature allows for deliberation, problem solving, understanding, and ongoing complex organization. It is difficult for the organization to continue its existence if it is rigid, unyielding, and in total control of the actor. Creativity and change are necessary, and freedom encourages them. The issue, of course, has to do with a working balance between order and freedom. In a democratic society, it is important to maximize freedom, and that means creating those conditions that encourage free thought and free action.

Freedom, however, if it exists at all, must be understood as relative and never absolute. No idea, act, or society can be completely free, but if it exists at all, we need to examine how it is possible. In this chapter there is an attempt to describe one of the most useful explanations.

## Summary and Conclusion

It should be obvious now that freedom is far more complex than what most of us learn from political leaders, the media, and everyday social interaction. It is a complex subject, and sociology tries to make our understanding much more than taken-for-granted patriotism. It is more than political dictatorship or democratic government, but it includes this element. It is more than conformity or nonconformity, but it includes this element, too. It is more than thinking or control of thinking, and it is more than acting or control of acting, but it includes these. A summary is in order to spell out briefly and to the point how freedom was discussed in this chapter.

1. Freedom was defined as the ability to control oneself, one's thinking, and one's actions.
2. Much of the sociological perspective emphasizes how society and various social forces control the human being.
3. Freedom can be divided into free thought and free action.
4. To the sociologist, thinking is the product of the social construction of reality. Control over thinking is highly social, arising from culture, language, structure, and knowledge.

5. To the sociologist, freedom to act is the second level of freedom.

6. Yet society is more than a prison. Through socialization and the development of language, self, and mind that allows for both thinking and directing our self as we act.

Freedom is far from automatic; it neither comes naturally nor simply comes to us as we grow older. Its likelihood depends on a society that allows and encourages it, on social conditions that do not oppress it, and on the continuous efforts of the individual to actively and intelligently pursue it.

Erich Fromm (1956: 48) writes that the most difficult kind of love is love for our own children because the whole purpose is to love them so that they can leave us and take control over their own lives, to freely choose their thoughts and actions. To work actively for the freedom of others takes intelligence, courage, and often great sacrifice. There are many who would tempt us to give up our freedom or who silently or forcefully take away whatever freedom we have. We need to recognize who they are, and how to liberate ourselves and others from them. Perhaps that is the real meaning of Henry David Thoreau's imprisonment discussed earlier in the chapter. Yes, he was imprisoned for his actions; and, yes, he lost his freedom of action, but he decided he had to give up his freedom of action so that others might live more freely. Whatever rights we have achieved in this society were fought for by others; whatever freedom we actually have came through many acts of others. Freedom should not simply become a slogan or a meaningless word; it should become a profoundly complex way of thinking and acting, never complete, always tenuous, always attacked, and always worth creating and defending.

## Questions to Consider

1. Is it possible to measure if anyone is free to any degree? Can freedom be proved?

2. Is language a prerequisite for freedom? Is it possible to be free without using language to think and control one's own thoughts and actions?

3. Do you agree that free thought is necessary for free action?

4. Why is it so important for humans to believe that they are free?

5. Is society really "our prison in history," as Peter Berger puts it?

6. How would you define freedom? Is it different from this chapter's definition?

7. Can you think of any belief that is entirely your own? What is its origin?

8. Is it possible for a man to think like a woman? Is it possible for someone who is white to think like someone who is black? Is it possible for a teacher to think like a student?

9. How would a judge in a court of law answer the question, Are human beings free?

10. How would a patriotic American answer the question, Are human beings free?

## References

The following works deal with the question of freedom in society by examining many of the social forces that act on individuals and limit their free thought or free action or both. Some will attempt to show the importance of language, self, and mind in creating some degree of freedom.

**Adams, Bert N., and R. A. Sydie.** 2002. *Contemporary Sociology Theory.* Thousand Oaks, CA: Pine Forge.

**Adams, E. M.** 1993. *Religion and Cultural Freedom.* Philadelphia: Temple University Press.

**Adler, Patricia A., and Peter Adler**, eds. 1996. *Constructions of Deviance: Social Power, Context, and Interaction.* 2nd ed. Belmont, CA: Wadsworth.

**Akers, Ronald L.** 1994. *Criminological Theories: Introduction and Evaluation.* Los Angeles: Roxbury.

**Anderson, Elijah.** 1999. *Code of the Street: Decency, Violence, and the Moral Life of the Inner City.* New York: Norton.

**Anderson, Eric.** 2005. *In the Game: Athletes and the Cult of Masculinity.* New York: State University of New York.

**Aronowitz, S., and W. DiFazio**. 1994. *The Jobless Future: Sci/Tech and the Dogma of Work.* Minneapolis: University of Minnesota Press.

**Aronowitz, Stanley.** 2003. *How Class Works: Power and Social Movement.* New Haven, CT: Yale University Press.

———. 2005. *Just around the Corner: The Paradox of the Jobless Recovery.* Philadelphia: Temple University Press.

**Aronson, Elliot.** 1998. *The Social Animal.* 8th ed. San Francisco: Freeman.

**Asch, Solomon E.** 1951. "Effects of Group Pressure upon the Modification and Distortion of Judgments," pp. 00–00 in *Groups, Leadership, and Men,* edited by Harold Guetzdow. New York: Carnegie.

**Baldwin, James.** 1963. *The Fire Next Time.* New York: Dial.

**Ballantine, Jeanne H.** 1997. *The Sociology of Education.* 4th ed. Englewood Cliffs, NJ: Prentice Hall.

**Banes, Colin, and Geof Mercer**. 2003. *Disability.* Malden, MA: Blackwell.

**Barlett, Donald L., and James B. Steele**. 1996. *America: Who Stole the Dream?* Kansas City, MO: Andrews & McMeel.

**Barnes, Sandra L.** 2005. *The Cost of Being Poor: A Comparative Study of Life in Poor Urban Neighborhoods in Gary Indiana.* New York: State University of New York Press.

**Becker, Howard S.** 1953. "Becoming a Marihuana User." *American Journal of Sociology* 59: 235–247.

**Bellah, Robert N., Richard Madsen, William M. Sullivan, Ann Swidler, and Steven M. Tipton**. 1985. *Habits of the Heart: Individualism and Commitment in American Life.* Berkeley: University of California Press.

**Berger, Bennett M.** 1995. *An Essay on Culture: Symbolic Structure and Social Structure.* Berkeley: University of California Press.

**Berger, Peter L.** 1963. *Invitation to Sociology.* Garden City, NY: Doubleday.

**Berger, Peter L., Brigitte Berger, and Hansfried Kellner**. 1974. *The Homeless Mind: Modernization and Consciousness.* New York: Vintage Books.

**Berger, Peter L., and Thomas Luckmann**. 1966. *The Social Construction of Reality.* Garden City, NY: Doubleday.

**Berrick, Jill Duerr.** 1995. *Faces of Poverty: Portraits of Women and Children on Welfare.* New York: Oxford University Press.

**Blumer, Herbert.** 1969. *Symbolic Interactionism: Perspective and Method.* Englewood Cliffs, NJ: Prentice Hall.

**Borchard, Kurt.** 2005. *The Word on the Street: Homeless Men in Las Vegas.* Reno, NV: University of Nevada Press.

**Bowles, Samuel, Herbert Gintis, Melissa Osborne-Groves**. 2005. *Unequal Chances: Family Background and Economic Success.* Brunswick, NJ: Princeton University Press.

**Brecher, Jeremy, and Tim Costello.** 1994. *Global Village or Global Pillage: Economic Reconstruction from the Bottom Up.* Cambridge, MA: South End Press.

**Brim, Orville G., Jr.** 1968. "Adult Socialization." In *Socialization and Society,* edited by John A. Clausen. Boston: Little, Brown.

**Brim, Orville G., and S. Wheeler**, eds. 1966. *Socialization after Childhood.* New York: Wiley.

**Danzinger, Sheldon, and Ann Chih Lin**, eds. 2000. *Coping with Poverty: The Social Contexts of Neighborhood, Work, and Family in the African-American Community.* Ann Arbor: University of Michigan Press.

**Davey, Joseph Dillon.** 1995. *The New Social Contract: America's Journey from Welfare State to Police State.* Westport, CT: Praeger.

**Derber, Charles.** 1996. *The Wilding of America: How Greed and Violence Are Eroding Our Nation's Character.* New York: St. Martin's.

**Diamond, Larry.** 1999. *Developing Democracy.* Baltimore, MD: Johns Hopkins University Press.

**Dohan, Daniel.** 2003. *The Price of Poverty: Money, Work, and Culture in the Next Mexican American Barrio.* Berkeley: University of California Press.

**Durkheim, Émile.** [1895] 1964. *The Rules of the Sociological Method,* translated by Sarah A. Solovay and John H. Mueller. New York: Free Press.

———. [1915] 1954. *The Elementary Forms of Religious Life,* translated by Joseph Swain. New York: Free Press.

**Edin, Kathryn, and Maria Kefalas.** 2007. *Promises I Can Keep: Why Poor Women Put Motherhood before Marriage.* Berkeley: University of California Press.

**Ehrenreich, Barbara.** 2006. *Bait and Switch: The (Futile) Pursuit of the American Dream.* New York: Metropolitan Books

**Ehrensal, Kenneth.** 2001. "Training Capitalism's Soldiers: The Hidden Curriculum of Undergraduate Business Education," pp. 00–00 in *The Hidden Curriculum,* edited by Eric Margolis. New York: Routledge/ Taylor & Francis.

**Eliade, Mircea.** 1954. *Cosmos and History.* New York: Harper & Row.

**Etzioni, Amitai.** 1991. "Too Many Rights, Too Few Responsibilities." *Society,* 28(2): : 41–48.

———. 2001. *The Monochrome Society.* Princeton, NJ: Princeton University Press.

**Ewen, Stuart.** 1976. *Captains of Consciousness.* New York: McGraw-Hill.

**Ezekiel, Raphael S.** 1995. *The Racist Mind.* New York: Viking.

**Faulks, Keith.** 2000. *Political Sociology: A Critical Introduction.* New York: New York University Press.

**Feagin, Joe R.** 1975. *Subordinating the Poor: Welfare and American Beliefs.* Englewood Cliffs, NJ: Prentice Hall.

**Festinger, Leon.** 1954. "A Theory of Social Comparison Processes." *Human Relations* 7: 117–140.

———. 1956. *When Prophecy Fails.* Minneapolis: University of Minnesota Press.

**Fine, Gary Alan.** 1987. *With the Boys: Little League Baseball and Preadolescent Culture.* Chicago: University of Chicago Press.

**Freie, John F.** 1998. *Counterfeit Community: The Exploitation of Our Longings for Connectedness.* Lanham, MD: Rowman & Littlefield.

**Freud, Sigmund**. [1930] 1953. *Civilization and Its Discontents.* London: Hogarth.

**Friedman, Thomas L.** 2005. *The World Is Flat: A Brief History of the Twenty-First Century.* New York: Farrar, Straus & Giroux.

**Fromm, Erich.** 1941. *Escape from Freedom.* New York: Holt, Rinehart & Winston.

———. 1956. *The Art of Loving.* New York: Harper & Row.

———. 1962. *Beyond the Chains of Illusion.* New York: Simon & Schuster.

**Fuller, Robert W.** 2003. *Somebodies and Nobodies: Overcoming the Abuse of Rank.* Gabriola Island, BC: New Society.

**Geertz, Clifford.** 1965. "The Impact of the Concept of Culture on the Concept of Man," pp. 00–00 in *New Views of the Nature of Man,* edited by John R. Platt. Chicago: University of Chicago Press.

**Giddens, Anthony.** 2000. *Runaway World: How Globalization Is Reshaping Our Lives.* New York: Routledge.

**Gilbert, Dennis, and Joseph A. Kahl**. 1997. *The American Class Structure in an Age of Growing Inequality: A New Synthesis.* 5th ed. Belmont, CA: Wadsworth.

**Glassner, Barry, and Rosanna Hertz**. 2003. *Our Studies, Ourselves: Sociologists' Lives and Work.* New York: Oxford University Press.

**Goffman, Erving.** 1959. *The Presentation of Self in Everyday Life.* Garden City, NY: Doubleday.

**Goldhagen, Daniel Jonah.** 1996. *Hitler's Willing Executioners: Ordinary Germans and the Holocaust.* New York: Knopf.

**Goldman, Alvin I.** 2002. *Pathways to Knowledge: Private and Public.* New York: Oxford University Press.

**Goode, Erich**, ed. 1996. *Social Deviance.* Boston: Allyn & Bacon.

**Goode, Erich.** 2002. *Deviance in Everyday Life: Personal Accounts of Unconventional Lives.* Prospect Heights, IL: Waveland.

**Guillaumin, Colette.** 1995. *Racism, Sexism, Power and Ideology.* New York: Routledge.

**Hall, John R., and Mary Jo Neitz**. 1993. *Culture: Sociological Perspectives*. Englewood Cliffs, NJ: Prentice Hall.

**Hammond, Phillip E.** 1992. *Religion and Personal Autonomy: The Third Disestablishment in America*. Columbia: University of South Carolina Press.

**Hare, Bruce R.**, ed. 2002. *2001 Race Odyssey: African Americans and Sociology*. Syracuse, NY: Syracuse University Press.

**Harris, Scott R.** 2006. *The Meaning of Marital Equality*. New York: University of New York Press.

**Hayes, Sharon.** 2003. *Flat Broke with Children: Women in the Age of Welfare Reform*. New York: Oxford University Press.

**Healey, Joseph F.** 1997. *Race, Ethnicity, and Gender in the United States: Inequality, Group Conflict, and Power*. Thousand Oaks, CA: Pine Forge.

**Herman, Edward S., and Noam Chomsky**. 1988. *Manufacturing Consent: The Political Economy of the Mass Media*. New York: Pantheon.

**Hertzler, Joyce O.** 1965. *A Sociology of Language*. New York: Random House.

**Hewitt, John P.** 2000. *Self and Society*. 8th ed. Boston: Allyn & Bacon.

**Hochschild, Arlie Russell, and Anne Machung**. 2003. *The Second Shift*. New York: Viking Penguin.

**Iceland, John.** 2003. *Poverty in America: A Handbook*. Berkeley: University of California Press.

**Janis, Irving L.** 1982. *Groupthink*. 2nd ed. Boston: Houghton Mifflin.

**Johnson, Heather B.** 2006. *The American Dream and the Power of Wealth: Choosing Schools and Inheriting Inequality in the Land of Opportunity*. New York: Routledge.

**Jones, Ron.** 1981. *No Substitute for Madness*. Covelo, CA: Island Press.

**Kelso, William A.** 1994. *Poverty and the Underclass: Changing Perceptions of the Poor in America*. New York: New York University Press.

**Kerbo, Harold R.** 1999. *Social Stratification and Inequality*. 4th ed. New York: McGraw-Hill.

**Kerbo, Harold R, and John A. McKinstry**. 1995. *Who Rules Japan: The Inner Circles of Economic and Political Power*. Westport, CT: Praeger.

**Lareau, Annette.** 2003. *Unequal Childhoods: Class, Race, and Family Life*. Berkeley: University of California.

**Liebow, Elliot.** 1967. *Tally's Corner*. Boston: Little, Brown.

———. 1993. *Tell Them Who I Am: The Lives of Homeless Women*. New York: Simon & Schuster.

**Lindesmith, Alfred R., Anselm L. Strauss, and Norman K. Denzin**. 1999. *Social Psychology*. 8th ed. Thousand Oaks, CA: Sage.

**Lofland, John.** 1966. *Doomsday Cult*. Englewood Cliffs, NJ: Prentice Hall.

**Lukes, Steven.** 1973. *Individualism*. New York: Harper & Row.

———. 2005. *Power: A Radical View.* 2nd ed. Oxford, UK: Palgrave Macmillan.

**Malcolm X and Alex Haley**. 1965. *The Autobiography of Malcolm X.* New York: Grove Press.

**Mannheim, Karl**. [1929] 1936. *Ideology and Utopia.* New York: Harcourt Brace Jovanovich.

**Marcuse, Herbert.** 1964. *One-Dimensional Man.* Boston: Beacon.

**Marger, Martin.** 1999. *Race and Ethnic Relations: American and Global Perspectives.* 5th ed. Belmont, CA: Wadsworth.

**Marx, Karl, and Friedrich Engels**. [1848] 1955. *The Communist Manifesto.* New York: Appleton-Century-Crofts.

**McCall, George J., and J. L. Simmons**. 1978. *Identities and Interactions.* New York: Free Press.

**McCall, Nathan.** 1994. *Makes Me Wanna Holler.* New York: Random House.

**McCord, Joan**, ed. 1997. *Violence and Childhood in the Inner City.* New York: Cambridge University Press.

**McNamee, Stephen J., and Robert K. Miller, Jr.** 2004. *The Meritocracy Myth.* Landham, MD:. Rowman & Littlefield.

**McNeill, William H.** 1993. "Epilogue: Fundamentalism and the World of the 1990s," pp. 00–00 in *Fundamentalisms and Society: Reclaiming the Sciences, the Family, and Education,* edited by Martin E. Marty and R. Scott Appleby. Chicago: University of Chicago Press.

**Mead, George Herbert.** 1925. "The Genesis of the Self and Social Control." *International Journal of Ethics* 35: 251–277.

———. 1934. *Mind, Self and Society.* Chicago: University of Chicago Press.

**Merry, Sally E.** 2005. *Human Rights and Gender Violence: Translating International Law into Local Justice.* Chicago: University of Chicago Press.

**Milgram, Stanley.** 1963. "Behavioral Study of Obedience." *Journal of Abnormal and Social Psychology* 67: 371–378.

**Miller, W. Watts.** 1996. *Durkheim, Morals and Modernity.* Montreal/London: McGill-Queen's University Press/UCL Press.

**Mills, C. Wright.** 1956. *The Power Elite.* New York: Oxford University Press.

———. 1959. *The Sociological Imagination.* New York: Oxford University Press.

**Morrison, Ken.** 1995. *Marx, Durkheim, Weber: Formations of Modern Social Thought.* Thousand Oaks, CA: Sage.

***New York Times.*** 2005. *Class Matters.* New York: Times Books.

**Newman, Katherine S.** 1999. *No Shame in My Game: The Working Poor in the Inner City.* New York: Knopf.

**Parenti, Michael.** 1998. *America Besieged.* San Francisco, CA: City Lights.

**Parrillo, Vincent N.** 2002. *Contemporary Social Problems.* 5th ed. Boston: Allyn & Bacon.

**Perrow, Charles.** 1986. *Complex Organizations.* 3rd ed. New York: Random House.

**Perrucci, Robert, and Earl Wysong**. 2003. *The New Class Society: Goodbye American Dream?* 2nd ed. Landham, MD: Rowman & Littlefield.

**Reich, Charles A.** 1995. *Opposing the System.* New York: Crown.

**Rose, Peter I.**, ed. 1979. *Socialization and the Life Cycle.* New York: St. Martin's.

**Rousseau, Nathan**. 2002. *Self, Symbols, and Society: Classic Readings in Social Psychology.* New York: Rowman & Littlefield.

**Ryan, William.** 1976. *Blaming the Victim.* Rev. ed. New York: Vintage.

**Sapir, Edward.** 1949. *Language: An Introduction to the Study of Speech.* New York: Harcourt, Brace, & World.

**Scaff, Lawrence A.** 2000. "Weber on the Cultural Situation of the Modern Age." Chapter 5 in *The Cambridge Companion to Weber,* edited by Stephen Turner. New York: Cambridge University Press.

**Schachter, Stanley.** 1951. "Deviation, Rejection, and Communication." *Journal of Abnormal and Social Psychology* 46: 229–238.

**Schlosser, Eric.** 2001. *Fast Food Nation: The Dark Side of the All-American Meal.* Boston: Houghton Mifflin.

**Schwartz, Barry.** 1994. *The Costs of Living: How Market Freedom Erodes the Best Things in Life.* New York: W. W. Norton.

**Shapiro, Thomas M.** 2003. *The Hidden Costs of Being African American.* New York: Oxford University Press.

**Shibutani, Tamotsu.** 1955. "Reference Groups as Perspectives." *American Journal of Sociology* 60: 562–569.

———. 1961. *Society and Personality: An Interactionist Approach to Social Psychology.* Englewood Cliffs, NJ: Prentice Hall.

———. 1986. *Social Processes: An Introduction to Sociology.* Berkeley: University of California Press.

**Smart, Julie.** 2001. *Disability, Society, and the Individual.* Gaithersburg, MD: Aspen.

**Spindler, George, Louise Spindler, Henry T. Trueba, and Melvin D. Williams**. 1990. *The American Cultural Dialogue and Its Transmission.* Bristol, PA: Falmer.

**Stirk, Peter M. R.** 2000. *Critical Theory, Politics, and Society: An Introduction.* New York: Pinter.

**Thoreau, Henry David**. [1849] 1948. "On the Duty of Civil Disobedience," pp. 00–00 in *Walden: On the Duty of Civil Disobedience.* New York: Holt, Rinehart & Winston.

**Tichenor, Veronica J.** 2005. *Earning More and Getting Less: Why Successful Wives Can't Buy Equality.* New Brunswick, NJ: Rutgers University Press.

**Titchkosky, Tanya.** 2003. *Disability, Self, and Society.* Toronto: University of Toronto Press.

**Tonry, Michael.** 2004. *Thinking About Crime: Sense and Sensibility in American Penal Culture.* New York: Oxford University Press.

**Travis, Jeremy, and Michelle Waul**, ed. 2003. *Prisoners Once Removed: The Impact of Incarceration and Reentry on Children, Families, and Communities.* Washington, DC: The Urban Institute.

**Trotter, Joe W., with Earl Lewish and Tera W. Hunter**, eds. 2004. *African American Urban Experience: Perspectives from the Colonial Period to the Present.* New York: Palgrave Macmillan.

**Tucker, Robert C.**, ed. 1972. *The Marx-Engels Reader.* New York: Norton.

**U.S. Bureau of the Census.** 1999. *Current Population Survey, March.* Washington, DC: U.S. Government Printing Office.

**Venatesh, Sudhir Alladi.** 2006. *Off the Books: The Undergound Economy of the Urban Poor.* Cambridge, MA: Harvard University Press.

**Waldinger, Roger David, and Michael I. Lichter**. 2003. *How the Other Half Works: Immigration and the Social Organization of Labor.* Berkeley: University of California Press.

**Walker, Beverly M.**, ed. 1995. *Construction of Group Realities: Culture, Society, and Personal Construction Theory.* New York: Praeger.

**Warriner, Charles K.** 1970. *The Emergence of Society.* Homewood, IL: Dorsey.

**Weber, Max**. [1905] 1958. *The Protestant Ethic and the Spirit of Capitalism,* translated and edited by Talcott Parsons. New York: Scribner's.

**Western, Bruce** 2006. *Punishment and Inequality.* New York: Russell Sage Foundation.

**White, Leslie A.** 1940. *The Science of Culture.* New York: Farrar, Straus & Giroux.

**Whorf, Benjamin Lee.** 1941. "Languages and Logic." *Technology Review* 43: 250–252, 266, 268, 272.

**Whyte, William Foote.** 1955. *Street Corner Society.* Chicago: University of Chicago Press.

**Wilson, William Julius**, ed. 1993. *The Ghetto Underclass: Social Science Perspectives.* Newbury Park, CA: Sage.

**Winant, Howard.** 1994. *Racial Conditions: Politics, Theory, Comparisons.* Minneapolis: University of Minnesota Press.

**Wolfe, Alan.** 2001. *Moral Freedom: The Impossible Idea That Defines the Way We Live Now.* New York: Norton.

**Wright, Erik Olin.** 1997. *Class Counts: Comparative Studies in Class Analysis.* Cambridge: Cambridge University Press.

**Wrong, Dennis H.** 1961. "The Oversocialized Conception of Man in Modern Sociology." *American Sociological Review* 26: 183–193.

**Yinger, Milton J.** 1982. *Countercultures: The Promise and Peril of a World Turned Upside Down.* New York: Free Press.

**Young, Alfred A.** 2004. *The Minds of Marginalized Black Men.* Princeton, NJ: Princeton University Press.

**Zimbardo, Philip.** 1972. "Pathology of Imprisonment." *Society* 9: 4–8.

# Why Can't Everyone Be Just Like Us?

## Value Judgments, Ethnocentrism, and Human Differences

### Concepts, Themes, and Key Individuals

- ❑ Values and value judgments
- ❑ Culture
- ❑ Ethnocentrism
- ❑ Deviance, oppression, ideology, and social conflict
- ❑ Functions and costs of ethnocentrism

The ancient Greeks lived in many small city-states. Each city-state was independent and had its own government, army, and economy. Some, such as Athens, were democracies; some, such as Sparta, were autocracies. Together, however, their citizens shared a heritage: They were all Greeks. Beyond the mountains and sea lived *other* peoples. Such people were strangers, barbarians. Their ways were different and less desirable. They were, in a word, "uncivilized." Like the Greeks, the Romans also saw the world divided into two: the civilization of Rome and the barbarian peoples. The medieval world divided people into heathen and Christian, and the European peoples who came to the Americas encountered many different cultures but called them all "Indian" and commonly described their ways as savage.

When I attended North High School, I honestly believed that somehow our school, our student body, our teachers, and our teams were better than others. My loyalty to North included a lingering belief that we were truly blessed over those who attended other schools. At all athletic events, I was sure that in controversial calls we were right and the other team somehow had the referees on its side.

Most Americans do, in fact, believe that the United States is the greatest nation in the world, and it is difficult for us to believe that there are other ways of living that are equally good or even better. When we looked at the former Soviet Union, we blamed its problems and shortcomings on an authoritarian regime and on government involvement in the economy, and we claimed that if only the Kremlin's ways could become similar to our ways, the people could have enjoyed what we enjoy. Indeed, when we look at other cultures, we tend to distinguish them according to how close they come to our own. We see some as primitive, some as developing, and some as developed and civilized.

I am talking here of several critical issues, all intimately related. It is important to examine these one at a time.

## The Meaning of Values

"Oompa, oompa, oompa-pa, my pa's better than your pa-pa." My religion is better than yours. My school. My major. My parents are. My morals. My life plans. My goals in life. My car. My friends.

Comparisons have something to do with values. Whenever we use terms such as *better, best, good, bad, superior, inferior, should,* and *should not,* we enter the complex world of values. The tip-off that people are discussing values is whenever they use or imply the word *should.* In that case, someone is always making a value judgment. The statement has to do with what should exist in the world rather than what actually exists. The title of this chapter is "Why Can't Everyone Be Just Like Us?" Although *should* is not in the question, it is certainly implied: "Others should be like us, so why aren't they?"

I vividly remember a conversation with two professors on a four-hour drive. They were singing the praises of higher education. "Everyone should get a college education," they said. "Knowledge is better than ignorance." I turned to them and boldly declared, "You're making a value judgment. Although I generally agree with you, there's no way any of us can prove that we're right. Only statements of fact can be proved." They disagreed, and we argued back and forth. I asked, "Why is knowledge better than ignorance?" Their answer: "Because it helps us succeed in the occupational world." "Well," I replied, "who says that we *should* succeed in the occupational world?" Such questions are often important and sometimes trivial, but they always involve assumptions of what we think life *should be like,* and thus they become questions of values.

Values are our commitments, and they reflect our image of what is good and what is not good in this world. Values are the standards against which people judge their own acts and the acts of others. They tell us "what goals people ought to seek, what is required or forbidden, what is honorable and shameful, and what is beautiful and ugly" (Shibutani, 1986: 68). If I really believe that having a family is important to a meaningful life, then that is a value to me. I live my life for my family, I vote on issues affecting my family, and I spend time and money on my family. Perhaps I even broaden this commitment to acting in favor of family life throughout the United States and even the world. As often happens, I become so committed to my family that I find it difficult to understand how others who do not have a family life similar to mine can possibly find happiness. I may also claim that this lack makes them immoral or selfish. I find threats to my family and family life in general to be important threats to my existence, and I support efforts to rid society of these threats.

For some of us, freedom is an important value. ("I should be free, all Americans should be free, and all people should be free"). Likewise, law and order might be a value—or religion, equality, artistic expression, education, a healthy body, physical beauty, tradition, individualism, friendship, helping others, living a moral life, making a lot of money, being a good citizen, and so on. These are all examples of what we regard as worthwhile. If I believe in them, they are my values; if you believe in them, they are yours. But there is no way either one of us can prove that ours are better than the other person's, for whenever we try to do this, we inevitably come up against more and more value judgments, none of which can be proved.

Our values can be contradictory. Americans can believe in both a segregated society and equal opportunity for all, or they can find themselves simultaneously worshiping individualism and group loyalty. I find myself attracted to tradition and progress at the same time, and sometimes I am torn between spending my time writing a book and listening to the concerns of my wife and children.

Most of our decisions in life involve choices we make among several values that we hold. We might value both freedom of expression and the rights of women. On the issue of pornography, we might have to choose between these values: "Yes, I value freedom of expression, but I don't think people have the right to produce pornography that denigrates women." Or on matters of civil rights: "Of course, I favor equality for all races. But I also believe that people should go to schools in their own neighborhoods." It is not always easy to turn our backs on one of our values so that we can work for another, but on occasion we must do just that. And, of course, this causes conflict in most of us whenever we recognize the contradiction.

## Values and Making Value Judgments

Judging other people is how values enter our social life. We like others from the value judgments we make: "He is a true individual." "She is a really ambitious person." "I respect the fact that he speaks up." "She is really pretty." We also dislike others based on value judgments we make: "He's dishonest." "She's stuck up." "They're immoral." "They're lazy." In each case, we create a measuring stick (a value) and use it to judge others. Judging, of course, is more than liking or not liking others. It is also deciding who should be punished, who should be promoted, whose death is called for, or who should live a happy life. It is deciding who must be changed, and whom we make war on.

When we ask, "Why can't others be just like us?" we are asking a question based on a yardstick we have somehow developed. It is a question that is at heart a statement of values, a statement that our ways are better than others' ways, and that to make a better world, others should become like us.

All of us probably do this type of judging on occasion. Some of us do it often. But value judgments are statements of preference, not fact. There is no way to prove that "my pa's better than your pa-pa" unless we specify what we mean by "better," and as soon as we do that, we are making a value judgment, which really is an *assumption* about what is preferable in fatherhood.

Several issues brought up in this chapter are at the core of many controversies in our society today. These issues come up time and time again in Washington, D.C., in the various states, in the media, and in our everyday interactions.

Many political, religious, and economic leaders argue that our values are not simply socially developed; they are either handed down by a supernatural being or they are freely arrived at by the individual. Many argue that our values are true and self-evident; it is unnecessary to even evaluate them. Finally, many argue that standing in judgment of others is not a problem—in fact, because our values are the right ones, those who act contrary to our values must be wrong.

For the sociologist, these ideas concerning values that arise from many leaders in society need to be examined more critically if we are to understand human beings. For the sociologist, these ideas have contributed to serious problems within societies and among societies. They are not ideas that are wrong necessarily, but they are ideas that need to be more carefully examined. As they stand, they are much too simple and misleading. In the end, they cause the serious thinker to confront and try to answer one of the most basic human questions of all: When is it necessary to take a stand and judge others, and when should we cease judgment and accept their differences? This is hardly ever an easy question to answer, except for the tiny few who always judge others and the tiny few who never judge. Let me briefly state how sociologists tend to approach values and making judgments about others.

1. *Values, like ideas and rules, are cultural.* We must always remember that our values are anchored in our social life. That does not mean they are wrong. They are our preferences, and people, because of their social life, learn to have different preferences.

2. *Human beings tend to believe their own values are true and right, they regularly defend them, and they often try to convince others to consider their goodness.* For example, I have learned in my social life that equality of opportunity and personal freedom are important. I am angry when these values are violated. I try to convince others to believe in these values. I sometimes condemn those who act contrary to them.

3. *Values are matters of preference, and it is impossible to prove that certain ones are the true ones for all to follow.* They are not statements of fact but commitments to what we think life should be. Most of us try hard to make our values seem factual, absolute, and true, but ultimately it is a matter of faith and commitment rather than proof. We might claim these are the true values because they are given to us by God; we might claim people can survive or find meaning only if they follow our specific values; and we might claim

that because most people follow these values they must be the right ones. Perhaps we need to make such claims for personal or social reasons, but for us to understand human action we must continue to recognize that value judgments remain value judgments and are not provable facts.

4. *All humans have values and all make judgments about other people based on those values.* The serious student must understand why this is necessary for us to do, and what some of the consequences can be when we make such judgments.

I have values, and I make judgments of others. My values and value judgments are socially constructed. Therefore, as a member of a group, I tend to believe in these values and tend to judge others accordingly. I want to. Sexism and racism are wrong to me. They violate my view of what the world should be. Oppression of people in any form violates what I hold to be right. These are the values I have decided to fight for; these are the values I use to judge other people's actions and to tell people they should change.

Not all of my values lead me to judge other people. Some I try only to apply to my own life and not judge others because they violate them. I believe in planning for the future, formal education, having a good family, and not wasting time—I try hard to understand why others might not agree with these values; if I am successful, I do not judge them because I know they are my values and not theirs.

Ethnocentrism is a tendency of a group of people to make value judgments about other people, value judgments that arise from their culture. It is a tendency to believe that our ways are right, and those without our ways, are less right. Ethnocentrism is a mixed blessing: it is almost inevitable, it often encourages and rationalizes inhumanity and oppression, but it also may contribute to society's stability by bringing people together around what they have come to believe is "truth."

## Meaning of Ethnocentrism

Consider what happens as people interact on a continuous basis. Each acts with the others in mind, people act back and forth, and each considers the acts of the others. The more they interact with one another, the less each has an opportunity to encounter outsiders. Cohabitation, going steady, being engaged, being married are cases in point. These relationships often mean cutting off regular interaction with other people. This sometimes means cutting off regular interaction with one's friends outside the relationship. Once one relationship takes over, time becomes short for others. It is difficult to have close friendships with many other people simply because a close friendship takes time and commitment; the more it takes, the less time we have to develop others.

Over time, continuous interaction develops a likeness among the actors, who communicate, share and discuss experiences, and adopt rules, ideas, and values in their relationship. They devise ways of dealing with the world they encounter. They develop a language that has a unique meaning to them. In short, they develop *culture.*

Differences with those outside the interaction are created and accentuated, communication does not occur regularly, and sharing ideas with them proves increasingly difficult. In time, actors develop a set of meanings, understandings, and values that are different from those of the outsiders.

What happens, of course, is that outsiders not only appear different but also come to be seen as strange—maybe deviant, ill, or evil. We make value judgments on the basis of what we are familiar with. We judge others on the basis of the world in which we interact. It is common for us to develop what sociologists call *ethnocentrism.*

*Ethnocentrism* means that people think their culture ("ethno") is central ("centrism") to the universe. It is a tendency to use what we have shared—values, ideas, and rules—as a starting point for thinking about and judging other people. We tend to think in terms of what we have learned in interaction, and it is hard for most of us to stand back and declare: "They are different. So what?

Ethnocentrism involves (1) the development of truths, values, and norms—culture—in social interaction; (2) the perception of others through the lenses of that culture; and (3) the judging of what others think and do according to that culture. Ethnocentrism involves a tendency to assume that one's own culture is right and that others, by definition, must be wrong. It is believing that what is, in fact, a social construction—arrived at through social interaction—is true and right. Ethnocentrism involves little or no evidence, but it assumes that one's socially constructed truths are correct, that one's socially developed rules are morally right, and that one's socially constructed values are better than others. This tendency is different from trying to prove the truth or falsehood of an idea through argument and evidence: It is assuming that others are wrong because they are different from what you believe is true, right, and worthwhile.

Think of ethnocentrism existing throughout our social existence, from the smallest group we join to the society we live in. Actually, we might even imagine the earth as a whole and think what would happen if we encountered a world with different beings. In every case, we see interaction, sharing, isolation, differences from outsiders, and the tendency to develop feelings of ethnocentrism. Not all individuals fall into this trap, but virtually every social organization does. Why? Why is there such a strong tendency for people in a social organization to make value judgments about people outside that organization and declare, "Why can't everyone be just like us?"

## The Reasons Ethnocentrism Arises

### *Social Interaction Encourages Ethnocentrism*

Ethnocentrism develops first simply because of the nature of interaction. We interact and share; we become organized, form a structure and institutions, share a culture, and thus tend to become isolated from others with whom we do not interact. Groups develop differences from one another, as do formal organizations, communities, and societies. *Without interaction with outsiders, differences become difficult to understand and difficult not to judge.* What is real to us becomes comfortable; what is comfortable becomes right. What we do not understand becomes less than right to us. Ethnocentrism is encouraged.

### *Loyalty to an Organization Encourages Ethnocentrism*

Ethnocentrism also develops, however, because of the nature of *social organization.* As we interact and become part of a society or a group, we generally come to feel something good about belonging to that group. We not only are American but also come to feel good about being American. We support our troops in the world; we tend to give our leaders any benefit of the doubt when there is conflict with other nations. Our identity becomes tied to what we feel good about. Life takes on meaning in that organization. We feel good that we belong to something. I am a Marine, a Xerox employee, an Elk, a New Yorker, a student at Harvard, a member of the National Association for the Advancement of Colored People. Belonging brings direction, comfort, and security. It brings a social anchor to our lives, giving meaning to what we do and more certainty to what we believe. Becoming part of a social organization (a small group of friends, a large society) encourages a *sense of loyalty,* and that loyalty encourages ethnocentrism. Loyalty means a commitment to something we regard as important and right. It brings a feeling of obligation to serve and defend. Criticism and threats to the organization are defended against. *It is easy to see alternative ideas, values, rules, and actions as threats to what we feel loyalty to rather than simply qualities that are different from ours. This is a basic cause of ethnocentrism.*

### Socialization Encourages Ethnocentrism

Socialization is the process by which the individual learns the ways of the organization and becomes a part of that organization. The defenders of any organization—a society, a community, a formal organization, or a group—teach the culture, institutions, structure, and loyalty; these, in turn, become part of what the individual knows and comes to believe in. These products of socialization appear to be more than simply socially derived; they become morally right, natural, true to the individual. In this way, socialization upholds and increases the ethnocentrism, the assumption that the ways of the organization have a special central place in the universe. It then becomes difficult to believe that societies that are different can be true and good. Truth and goodness in others is judged by our standards. For most of us, this is the only standard that we have come to know through socialization.

### The Creation of Deviance Encourages Ethnocentrism

Wherever there is culture, there will be individuals who disagree with and violate that culture. Lines are drawn, and people are punished. Punishment shows all members of society that individuality can go only so far. It shows them the consequences of violating rules.

It is relatively easy to recognize that ethnocentrism usually leads to condemnation and punishment of others. However, it is also important to understand that, as we condemn others, we reaffirm the rightness of our culture, making ethnocentrism more legitimate and even increasing its importance. To sociologists, deviants are those who are perceived as violators of society's rules, truths, and values. They are "outsiders" in that they are placed outside of what people know is true and right. They are barbarian, uncivilized, savage, evil, criminal, terrorists, insane. Each society creates its own outsiders by drawing lines: "Over this line there is something wrong with you." The lines shift but always exist. Although individual differences can sometimes be tolerated or even encouraged to some extent, allowing widespread individuality is to admit that there is nothing special about the culture in which we are all supposed to believe. There is a danger that tolerance will legitimate alternatives to our culture, that people will become too critical about what we believe, take for granted, and hold as absolute. The attempt to draw lines, the identification of certain people as violators, the punishment of and lasting stigma associated with these violators all result in reaffirmation of our culture, reinforcement of the sacredness of our rules, and increased ethnocentrism. In a basic sense, punishment of those who violate culture creates a greater certainty that "we are indeed right."

### *Dominance and Oppression Create Ethnocentrism*

The trade in African slaves that prospered from the seventeenth to the early nineteenth centuries was the product of people who realized that there was a fortune to be made by uprooting, transporting, and oppressing large numbers of people without any concern for their own desires, plans, values, or ways of life. Like most of the rest of humanity, slave traders and slave owners probably believed in God, and they probably considered themselves good, upstanding citizens of their world. It is too easy for us to dismiss them as insane or evil. How, in fact, did they live with themselves? Did they have consciences? Did they consider themselves to be moral people?

Cases of inhumanity exist in every people's history. The United States systematically destroyed Native Americans. The Germans murdered millions of people who were defined as less human. Southeast Asia, Yugoslavia, and parts of Africa in the last part of the twentieth century and the early years of the twenty-first were filled with more examples of one group of people systematically and intentionally killing others whom they defined as different. The war in Sudan that is a campaign to murder the people of Darfur for the intentional purpose of ethnic cleansing is a current tragic example.

What is the link between such oppression and ethnocentrism? Without question, ethnocentrism sometimes encourages war, systematic murder, slavery, exploitation, and inequality. In addition, it is also true that ethnocentrism is the result of such acts. Racism did not precede (and cause) slavery; it is clear from the historical record that it was the existence of slavery that influenced the development of racism. Slavery was developed for economic gain, not because one group was seen to be inferior. A racist philosophy, inspired by ethnocentrism, developed to try to justify and protect the institution.

Extend this argument to any instance of inhumanity. Where people oppress others, there normally needs to be a justification for their actions to convince themselves and others that what they do is all right. Some form of ethnocentrism is generally the result. It is all right to oppress because "what they are" is less worthy than "what we are." "God decided that our people should conquer and control the world. The sacrifice of others for our benefit is both necessary and right." The "old boys' network" develops a rationale for its treatment of women that tells them that their own ways are superior; the employer who exploits cheap labor comes to believe that he or she is helping "those people" who do not need the same income as "people like us"; and the conquerors who grab the land and imprison or destroy those who owned it explain that "they didn't use it the right way anyway."

Recognize, then, that ethnocentrism is an ideology, a way of thinking that people use to justify to themselves and others the oppression of people unlike themselves. Where oppression exists, ethnocentrism is encouraged.

### *Social Conflict Encourages Ethnocentrism*

Social conflict is an inherent part of all social life. Wherever there are differences or wherever there is scarcity, there is conflict—not necessarily violent conflict, but at least a struggle over whatever is scarce. Interorganizational conflict (conflict between organizations) normally encourages ethnocentrism. War between societies is the best example, but less violent competition between companies, teams, or communities also reveals this tendency. Those with whom we do battle are portrayed as less worthy and deserving of our contempt:

> We tend to impute to our enemies the most foul motives, often those that we have trouble avowing ourselves: [We tend to believe that] the enemy is inherently perfidious, insolent, sordid, cruel, degenerate, lacking in compassion, and enjoys aggression for its own sake. Everything he does tends to be interpreted in the most unfavorable light. (Shibutani 1970: 226)

At the same time, when we are involved in conflict, we tend to describe our own motives and our own ways as noble.

> [We maintain that] we seldom engage in wars because of greed. We fight for freedom and justice or in defense against unwarranted aggression. We are strong, courageous, truthful, compassionate, peace-loving, and self-sacrificing. We respect the independence of others and are loyal to our allies. (Shibutani 1970: 226)

In conflict with others, we tend to idealize our own ways. We selectively see who they and we are. We exaggerate their faults and exaggerate our own virtues. Enemies are transformed from human beings to objects without rights; it becomes increasingly difficult to see the world from their perspective. What we do against the enemy becomes more acceptable to us, because we are able to rationalize the defense of goodness against evil. Conflict leads us to increase our ethnocentrism, and increasing our ethnocentrism serves to justify and increase conflict. The two build on each other; over time, the world increasingly appears to be a struggle between good and evil. The 1990–91 crisis in the Persian Gulf, for example, became a struggle between the "forces of good" as represented by President George H. W. Bush and the "forces of evil" as represented by President Saddam Hussein. In 2002, George W. Bush reminded us that there are "evil nations" outside of the civilized world, and because of the horrible tragedy committed by those who destroyed the World Trade Center in New York and attacked the Pentagon in Washington, D.C., intentionally destroying the lives of thousands of people, it was relatively easy to label the other side "terrorists," "barbarians," and "murderers" so that we, of course, would be justified in destroying the other side. And, of course, those who became the victims of our battle reaffirmed the rightness of our cause.

Georg Simmel (1908) showed how conflict between organizations encourages both ethnocentrism and a tendency to silent internal dissent. People become more united in belief, more intolerant of people who question, and increasingly aggressive toward the other side. As Shibutani (1970) writes, in war "moderate and reasonable men are virtually immobilized, and the public gets a constant repetition of a single point of view" (p. 228). Criticism of policy is perceived as disloyalty, just as the social conflict with outsiders is perceived to be a threat to everything that is right. Social conflict such as war brings out the rightness of our cause, our ways, and our truths. Judgment of others is more likely than understanding.

### *Summary*

Where social organization exists over time, ethnocentrism is commonplace, probably inevitable. Interaction itself encourages ethnocentrism by limiting our interaction to a relatively few people. Feelings of loyalty to organization, encouraged by leaders, bring ethnocentrism. Socialization by various representatives from family to society, by teaching that social patterns are more than social—indeed, even sacred and universally true—contributes to ethnocentrism. Regularly using culture as a standard of judging people, stigmatizing others as deviant, and punishing violators of organizational patterns reinforce and justify a feeling of rightness in the ways of the organization. Oppressing and exploiting others encourage ethnocentrism because ethnocentrism is used to dehumanize and to justify victimization. War encourages ethnocentrism because leaders inevitably paint the enemy as evil and our own ways as just and good.

It is difficult for people to exist in any form of social organization for a period of time without slipping into the view that their world, their truths, their rules, their values are normal, true, and right. Once this happens, it almost inevitably brings a standard that is used to judge outsiders. This is the essence of ethnocentrism. It is difficult for people in any group to simply accept the differences of those outside that group. The temptation is to see that "either they or we are wrong, and since I know I'm right, they must be wrong."

## Human Differences

"Why can't everyone be just like us?" The question usually implies a value judgment. ("After all, our ways are better!") The question therefore is itself ethnocentric. ("Our ideas, values, norms are right; those that differ from ours are less right.") However, let us ask the question without making a value judgment, without being ethnocentric. Let us become more scholarly and objective and phrase the question differently: "Why are people different from us?" "Why are societies different?" "Why are communities, groups, and formal organizations different from one another?"

### *Social Interaction and Human Differences*

Previously, we discussed social interaction as an important source of ethnocentrism because many human differences are traceable to such interaction. Recall that social interaction pulls some people together while they simultaneously become separated from others. To some, interaction brings familiarity, interdependence, and the social patterns of culture, structure, and social institutions. People become increasingly used to one another as they interact. Over time, their social world becomes part of them, and it appears to be a natural part of the universe.

It is this interaction and the patterns that result that distinguish groups of people from one another. My ideas, values, morals, and traditions are different from yours because the groups within which we were formed are different. I have been influenced by different social organizations. Drugs and alcohol are not part of my life because my interaction has not taken me in that direction. The religious beliefs I hold are traceable to my interaction, and so are my interests and talents. My life is different from yours in part because I grew up in Minneapolis and moved to Moorhead, and you stayed in Minneapolis. My life is different from yours because I interact with sociologists and your contacts are different.

The real meaning of social organization is that it brings commonality, communication, and cooperation with those inside, and it also brings differentiation from, lack of communication with, and much less cooperation with those outside. Interaction and organization bring internal unity and external differences. *So long as there is interaction and so long as that interaction does not include everyone at one time, it is impossible for all of us to be the same.*

### *Social History and Human Differences*

No two societies (or groups or formal organizations or communities) develop in the same way. *All have a different history.* The unique aspects of their development will produce differences between those inside and those outside society, making it impossible for "them" to be like "us." Societies may appear to be alike in that all have important charismatic leaders in their history—a Lenin, a Luther, a Muhammad, a Napoleon, or a Gandhi—but each leader will have brought a unique set of changes, unlike the leaders in other societies. Each society will have a mixture of tradition and modernization, and that mixture will always be unique. Each may depend heavily on one major religion, and in that sense they will be similar, but each religion will be different in several basic ways. Even when two have the same religion (Catholicism, for example), each will, in fact, be different, because the religion will exist within a larger social context. Iran, Syria, Egypt, and Indonesia may all be Muslim, but the life, ideas, values, and even the religion of the people will differ considerably because of their different histories.

All social organizations have unique histories and thus create different social patterns from all others. I belonged to both a poker group and an investment group in Minneapolis. When I moved to another city, I helped form these groups anew. The groups in Minneapolis still exist; in Fargo–Moorhead they also exist. Even though I tried to form the same groups that existed in Minneapolis, they evolved much differently. Why can't the groups in Fargo–Moorhead be just like the ones in Minneapolis? Because they differed in their histories: their experiences, problems, solutions, and social patterns.

"Why is it so difficult for African Americans in the United States to make it economically, politically, and educationally? Other minorities have. Other immigrant groups have. The Jewish people have. The Japanese Americans are. What is different about the African Americans?" The answer to this question is complex, but part of it is found in the different social histories. It lies in the history of the United States—in the force that brought Africans to the Americas, in the institution of slavery, in the Civil War, in post–Civil War conflict and domination, in the immigration of large numbers of whites in the late nineteenth and early twentieth centuries, and in the migration of African Americans from Southern rural areas to Northern urban areas during and after World War I. It lies in the nature of the historical relationship between whites and African Americans—the patterns of segregation, poverty, exclusion, and domination that prevailed for hundreds of years. It lies in our interaction patterns in a highly segregated society, which led to separate communities without open interaction and communication, encouraging different and sometimes clashing social patterns. It lies in a heritage of mistrust and hopelessness, fostered by discrimination in every area of American life. All this history is important in order to understand racism today. The accumulation of economic, educational, social, and political problems developed through hundreds of years does not simply disappear once we realize that rights have been violated in our democratic society. To argue that "I haven't oppressed others; I haven't had slaves; I have no responsibility to sacrifice my advantages" ignores a major misunderstanding of why we face great racial and ethnic inequalities. Today African Americans are different from every other minority in our history; so, too, are Jewish Americans, Japanese Americans, Native Americans, and Mexican Americans. There is no reason to believe that these groups are the same simply because they are or were disadvantaged.

### *Problems and Social Patterns*

People therefore differ from one another because their interaction separates them and their unique histories create different social patterns. Groups, formal organizations, communities, and societies also develop differently *because the problems they encounter are different.* Organizations develop structure, culture, and institutions that work. China cannot be like the United States because the problems it must solve are entirely different from those in the United States and call for different social patterns. For example, the problem of unity and social order has always plagued China. China has really been many societies, not one, and there has been a strong tradition of division. The history of China is one of separate feudal empires, fighting warlords, decentralized governments, and decentralized economies. In contrast, the United States, although it began as separate states, has a stronger tradition of unity, fostered by the Revolutionary War and the founding of a society separate from England, forced by the Civil War, and encouraged by transportation, communication, and economic systems that developed rapidly after the Civil War. China also has been conquered by Japan and attacked by the Soviet Union and has developed a mistrust of its neighbors. This fact influences its ways. The United States, on the other hand, has never lost wars to its neighbors and has not developed this same fear. Finally, the massive population problems created out of a long history of loyalty to family and tradition as well as a long history of widespread poverty have made China a different society from the United States.

Given these different problems, how can we think that U.S. and Chinese societies can be alike? How can we imagine that what works here is going to work there? Private enterprise may be a great institution in the United States, but it is difficult to transplant to a society with different problems and one that has traditionally valued kinship and community over individualism.

Baseball teams that win pennants cannot be like baseball teams that are trying to win their first game. Universities that graduate those who fill elite positions cannot be like universities that try to offer some education to anyone and everyone who wants it. Communities that have serious pollution problems cannot be like communities that must solve the problem of unemployment.

We are often tempted to compare ourselves with other societies, bragging about our progress or even complaining about something we would like improved. Many of us in education yearn for the educational system of a Great Britain or a Singapore. However, the purpose of our educational system has always been different from those of other societies. We have tried to build a high school system that equalizes opportunity as much as possible and a university system that appeals to the needs of the entire population. Our resulting institutions have therefore been different. For good or for bad, ours tries to be an open system in which we give the individual many opportunities for success; until we radically change the purpose of our schools, it is impossible to build a school system similar to those in other societies. If we accepted only the most academically talented in our high schools and then closed the universities to all others, then we could develop a system of education similar to those in other societies, but the whole purpose of education would change and with it the whole nature of our society. Institutions do not develop in a vacuum. Our ways have developed around values and problems that we have designated as important.

Simply put, organizations differ from one another for three reasons: (1) interaction isolates and differentiates them; (2) their histories are unique; and (3) the problems with which their social patterns must deal are different, and this influences what patterns develop.

A fourth reason should be pointed out. Remember that earlier in this chapter we explored ethnocentrism, the tendency for us to regard our ways as right and others' ways as less attractive. *Ethnocentrism enters into why we are all different.* As we begin to be different from others, we fight for what we are, we defend the ways that we are used to. We are reluctant to give up what we have in order to become "like them," and we do what we can to protect our ways. Who wants to be like the strangers anyway? If they try to force us, we will use force. If they try to convert us, we will pull back into our community. Not only do we try to maintain our differences, but also conflict with others actually encourages us to hold on to our differences, to maintain our separate identities as much as we can.

## Summary and Conclusion

You and I exist in a social context. Where we happen to live our lives and with whom we live will influence who we are, what we do, and what we believe. This, in turn, will make you and me different from each other. The intensity of our interaction, the history, problems, and patterns of our organizations, together with the feelings of ethnocentrism that inevitably arise in organizations, will keep us different. Although it might seem someday that you are becoming more and more like me (or vice versa), we should expect that differences will remain and that they will always be substantial.

It is easy to forget the role of social organization in creating the differences between people. In the 1990s, we were too often attracted to racial or biological differences as explanations. It is too easy to think real differences are caused by how people look physically; it is too easy to equate belief and behavior with physical appearance. Biological differences, although they are sometimes important for understanding individual differences, are much less relevant to understanding differences between groups or societies.

We can never have a world where all agree and cooperation is perfect. When it really comes down to it, why should we want that anyway? Human differences are not necessarily bad, and a strong case can be made that they are good. Diversity encourages a dynamic approach to understanding anything in the universe.

It encourages us to evaluate who we are, how we live, and what we believe. Diversity brings alternatives to what we know, new solutions to problems we encounter, and new meanings to our lives. It can teach us respect for differences and humility concerning our own views of reality. It can bring a people a much richer democracy because it can teach them mutual respect rather than simply accepting what the majority wants.

And ethnocentrism? Is that good or bad? It depends on our values, of course. It seems that ethnocentrism may contribute to social solidarity and social order. It helps bind us, and it creates in us a commitment to society. It makes it easier to follow the rules, because the rules seem right. Ethnocentrism makes us feel good about who we are and more certain about what we believe. It gives us an anchor; it helps us decide what is and is not good in the world. It encourages our community to retain its unique qualities. Some ethnocentrism is undoubtedly necessary for the continuation of society.

On the other hand, ethnocentrism is costly. From the standpoint of society, it discourages innovation and change as well as the solution of serious problems. People become opposed to change when it is perceived to threaten qualities in society that they cherish. There is a tendency to ignore serious social problems because their solution is not worth giving up what we cherish. In fact, ethnocentrism discourages us from finding creative approaches to solving problems because we are also generally committed to our particular way of solving our problems.

From the standpoint of the individual, ethnocentrism gets in the way of important values we often express. It hinders our understanding of other people, for it makes us too quick to judge those who are different. It encourages narrow-mindedness and an unwillingness to recognize many human differences for what they are. Ethnocentrism not only stands in the way of understanding others but also hampers us in understanding ourselves because we never can appreciate the fact that we, our society, and our society's rules and truths are, to a great extent, part of a social reality. Ethnocentrism tends to confuse culture with truth; it makes us feel that what we believe is true and right rather than socially developed and open to criticism.

Finally, ethnocentrism too often encourages and justifies inhumanity. It is used by political opportunists to gain support for persecuting and warring against others, to justify stealing from and enslaving others. It leads to persecuting minorities and destroying individuals whose only sin is that they are different from the rest of us.

The dilemma is that ethnocentrism fosters the continuation of society and the security of its members, yet it undermines important qualities that many regard as central to a democratic society.

## Questions to Consider

1. Most beliefs can be examined by evidence to determine the extent to which they are accurate. Can the same be done with values?
2. If we put culture aside, what exactly should be the standards by which humans judge one another?
3. Why do we have such a difficult time accepting human differences without judging those differences to be good or bad?

4. How important is ethnocentrism to oppression and war?

5. Exactly why are people who live in China today different from people who live in the United States?

6. How would someone who is active in the Ku Klux Klan answer the question, Why can't everyone be just like us?

## REFERENCES

The following works deal with human values and their development in society, with ethnocentrism, or with how differences between groups arise.

**Adler, Patricia A., and Peter Adler,** eds. 1996. *Constructions of Deviance: Social Power, Context, and Interaction.* 2nd ed. Belmont, CA: Wadsworth.

**Altheide, David.** 2006. *Terrorism and the Politics of Fear.* Landham, MD: Rowman & Littlefield.

**Anderson, Elijah.** 1990. *Streetwise: Race, Class, and Change in an Urban Community.* Chicago: University of Chicago Press.

———. 1999. *Code of the Street: Decency, Violence, and the Moral Life of the Inner City.* New York: Norton.

**Barton, Bernadette.** 2006. *Stripped: Inside the Lives of Exotic Dancers.* New York: New York University Press.

**Becker, Howard S.** 1973. *Outsiders.* Enlarged ed. New York: Free Press.

**Berger, Bennett M.** 1995. *An Essay on Culture: Symbolic Structure and Social Structure.* Berkeley: University of California Press.

**Berger, Peter L., and Thomas Luckmann.** 1966. *The Social Construction of Reality.* Garden City, NY: Doubleday.

**Bowles, Samuel, Herbert Gintis, and Melissa Osborne-Groves.** 2005. *Unequal Chances: Family Background and Economic Success.* Princeton, NJ: Princeton University Press.

**Chang, Iris.** 1997. *The Rape of Nanking: The Forgotten Holocaust of World War II.* New York: Basic Books.

**Charon, Joel M.** 2009. "An Introduction to the Study of Social Problems," pp. 1–12 in *Social Problems: Readings with Four Questions,* edited by Joel Charon and Lee Vigilant. 3rd ed. Belmont, CA: Wadsworth Cengage Learning.

**Charon, Joel M., and Lee Garth Vigilant.** 2009. *Social Problems: Readings with Four Questions,* 3rd ed. Belmont, CA: Wadsworth Cengage Learning.

**Cohen, Mark Nathan.** 1998. *Culture of Intolerance: Chauvinism, Class, and Racism in the United States.* New Haven, CT: Yale University Press.

**Crothers, Lane.** 2003. *The American Militia Movement from Ruby Ridge to Homeland Security.* New York: Rowman & Littlefield.

**Davis, Mike.** 2000. *Magical Urbanism: Latinos Reinvent the U.S. City.* New York: Verso.

**Diamond, Larry.** 1999. *Developing Democracy.* Baltimore, MD: Johns Hopkins University Press.

**Dohan, Daniel.** 2003. *The Price of Poverty: Money, Work, and Culture in the Next Mexican American Barrio.* Berkeley, CA: University of California Press.

**Durkheim, Émile**. [1893] 1964. *The Division of Labor in Society,* translated by George Simpson. New York: Free Press.

———. [1915] 1954. *The Elementary Forms of Religious Life,* translated by Joseph Swain. New York: Free Press.

**Edin, Kathryn, and Maria Kefalas**. 2007. *Promises I Can Keep: Why Poor Women Put Motherhood Before Marriage.* Berkeley: University of California Press.

**Eliade, Mircea.** 1954. *Cosmos and History.* New York: Harper & Row.

**Emerson, Michael O., and Rodney M. Woo**. 2006. *People of the Dream: Multiracial Congregations in the United States.* Princeton, NJ: Princeton University Press.

**Erikson, Kai T.** 1966. *Wayward Puritans: A Study in the Sociology of Deviance.* New York: Wiley.

———. 1976. *Everything in Its Path.* New York: Simon & Schuster.

**Ezekiel, Raphael S.** 1995. *The Racist Mind.* New York: Viking.

**Farley, John E., and Gregory D. Squires**. 2005. "Fences and Neighbors: Segregation in 21st-Century America" *Contexts* 4(1): 33–39.

**Farley, John E.** 1999. *Majority-Minority Relations.* 4th ed. Upper Saddle River, NJ: Prentice Hall.

**Feagin, Joe R., and Hernan Vera**. 1995. *White Racism: The Basics.* New York: Routledge.

**Featherstone, Mike**, ed. 1990. *Global Culture: Nationalism, Globalization, and Modernity.* London: Sage.

**Foner, Nancy.** 2005. *In a New Land: A Comparative View of Immigration.* New York: New York University Press.

**Friedman, Thomas L.** 2002. *Longitudes and Attitudes: Exploring the World after September 11.* New York: Farrar, Straus & Giroux.

**Geertz, Clifford.** 1965. "The Impact of the Concept of Culture on the Concept of Man," pp. 93–118 in *New Views of the Nature of Man,* edited by John R. Platt. Chicago: University of Chicago Press.

———. 1984. "Distinguished Lecture: Anti Anti-Relativism." *American Anthropologist* 86: 263–278.

**Giddens, Anthony.** 2000. *Runaway World: How Globalization Is Reshaping Our Lives.* New York: Routledge.

**Goffman, Erving.** 1963. *Stigma: Notes on the Management of Spoiled Identity.* Englewood Cliffs, NJ: Prentice Hall.

**Goldberg, Michelle.** 2006. *Kingdom Coming: The Rise of Christian Nationalism.* New York: W.W. Norton

**Goldhagen, Daniel Jonah.** 1996. *Hitler's Willing Executioners: Ordinary Germans and the Holocaust.* New York: Alfred A. Knopf.

**Goode, Erich.** 1984. *Deviant Behavior: An Interactionist Approach.* 2nd ed. Englewood Cliffs, NJ: Prentice Hall.

———. 2002. *Deviance in Everyday Life: Personal Accounts of Unconventional Lives.* Prospect Heights, IL: Waveland Press.

———, ed. 1996. *Social Deviance.* Boston: Allyn & Bacon.

**Haenfler, Ross.** 2006. *Straight Edge: Hardcore Punk, Clean-Living Youth, and Social Change.* New Brunswick, NJ: Rutgers University Press.

**Hagan, William Thomas.** 1993. *American Indians.* Chicago: University of Chicago Press.

**Hall, John A., and Charles Lindholm**. 2000. *Is America Breaking Apart?* Princeton, NJ: Princeton University Press.

**Hallinan, Maureen T.** 2005. *The Socialization of Schooling.* New York: Russell Sage Foundation.

**Hancock, Angie-Marie.** 2004. *The Politics of Disgust: The Public Identity of the Welfare Queen.* New York: New York University Press.

**Hedges, Chris.** 2002. *War Is a Force That Gives Us Meaning.* New York: Public Affairs.

**Henslin, James M.** 2000. *Social Problems.* 5th ed. Upper Saddle River, NJ: Prentice Hall.

**Herskovits, Melville Jean.** 1972. *Cultural Relativism,* edited by Frances Herskovits. New York: Random House.

**Hertz, Rosanna.** 2006. *Single by Chance, Mothers by Choice: How Women are Choosing Parenthood without Marriage and Creating the New American Family.* Oxford, UK: Oxford University Press.

**Hervieu-Leger, Daniele.** 1998. "Secularization, Tradition and New Forms of Religiosity: Some Theoretical Proposals," pp. 28–44, in *New Religions and New Religiosity,* edited by Eileen Barker and Margit Warburg. Cambridge, England: Cambridge University Press.

**Hochschild, Jennifer L.** 1995. *Facing Up to the American Dream: Race, Class, and the Soul of the Nation.* Princeton, NJ: Princeton University Press.

**Hostetler, John A.** 1980. *Amish Society.* Baltimore, MD: Johns Hopkins University Press.

**Hull, Kathleen E.** 2006. *Same-Sex Marriage: The Culture Politics of Love and Law.* Cambridge: Cambridge University Press.

**Ignatieff, Michael**, ed. 2004. *The Lesser Evil: Political Ethics in an Age of Terror.* Princeton, NJ: Princeton University Press.

**Jacobs, Aton K.** 2006. "The New Right, Fundamentalism, and Nationalism in Postmodern America: The Marriage of Heat and Passion." *Social Compass* 53(3): 357–366.

**Jankowski, Martin Sanchez.** 1991. *Islands in the Street: Gangs and American Urban Society.* Berkeley: University of California Press.

**Jones, Ron.** 1981. *No Substitute for Madness.* Covelo, CA: Island.

**Katkin, Wendy F., Ned Landsman, and Andrea Tyree**, eds. 1998. *Beyond Pluralism: The Conception of Groups and Group Identities in America.* Urbana: University of Illinois Press.

**Keiser, R. Lincoln.** 1979. *Vice Lords: Warriors of the Street.* New York: Holt, Rinehart & Winston.

**Keller, Suzanne.** 2003. *Community: Pursuing the Dream, Living the Reality.* Princeton, NJ: Princeton University Press.

**Kephart, William M.** 1991. *Extraordinary Groups: The Sociology of Unconventional Life-Styles.* 4th ed. New York: St. Martin's Press.

**Kerbo, Harold R.** 1999. *Social Stratification and Inequality.* 4th ed. New York: McGraw-Hill.

**Koonigs, Kees, and Dirk Druijt**, eds. 1999. *The Legacy of Civil War, Violence and Terror in Latin America.* London: Zed.

**Kosmin, Barry A., and Seymour P. Lachman**. 1993. *One Nation under God: Religion in Contemporary American Society.* New York: Harmony.

**Kraybill, Donald B., and Marc A. Olshan**, eds. 1994. *The Amish Struggle with Modernity.* Hanover, NH: University Press of New England.

**Lareau, Annette.** 2003. *Unequal Childhoods: Class, Race, and Family Life.* Berkeley: University of California.

**Lee, Robert G.** 1999. *Orientals: Asian Americans in Popular Culture.* Philadelphia: Temple University Press.

**Lewellen, Ted C.** 1995. *Dependency and Development: An Introduction to the Third World.* Westport, CT: Bergin & Garvey.

**Lewis, Bernard.** 2002. *What Went Wrong? Western Impact and Middle Eastern Response.* New York: Oxford University Press.

**Liebow, Elliot.** 1993. *Tell Them Who I Am: The Lives of Homeless Women.* New York: Simon & Schuster.

**Lofland, John.** 1966. *Doomsday Cult.* Englewood Cliffs, NJ: Prentice Hall.

**Lorber, Judith.** 2005. *Breaking the Bowls: Degendering and Feminist Change.* New York: Norton.

**Lynch, Michael J.** 2000. *The New Primer in Radical Criminology: Critical Perspectives on Crime, Power, and Identity.* Monsey, NY: Criminal Justice Press.

**Madsen, Richard**, ed. 2002. *Meaning and Modernity: Religion, Polity, and Self.* Berkeley: University of California Press.

**Mann, Michael.** 2004. *The Dark Side of Democracy: Explaining Ethnic Cleansing.* New York: Cambridge University Press.

**Manning, Christel J.** 1999. *God Gave Us the Right: Conservative Catholic, Evangelical Protestant, and Orthodox Jewish Women Grapple with Feminism.* New Brunswick, NJ: Rutgers University Press.

**Marger, Martin.** 1999. *Race and Ethnic Relations: American and Global Perspectives.* 5th ed. Belmont, CA: Wadsworth.

**McCall, George J., and J. L. Simmons**. 1978. *Identities and Interactions.* New York: Free Press.

**McCall, Nathan.** 1994. *Makes Me Wanna Holler.* New York: Random House.

**McCord, Joan**, ed. 1997. *Violence and Childhood in the Inner City.* New York: Cambridge University Press.

**McNamee, Stephen J., and Robert K. Miller Jr.** 2004. *The Meritocracy Myth.* Landham, MD.: Rowman & Littlefield Publishers.

**McNeill, William H.** 1993. "Epilogue: Fundamentalism and the World of the 1990s," pp. 558–574, in *Fundamentalisms and Society: Reclaiming the Sciences, the Family, and Education,* edited by Martin E. Marty and R. Scott Appleby. Chicago: University of Chicago Press.

**Mendelsohn, Everett.** 1993. "Religious Fundamentalism and the Sciences," pp. 23–41, in *Fundamentalisms and Society: Reclaiming the Sciences, the Family, and Education,* edited by Martin E. Marty and R. Scott Appleby. Chicago: University of Chicago Press.

**Mills, Nicolaus.** 1997. *The Triumph of Meanness: America's War against Its Better Self.* Boston: Houghton Mifflin/Sage Foundation.

**Nash, Kate.** 2000. *Contemporary Political Sociology: Globalization, Politics, and Power.* Malden, MA: Blackwell.

**Neuman, W. Lawrence.** 2004. *Basics of Social Research: Qualitative and Quantitative Approaches.* Boston: Pearson.

**New York Times**, ed. 2005. *Class Matters.* New York: Times Books.

**Newman, Katherine S.** 1999. *No Shame in My Game: The Working Poor in the Inner City.* New York: Knopf.

**Parenti, Michael.** 1998. *America Besieged.* San Francisco: City Lights.

**Parrillo, Vincent N.** 2002. *Contemporary Social Problems.* 5th ed. Boston: Allyn & Bacon.

**Perrucci, Robert, and Earl Wysong**. 1999. *The New Class Society.* Lanham, MD: Rowman & Littlefield.

**Pfuhl, Erdwin H., and Stuart Henry**. 1993. *The Deviance Process.* 3rd ed. New York: Aldine de Gruyter.

**Phillips, Kevin P.** 2002. *Wealth and Democracy: A Political History of the American Rich.* New York: Broadway.

**Rashid, Ahmed.** 2002. *Jihad: The Rise of Militant Islam in Central Asia.* New Haven, CT: Yale University Press.

**Rokeach, Milton.** 1969. *Beliefs, Attitudes, and Values.* San Francisco: Jossey-Bass.

**Rotenberg, Paula S.** 2002. *White Privilege: Essential Readings on the Other Side of Racism.* New York: Worth.

**Rouner, Leroy S., and James Langford**, eds. 1996. *Philosophy, Religion, and Contemporary Life: Essays on Perennial Problems.* Notre Dame, IN: University of Notre Dame Press.

**Rousseau, Nathan**. 2002. *Self, Symbols, and Society: Classic Readings in Social Psychology.* New York: Rowman & Littlefield.

**Rubington, Earl, and Martin S. Weinberg**. 1987. *Deviance: The Interactionist Perspective.* 5th ed. New York: Macmillan.

**Schwartz, Barry.** 1994. *The Costs of Living: How Market Freedom Erodes the Best Things in Life.* New York: Norton.

**Seligman, Adam.** 1995. *The Idea of Civil Society.* Princeton, NJ: Princeton University Press.

**Shapiro, Thomas M.** 2003. *The Hidden Costs of Being African American.* New York: Oxford University Press.

**Shibutani, Tamotsu.** 1955. "Reference Groups as Perspectives." *American Journal of Sociology* 60: 562–569.

———. 1970. "On the Personification of Adversaries." *Human Nature and Collective Behavior,* edited by Tamotsu Shibutani. Englewood Cliffs, NJ: Prentice Hall.

———. 1986. *Social Processes: An Introduction to Sociology.* Berkeley: University of California Press.

**Simmel, Georg**. [1908] 1955. "Conflict," pp. 00–00 in *Conflict and the Web of Group Affiliations.* Translated by Kurt H. Wolff. New York: Free Press.

**Smart, Julie.** 2001. *Disability, Society, and the Individual.* Gaithersburg, MD: Aspen.

**Staub, Ervin.** 1989. *The Roots of Evil: The Origins of Genocide and Other Group Violence.* New York: Cambridge University Press.

**Stirk, Peter M. R.** 2000. *Critical Theory, Politics, and Society: An Introduction.* New York: Pinter.

**Suarez-Orozco, Carola.** 2001. "Immigrant Families and Their Children: Adaptation and Identity Formation," pp. 129–139 in *The Blackwell Companion to Sociology,* edited by Judith Blau. Oxford, UK: Blackwell.

**Sullivan, Maureen.** 2004. *The Family of Woman: Lesbian Mothers, Their Children, and Undoing of Gender.* Berkeley: University of California Press.

**Sullivan, Oriel.** 2006. *Changing Gender Relations, Changing Families: Tracing the Pace of Change over Time.* New York: Rowman & Littlefield.

**Sumner, William Graham**. [1906] 1940. *Folkways.* Boston: Ginn.

**Suro, Roberto.** 1999. *Strangers among Us: Latino Lives in a Changing America.* New York: Knopf.

**Titchkosky, Tanya.** 2003. *Disability, Self, and Society.* Toronto: University of Toronto Press.

**Toennies, Ferdinand**. [1887] 1957. *Community and Society,* translated and edited by Charles A. Loomis. East Lansing, MI: Michigan State University Press.

**Tonry, Michael.** 2004. *Thinking about Crime: Sense and Sensibility in American Penal Culture.* New York: Oxford University Press.

**Turner, Bryan S.** 1983. *Religion and Social Theory: A Materialist Perspective.* London: Heinemann.

**Venatesh, Sudhir Alladi.** 2006. *Off the Books: The Underground Economy of the Urban Poor.* Cambridge Mass: Harvard University Press.

**Waldinger, Roger David, and Michael I. Lichter**. 2003. *How the Other Half Works: Immigration and the Social Organization of Labor.* Berkeley: University of California Press.

**Walker, Henry A., Phyllis Moen, and Donna Dempster-McClain**, eds. 1999. *A Nation Divided: Diversity, Inequality, and Community in American Society.* Ithaca, NY: Cornell University Press.

**Ward, Keith.** 2000. *Religion and Community.* New York: Oxford University Press.

**Weber, Max**. [1905] 1958. *The Protestant Ethic and the Spirit of Capitalism,* translated and edited by Talcott Parsons. New York: Scribner's.

**Western, Bruce** 2006. *Punishment and Inequality.* New York: Russell Sage Foundation.

**White, Leslie A.** 1940. *The Science of Culture.* New York: Farrar, Straus & Giroux.

**Whorf, Benjamin Lee.** 1956. *Language, Thought, and Reality.* New York: John Wiley.

**Whyte, William Foote.** 1955. *Street Corner Society.* Chicago: University of Chicago Press.

**Wilson, William Julius**, ed. 1993. *The Ghetto Underclass: Social Science Perspectives.* Newbury Park, CA: Sage.

———. 1997. *When Work Disappears: The World of the New Urban Poor.* New York: Knopf.

———. 1999. *The Bridge over the Racial Divide.* Berkeley: University of California Press.

**Yinger, Milton J.** 1982. *Countercultures: The Promise and Peril of a World Turned Upside Down.* New York: Free Press.

**Zellner, William W.** 1995. *Countercultures: A Sociological Analysis.* New York: St. Martin's.

**Zinn, Maxine Baca, and D. Stanley Eitzen**. 1999. *Diversity in Families.* Boston: Allyn & Bacon.

**Zweig, Michael.** 2000. *The Working Class Majority: America's Best Kept Secret.* Ithaca, NY: Cornell University Press.

# 7 Why Is There Misery in the World?

## Society as an Important Source of Human Problems

### Concepts, Themes, and Key Individuals

- Human misery
- Social inequality
- Social forces and social problems
- Destructive social conflict
- Socialization and alienation
- Society as cause of human misery

In his book *Beyond the Chains of Illusion* (1962), Erich Fromm describes three events that inspired him to become a social scientist. The first was the suicide of a dear friend right after the death of her father. The second was World War I, a war fought by "civilized" nations against one another, each claiming justice on its side. The third was the mass murder of the Jewish people during World War II by one of the most advanced societies in the world. These three events pushed Fromm to try to understand human beings in order to create a more just world.

Events such as suicide, genocide, and war cry out for explanation for at least two reasons: Their causes are difficult to understand, and their costs in human misery beg for a solution. All of Fromm's work was an effort to understand the actions of human beings and the reasons why there is so much misery and injustice in the world. Fromm's quest is similar to those of many other great thinkers and should be important to all of us.

Sociology has always attracted scholars driven by a desire to make sense of misery and to bring justice to the world. Karl Marx, reacting to horrible conditions of poverty and the accumulation of wealth by a few people, was inspired by a vision of equality for all. Émile Durkheim, reacting to conditions of rapid social change and rising individualism, sought a world of people bound together through a shared sense of morality. American sociologists, reacting to problems of migration, urbanization, poverty, and social inequality, were inspired to create a practical science applied to serious social problems. Indeed, like Fromm, many sociologists begin their intellectual journey because of their desire to improve the human condition. Auguste Comte, the nineteenth-century founder of sociology, believed that he was founding an academic discipline that would save humanity by studying and solving the problems that plague humankind. Comte undoubtedly exaggerated what sociology could do, but there is still a faith in most of us that sociological knowledge can make a substantial contribution to improving the world.

Strangely enough, it is difficult to define *misery.* "Unhappiness" and "suffering" come close, but unhappiness is less acute, and both terms imply a more temporary state. Everyone is unhappy sometimes; everyone suffers occasionally. Perhaps misery is best understood as a *state of chronic suffering and unhappiness.* Of course, if misery can come to anyone at random and if all people have an equal chance to experience a life of misery, then there is no reason to try to identify objective conditions that might create it. However, almost every social scientist, most journalists and religious leaders, and those who work closely with people in need of help would agree that some conditions in the world foster misery, and these conditions must be altered if misery among people is to be lessened. It may be that a poor person lives a much fuller life than a wealthy person, and that the wealthy person may actually live a life of misery, but it is far more likely to find great suffering in those who experience the horrors that accompany bare survival. War may create heroes, and abuse by a parent may create individuals who spend their lives helping those who are abused, but an honest objective observer would still accurately report that war and abuse create far more misery than peace and love. It is important, therefore, to recognize that although misery is, in part, a subjective feeling caused by many complex conditions too numerous to name, conditions of misery do exist in society, and it is worthwhile to identify what these are, why they exist, and why they create suffering for so many people. To argue that anything might cause misery is to ignore identifying and trying to change anything in society so that people can live better lives. Although we might disagree on what conditions need to be altered, and although we might also disagree over how best to change these conditions, ultimately it is important to identify and understand conditions. To the sociologist, these conditions are located in society itself. They are social conditions.

Sociologists do not concentrate on why people feel misery so much as on the conditions that encourage misery. It should be no surprise that sociologists focus on *social* conditions. Poverty is one such condition. In every society, poverty creates life problems for people: oppression, chronic economic hardship, victimization, lack of opportunity in society, and an inability to protect one's self and family from disease, starvation, and crime.

Sociology is not the only perspective that helps us understand misery in the world. Psychologists and psychiatrists test and treat people who are schizophrenic, paranoid, suicidal, and manic-depressive, and who lack self-worth and self-control. They have identified some important clues to why misery exists, including chemical imbalance, genetic predisposition, early childhood training, trauma, personality development, and failures in school and friendships. Religious leaders normally look to spiritual causes and call for spiritual solutions. Misery exists, they often contend, because our choices are not right, our values are poor, our actions are immoral. They often seek to understand why so many of us live without religious or ethical principles.

Misery is a religious question because, for thinking people, it gets at the heart of what God is and what the meaning of life is. For many contemporary religious leaders, the question is, How can a just God allow a world in which so much misery exists? Look at the wars of the last century. Look at the Holocaust during World War II and the murder of millions afterward. Look at the hunger in the world, the epidemics that destroy many thousands and even millions of innocents. Look at the innocent people who are murdered every day, the muggings, the drug dealers, the neighborhoods to which many people are condemned. What is the role of God? Is God truly loving? Is God the cause in any way? Is it our turning from God? How do we explain misery?

The story of Job in the Old Testament is often used by religious people to investigate misery and the role of the supernatural. For example, Rabbi Harold Kushner (1981) uses this story to illustrate his—and many other people's—moral problem. Job is described as a just man. This man has a wonderful life, but he loses everything. How can a just and all-powerful God allow this to happen? What did Job do to deserve what happens to him? How is it possible, Kushner asks, for an all-powerful and all-just God to curse this just man, Job, with misery? Kushner concludes that one cannot believe all three ideas at the same time—two of them, but not all three. Either Job is not really a just man, or God is not a just God, or God is not all-powerful. For example, if Job is not just, then an all-powerful, just God makes some sense—punishment seems rational. Or if an all-powerful God is really not just, it makes some sense that this God will bring misery to anyone, even someone like Job who is just. The description of Job is that he is a very just man. Therefore, God must be either all-powerful or all-just, but not both. Kushner is not willing to simply say that everything has a purpose. Instead, he needs to know why an all-powerful, all-just God punishes those who are just. His answer is that God is not all-powerful, that God does not determine events. He argues that there are many events in the world that are caused by natural forces rather than an all-powerful God. Misery is not an act of God; it occurs within the natural order that exists. Misery occurs to both good and bad people. Disease strikes the one who catches the germ, not the one who is evil. The earthquake destroys the property that happens to be in a vulnerable place, not the property owned by an evil person. Neither the germ nor the earthquake distinguishes between the just and the unjust. The American Civil War occurred because historical human social causes together created the conditions that led to war. So it also was with World War I, World War II, Korea, Vietnam, Afghanistan, and Iraq.

To understand disease, war, and earthquakes, scientists turn to nature. To understand human misery, we must identify natural—and to a great extent, social—causes. Misery occurs because of certain conditions present in nature, society, and the individual. Scientists assume that their purpose is to discover the conditions that create misery. Sociologists, in particular, focus on those conditions that are social.

Durkheim cautioned us long ago that if we are ever going to understand *social matters,* then we need to look at *social causes.* If poverty, violence, crime, oppression, and meaningless work are *social* matters (and they are), then we must look to a *social* explanation. If suicide, drug abuse, and fear are widespread in *society* (and they are), then we must examine the nature of that society to understand these seemingly individual acts. Sometimes it is the breakdown of society that is responsible; sometimes it is actually the successful operation of society that is to blame.

Sociologists look first to *social inequality* as the source of social problems underlying misery. Poverty, oppression, exploitation, and lack of hope and self-worth bring misery to many people, and these are linked to social inequality. Inequality also produces institutions—for example, public schools, private health care, and a criminal justice system that favors those who can pay—that cannot and do not work for large numbers of people, resulting in miserable conditions for many. Finally, societies built on inequality will produce people who, no matter what they have, feel misery even if they do not live in obviously deprived conditions.

Widespread *destructive social conflict* and the breakdown of social order are the second source of human misery from a sociological perspective. Society to a great extent is a cooperative order. It is built on trust and agreement. Conflict is necessary in and contributes to society, but conflict sometimes becomes destructive, disrupts and destroys life, and creates chaos. For many people, it brings fear and a feeling of vulnerability to the whims of others.

Third, sociologists focus on *socialization*. Human beings are socialized, or taught the ways of society, from birth to death. In many complex ways, however, socialization creates misery for people. For some, socialization is inadequate, and the individual lacks proper social and emotional support or does not learn the self-control needed for successful problem solving. For some, socialization leads them down the road to a life of misery. For some, socialization teaches moral rules that encourage exploitation and destruction of others. For still others, socialization creates unrealistic expectations, so that no matter how successful one is, one cannot overcome a feeling of misery.

Finally, several important traditions in sociology look to *alienation* as a cause of misery. Alienation is the separation of people from one another, from meaningful work, and from one's self (that is, a sense of ownership over our most fundamental possession, our self). Conditions in society create alienation of the individual, and this alienation is a fourth cause of human misery.

The accompanying figure summarizes these four general causes of misery. Each cause will be discussed in more depth in the rest of the chapter.

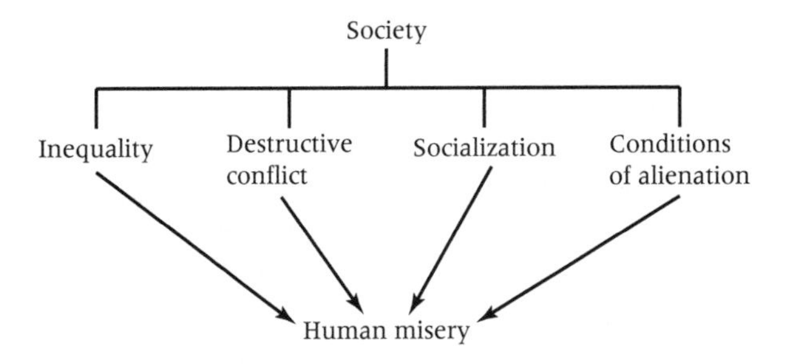


## The First Cause of Misery: Social Inequality

Inequality brings misery to many people. In the struggle to succeed, some people lose out and end up dependent, powerless, and/or exploited. Where they end up has implications for their families. Often, their lives are without much hope, and they struggle simply to survive. Often, they turn their anger inward and become self-destructive, or they turn their anger toward their family, neighbors, strangers, or those who exploit them. Inequality also encourages many who succeed to exploit those who do not succeed and to narrow their vision in the world in order to create and justify an even more secure position in society. Their success, they often find, depends in part on the efforts of those who are less successful.

Almost every society is built on three types of inequality—economic, political, and social—and in every case a belief system arises to justify that inequality and protect those who are successful. In the United States, we read and are told in no uncertain terms: "Work hard, and you, too, can achieve a life of privilege. You, too, can rise above all the rest. You, too, can have material success." "Competition brings out the best in the human being, and without it we would not succeed as individuals or as a society." "Promising great rewards to those who work hard and act smart is basic to building a good society." "In order to encourage people to take chances and in order to get them to take difficult and responsible jobs, we must give them hope for great material rewards." "People have a moral right to keep whatever they make." "Those who win deserve it, and they are not responsible for those who do not win." Always such ideas contain at least a kernel of truth, but they are usually exaggerations of reality, and they are meant to defend a system rather than to understand it. They are ideas that sociologists call *ideology*.

Inequality means that in social relationships, from dyads to societies, some people are somehow favored over others. Positions matter, and they matter a lot because they place some actors above or below others. Gender, race, class position, as well as positions in business and in government, rank the actors who fill them, but there are consequences. In a social world where positions are unequal, some win and some lose. All students cannot receive an A in a college class based on the normal curve. All people cannot be millionaires in a capitalist society. Indeed, one main reason there are millionaires in the first place is that some people are able to get others to work for them at a rate far lower than what they (the millionaires) receive. Some receive a little less (the upper middle class); most receive quite a bit less (the working class); and many receive almost nothing (the poor).

When we relate inequality to social class, we must note two important aspects. First, in a class society, one is advantaged or disadvantaged based on what economic resources one accumulates. Second, one tends to pass down advantages or disadvantages to children through educational opportunities, social contacts, or direct inheritance. The net effect is that, over time, great inequalities arise in the distribution of resources within society and among societies in the world. Some people live in splendor, some are affluent and secure, many can barely pay their bills, and many live on the verge of starvation. A sensitive tourist in any big city in the United States cannot help but see people without food, adequate clothing, shelter, or dignity crying out for help, while others casually walk into art galleries and pay $100,000 for a single painting.

Social inequality is linked to misery in seven ways: It produces poverty, contributes to crime, forces some people to work at bad jobs, facilitates the exploitation of some by others, creates low self-esteem and loss of hope, contributes to high levels of stress throughout society, and forms institutions that produce and maintain misery.

### *Consequences of Inequality: Poverty*

The critical question is, What happens to those left behind in society? Poverty is the result for many. From 13 percent to 25 percent of the American population lives in poverty, depending on how poverty is defined. Definition is partly a political question: By law, government determines a poverty level. In 1996, poverty was defined as a nonfarm family of four that earned less than $16,036 per year. Approximately

13.7 percent of the population was in this bracket (U.S. Bureau of the Census, 1999). Other definitions of poverty would include as much as 25 percent of the population.

Poverty is associated with a range of disadvantages that profoundly affect how people lead their lives: For those who are poor, work that is secure, safe, and productive is far less available. Lack of adequate health care services and poor physical and mental health are important problems. Neighborhood choice, educational opportunities, geographical mobility, and protection under the law are far more limited than for others in the population. Problems associated with poor neighborhoods, family stress and disorganization, and economic survival are more evident among the poor. Poverty strains people's ability to solve everyday problems; any planning for the future must be sacrificed to everyday survival. Many of society's most serious social problems can be traced to poverty, not only because the poor are deprived of what everyone else enjoys, but also because victims of poverty often fall prey to anger, crime, violent conflict, family disorganization, and political instability. And, of course, social problems also arise when the poor do not define their future as hopeful. Conditions of poverty tend to encourage this.

It is tempting to look down on the poor from more privileged heights and complain that "it's their fault" and that "we must take care of ourselves." Such attitudes help to perpetuate the unequal distribution of resources; redistribution becomes a major challenge that most of us are not motivated to take on. The fact is that poverty is built into a society of inequality. It results from a system in which some people succeed at the expense of others, a system in which some people are born into situations in which opportunities are fixed against them, or a system in which social change favors some and leaves others behind. Inequality is structured; it is not simply a result of personal effort.

### *Consequences of Inequality: Crime*

In a society of great inequality, people are socialized to judge themselves and others on the basis of material success: "I am good because I have achieved much in this competitive game of life." "Others have more than I, but maybe I can get there, too. Others have less, and I'm fortunate." Material success is a value many people share. A society of inequality bestows dignity on those who rise to the top and withholds dignity from those who remain below.

The game of life is fixed, however. Opportunities are never equal. We are all born into privilege or lack of it; class is largely an inherited rank. We learn what success means in society, and the choice becomes clearer and clearer to those in inferior positions: "Either accept a lowly position or work to change your position. And if you work to change your position, there is another choice: Work extra hard in a system that favors others, or go outside the legitimate system to make it." Many poor people accept their position and struggle simply to survive. Many work extra hard to make it in the legitimate order. But others see no reason to follow laws that seem to work against them in the competitive order, laws made by those who most benefit from that order. Stealing, prostitution, selling illegal drugs, and violent crime become attractive options. Some will overcome poverty through crime; the vast majority will not. Those who do not will remain poor, increasingly victimized by the welfare, court, medical, and prison systems that attempt to exercise control over their lives to ensure that they are not threats to the rest of society. Over time their misery worsens.

The poor are by no means the only ones who break the law and try to achieve success outside legitimate means. Crime exists at all levels of society because of the widespread inequality and the passion of people to improve their rank. It exists because the rich try to stay rich or get richer. And although those who succeed in improving their rank illegally may or may not overcome their misery, those who are caught and punished will have to deal with additional problems. Politicians who take bribes, stockbrokers who deal illegally, accountants who cook the books, and employers who do not protect employees from hazardous waste are all examples. Almost always, however, their situation will not come close to the misery of the unsuccessful lawbreaker among society's poor because the court system gives harsher treatment to the poor. The wealthy are more able to escape prison through paying fines, hiring expensive lawyers, and convincing the courts that they are not a danger to society.

When we think of crime, most of the time it is not the perpetrators but the victims whom we think of as living lives of misery. Those who are the victims, however—those who are preyed on by the lawbreaker—are also disproportionately the poor. They are the ones whose neighborhoods are infested with gangs and drugs. Organized crime infiltrates poor neighborhoods by providing illegal goods (handguns, stolen goods) and services (prostitution, fencing).

The poor are in close proximity to those who engage in crime, they are the ones who are most open to exploitation, and they are the ones least likely to be protected by the legal system.

This point should not be missed: Inequality is largely the cause of crime in society. Almost everything in society teaches us that being more successful materially is what makes life worthwhile, causing some people to see crime as the easiest way to achieve that success. Street crime, drug dealing, price-fixing, and bank robbery are all consequences of a society that emphasizes material success. And such crime has important consequences for all of us; we are all victims, because crime brings disorder to society, as well as fear and distrust to our everyday existence.

### *Consequences of Inequality: Bad Jobs*

But inequality fosters more than poverty and crime. It generates tedious, low-paying, dangerous, and insecure work for many. The work that many people do offers few material rewards; it traps them in a life of bare *survival.* Misery exists, in part, because miserable work exists; those who have little choice in the matter must take it or die. Consider mining, for example. For many generations, people in Appalachia have taken low-paying, tedious, physically demanding, dangerous, and insecure jobs as miners. Why? Because "someone has to do them" and because those who own the mines can profit only if those who work for them remain poor. If the workers become materially successful and expect decent wages, the rich will find that mines are no longer profitable and will have to close them down. So the poor must choose between no work and bad work. The same is often true for the women who clean middle-class homes or who care for middleclass children; their job security also depends on low wages. In other words, it is their "willingness" to take low-paying jobs that guarantees their continued work. Bad and low-paying jobs will always be a part of a society in which some succeed at the expense of those who need these jobs to survive.

Bad jobs are also the most insecure. Those at the bottom of the employment ladder are in unskilled occupations: those most likely to be replaced by machines, by labor in other societies, or simply by other workers willing to work for less. In times of depression, their jobs are the first to go, and they are the ones most likely to experience long-term unemployment.

Work is a major part of a human being's life. Miserable work contributes significantly to a miserable existence. Along with poverty and crime, it is a product of a society of inequality. Exploitation is another product.

### *Consequences of Inequality: Exploitation*

Inequality, Marx believed, causes misery in yet another way: It can always be translated into power (Marx and Engels 1848). Where inequality of any kind exists, unequal power is inevitable. It does not matter if that inequality is based on economic resources or political, occupational, gender, racial, or religious position in society. Where power is unequal, exploitation (selfish use by others) is likely, because those in need must depend on what others demand of them. "Do what I say or I will replace you." In general, the poorer a person is, the easier it is to be replaced because there is little protection. Marx emphasizes economic exploitation: Those who own the means of production (factories, for example) are extremely powerful, so they are able to exploit all who must depend on them for work.

Sociologists have gone beyond Marx in their analysis of inequality and power, however. Physical and mental abuse is far more likely in a situation of dependence and exploitation. Women and children are victims of domestic abuse in part because of physical inequality but also because they are made relatively powerless through relying on the man's paycheck, and because they fear the consequences of challenging or leaving him. And in many societies in the world as well as in many communities in the United States, governments do not adequately protect children and women from abuse. The dominant man controls, threatens, and exploits the dependent woman or child. Nonwhites are exploited by whites, the defenseless by the violent, the small business by the large corporation. Inequality of all kinds means dependence, and dependence facilitates exploitation.

Exploitation of the powerless characterizes almost every society. Our own history, which we too often idealize, has been one in which African Americans were enslaved; Asian Americans, immigrants from Southern and Eastern Europe, and Mexican Americans were used as cheap labor; and Native Americans were victims of our desire for good land. Most European societies discriminated against and exploited Jewish people, and most persecuted and exploited Christians who were not part of the dominant denomination. Fear, anger, physical expulsion, execution, extermination, and denial of rights and privileges enjoyed by the dominant group are but some of the instances of misery brought on by such systems of inequality. And, of course, where such inequality still exists, exploitation and misery continue.

Almost every society has also had a system of inequality based on gender. Where gender inequality is extreme, sexual exploitation is regarded as normal and legitimate, the destruction of infants because they are female is an accepted practice, the brutal practice of the circumcision of women is required, and the physical abuse of wives by husbands is a right. Gender inequality closes off women from equal participation in the political, economic, and social orders. It denies them the educational opportunities and legal rights that men enjoy. Men gain privilege at the expense of those in a less powerful position.

### *Consequences of Inequality: Lack of Self-Worth*

As people are exploited, as they work at bad jobs, as they barely survive in poverty, and as they engage in and are victims of crime, their views of themselves are formed. People in lower positions are affected; their self-respect is damaged. Those who are looked down on have a difficult time escaping poor self-images. Those without honor in the eyes of others have a much more difficult time finding honor in themselves. The poor are normally defined as undeserving and lazy, nonwhites are seen as less capable than whites, and women are defined as submissive, passive, inferior intellectually, and sexual objects for men's gratification. It is clear that these beliefs are used to justify the system of inequality and the right to discriminate against the exploited groups. It is also clear that these beliefs tend to "objectify" the dispossessed. Those people may be human, but they are somehow different from the rest of us, objects rather than real human beings with feelings, and this status makes their misery easier for us to accept.

Of course, such beliefs also contribute to the misery of those who are in the dominated groups because many come to believe what they are taught about themselves. Misery exists, in part, because people disparage themselves—and they do so largely because others define them as less worthy and they see that those who are similar to them are also defined in that way. If those who are in powerful positions are able to justify their status by claiming that they "worked hard" or that they are "superior," "smarter," or "more talented," then what does that imply about those who are in lower positions? That is part of the misery of the oppressed: Being told and then coming to believe that they are somehow to blame for their position or that their gender, color, or religion automatically makes them less worthy.

Great miseries result from lack of self-worth: anger and hatred toward others, mental illness, alcoholism, drug abuse, and suicide. Not all such conditions can be traced to the struggle for material gain in a society that measures dignity by one's level of material success, but many can. Misery takes a toll, and for some in society the toll is great. In every one of these problems, it is the poor who suffer the most, but all who think they fail to measure up are vulnerable.

### *Consequences of Inequality: Stress*

A society in which competition, material success, and extravagance is so important produces a lot of stress throughout its population. This is all too clear in the college community. In a society of such great inequality, much of our dignity as human beings is tied to economic success. We are taught to make it in a race we all must run, a "rat race," as some would call it. "Whoever ends up with the most toys wins." "Whoever is left behind deserves the misery they receive." We see a stratified society all around us, some above us, and some below us. Culture creates in most of us a commitment to winning and achieving more than others. Those of us who join in this game develop a fear of losing whatever we have, and some hope that we can even improve our position. Both fear and hope stimulate us to work hard, but fear that is realized or hope that is continuously frustrated brings misery. Misery can come to those at the top as well as to those on the bottom, because it arises in part from one's perceived lack of success in the system of inequality. Of course, the misery of the poor is compounded by the ever-present problem of bare physical survival on top of the stress related to falling lower in the structure.

The fight to stay even or do better brings with it the temptation to commit crime, and this, in turn, may bring more misery both to those who commit crimes and to those who are the victims. The simplest illustration is drugs. Crime associated with drugs is the work of both the poor and the wealthy; all are interested in improving rank in society. For many of the poor, this seems to be the only way to make it. For many among the wealthy, it is a chance that can quickly create more fame and fortune. But it ultimately is the poor who are hit the hardest. They tend to be the most likely victims, and they are the ones most exploited and more easily sent to prison for their crimes.

The study of the homeless should remind all of us that no one is immune from the bottom. We go to school, get a job, work hard, buy our home, pay the mortgage, and hope to live happily ever after. However, any of us can suddenly find ourselves out of work or broke in our businesses. Our companies and communities can close up overnight, and house values can plummet as international economic forces play themselves out. Our stock values can plunge and suddenly our life plans are altered. Or we can marry, have children, depend on a spouse for our survival, and suddenly find ourselves out of a marriage and in poverty. Or we can retire and suddenly find it impossible to survive on the little savings we have accumulated.

We share as a society a belief that wealth is a sure sign of success. No matter who we are, we have a lot to lose if we fail and much to gain if we succeed. Such a world may encourage hard work, but for many people the cost is a life filled with great fear.

### *Consequences of Inequality: Institutions That Produce and Maintain Misery*

Inequality produces institutions that do not serve all human beings satisfactorily. Those who create them believe that if we are to survive as a society all of us must follow these institutions. But all institutions are fixed—not only in our own society but also in every society where inequality exists. Institutions tend to benefit those at the top of society; rarely are they created to help those at the bottom. Indeed, institutions often work in such a way that they perpetuate the structure and create the conditions that keep people in the place they were born. Institutions themselves create misery by systematically protecting and giving benefits to those who succeed—benefits that others do not have. Segregation, neighborhood schools, and divorce benefit some at the expense of others. Capitalism, the corporation, regressive taxes, private health care, and a legal system based on the amount one can pay continue to keep some people rich and others poor. *Few of the institutions of society are meant to solve the problems of human misery, except when misery touches the lives of those who are powerful.*

Our system of private medicine and private health insurance primarily takes care of the needs of those who can afford it. Public and private education, supposedly set up for the purpose of helping people raise themselves up in the system of inequality, normally functions to keep them approximately where they began. Our system of law, our political party system, and our court and prison system protect and benefit primarily those high in the class system and, at the very least, function to keep stable the system of inequality that prevails in society. For fifty years, Eastern European societies built political, legal, economic, and educational systems that clearly came to benefit the politically powerful at the expense of everyone else. Throughout American history, we can identify a disproportionate number of institutions that benefit the wealthy and middle classes at the expense of the poor and working classes, whites at the expense of nonwhites, and men at the expense of women. Our efforts to correct misery in society have never equaled our commitment to a society built on the principle of inequality, so our institutions have changed slowly, usually when less powerful groups have organized and demanded that the powerful make changes. Democracy and justice mean something only when all people are respected, all people's freedom matters, and all people benefit from the particular institutions that exist.

### *Misery and Social Inequality: A Summary*

Social inequality is not the only cause of misery in society, but it goes a long way in explaining much of it. To the sociologist, it is almost always the first cause explored. For many, sociology is the study of class, racism, and sexism precisely because of the social problems they create. Such inequalities are believed to arise from the social patterns themselves, and thus an integral part of the society we live in. Such inequalities and many other types constitute the origin of much of the poverty, crime, stress, bad jobs, exploitation, low self-esteem, loss of hope, and inadequate institutions that are fixed to favor some over others and ultimately to perpetuate the misery of many.

Social problems will always be with us. Poverty, exploitation, stress, and crime cannot be eradicated. But these conditions can be altered and limited. Misery exists, but many of us recognize that a just and democratic society must work at these problems in order to minimize misery.

## The Second Cause of Misery: Destructive Social Conflict

*The second cause of misery in the world from a sociological perspective is destructive social conflict or violence.* When conflict becomes destructive, people suffer, lives are destroyed, and real problems go unsolved.

### *The Meaning of Destructive Conflict*

Not all conflict is destructive. Indeed, it is important to see most conflict as inevitable, necessary, and constructive. Conflict means that human beings in social interaction struggle with one another over something they value but cannot all achieve. Conflict is interaction in which actors use power—try to impose their will—in relation to one another. Competition is one form of conflict: It is conflict that takes place within clearly specified rules. Whenever actors try to persuade one another, whenever they fight one another for a cause they believe in, conflict arises. Whenever we try to achieve our goals and others are involved, there will invariably be some struggle, and usually negotiation and compromise result. The result of conflict is usually positive. Both parties get something, organizations change, people's interests are heard, and problems are identified and dealt with. Constructive conflict is a fact of life; instead of causing misery, it is one way in which misery can be recognized and alleviated.

Destructive conflict is something else. Wars are fought, and people are killed or made homeless, their lives left in ruin. Riots cause physical harm, killing, and the destruction of property by both rioters and the authorities. Terrorism by government, groups, and individuals preys on the general population. Spouse and child abuse physically harm people in the short run and cause destructive emotional effects in the long run. There are always victims in destructive conflict.

*Destructive conflict is characterized by intense anger and the desire to destroy or hurt one's opponent. Such conflict often escalates and becomes increasingly violent, inflicting physical and emotional harm on the victims. It ends in harm to others while ignoring the real issues between people that create the conflict in the first place.*

### *The Causes of Destructive Conflict*

Why does conflict become destructive? Why does something that has every potential for contributing to human welfare become a source of human misery?

*Destructive conflict partly arises when constructive conflict is discouraged or ignored, and real differences and problems are neither faced nor resolved.* The more powerful refuse to recognize the struggles of the less powerful. The less powerful may fear to express interests that might bring conflict out in the open. Or conflict sometimes seems irreconcilable to the parties involved, and thus there seems little point in trying to resolve it constructively. Often, people run from constructive conflict out of fear that it might escalate into highly aggressive and even violent confrontation (family conflict is an example). If conflict is repressed over time, however, it can grow more intense, more irrational, and more emotional. The goals that each party originally sought can be lost, replaced by hostility rather than goal-directed efforts to negotiate and resolve real differences.

*Social inequality is also an important source of destructive conflict.* Violent revolutions arise from inequality, often begun by people who are rising in the social order yet are still left out. Much violent crime arises from the frustration and anger fostered by inequality. Many wars are a result of one nation attacking another because it has superior resources and wants something that the other has. Often, aggression arises because inequalities within society are not faced, problems are externalized, and leaders try to ignore real conflict within the nation by creating a common outside enemy. Individuals who are deprived in a system of inequality become frustrated, angry, and sometimes violent offenders against family members, against strangers, against the successful, and ultimately against themselves. Destructive conflict also occurs because those in powerful positions find violence an attractive means for achieving their will when others question them: parents, teachers, political leaders, and gangsters are only a few examples.

*Destructive conflict also occurs because many of us have learned to use violence in dealing with the problems we face.* In trying to achieve our will in relation to others, we learn to use violent confrontation. The use of violence is sometimes cultural and institutional. American society is more violent than some societies and less violent than others. Our political leaders, through what they say and do, tell us that it is all right to resolve problems through violent, destructive conflict. In how parents act toward children, they too express this message, and movies, television, and even music reinforce it. Cartoon characters, superheroes, and men who must prove their manhood through aggressive violence are important examples. Several themes in our history teach us that destructive conflict is necessary and even good: the winning of the frontier, Western vigilantism, wars of expansion, slavery, and violent oppression of minorities are examples. Of course, there are also values, principles, and institutions in American society that limit violent destructive conflict: for example, participating in the democratic process through voting, a spirit of compromise and negotiation in politics, a reliance on law, and respect for individual rights.

To the extent that conflict is repressed and people are not encouraged to negotiate openly and constructively in their own interests, destructive conflict is encouraged. To the extent that culture encourages individuals and groups to use violence, it is encouraging destructive social conflict and human misery. Whenever government, family, or leaders legitimate the use of violence, they are displaying to others that violence is one way in which problems can be solved. *Violence does not simply happen. It is built into several social patterns in society.*

### *Destructive Conflict as a Cause of Misery*

*Destructive social conflict hurts the victim of violence.* It does not matter where it comes from: parents, police, lawbreakers, labor unions, management, or government. It is meant to hurt or destroy the other, and it often does. It produces anger in those who are victims. It may bring quiet anger, and often that anger becomes chronic and deep-seated. It may be expressed, or it may simply fester. Though it is often aimed at others, it can also be aimed at oneself in the form of self-destructive behavior: abuse, gambling addiction, or suicide, for example.

*There is every reason to believe that even if one wins in destructive conflict, misery is far from being eliminated.* The cycle of destructive conflict is difficult to halt. *Winning brings anger on the other side and the possibility of retaliation in the future.* Winning causes one (and others) to believe that destructive conflict is the way to achieve what one wants in situations, thus encouraging its continued use. Psychologically, it encourages more—rather than fewer—aggressive feelings in the perpetrator. Instead of making one feel good, aggression tends to make one more angry and destructive toward the victim. That is because aggressors justify violence through dehumanizing the victim, convincing others as well as themselves that "the victim deserved it." Such a belief system encourages further aggression.

Violence and destructive conflict often occur outside the legitimate order. They are not what we normally expect from one another. They are actions that flout convention. Social interaction depends on convention and the underlying idea that those around us—even strangers—will follow that convention. Interaction thrives on trust, and *one of the real victims of destructive conflict is trust.* Through violent conflict the world of the predictable and familiar becomes a world of disorder and unpredictability, without rules for people to put their faith in. A world of distrust, rule breaking, and unpredictability makes life miserable for many. They become victims of the strongest and most violent; they become afraid of the world that used to be taken for granted.

Misery in society can be traced to many root causes. We have thus far focused on two: social inequality and destructive conflict. A theme weaves itself through both of these causes: *that which seems to be an integral part of society, even necessary for its continuation, also brings misery to many people within society.* Inequality, so much a part of our society—and even what some would call a strength—brings misery to large segments of the population, and that misery can eventually turn on the rich and powerful and threaten the continuation of social institutions that most people have come to take for granted. Social conflict, a necessary and productive part of all societies, can become violent and destructive, and destructive conflict destroys victims, harms perpetrators, and undermines society itself.

## The Third Cause of Misery: Socialization

All sociologists recognize the significance of socialization on the kinds of individuals we become. Socialization influences our choices and teaches us the rules, values, and ideas by which we control ourselves and see the world. It places us into our positions within social structure, and it introduces us to the institutions that govern society. If successful, it encourages love, responsibility, emotional health, and ability to face and deal with problems that arise. Socialization finishes what nature has begun by forming us into social beings and by giving us language, self, and mind.

### *Inadequate Socialization: Problems of Self-Love and Self-Control*

Some people are born into situations in which socialization is insufficient or inadequate. Early interaction within families is sometimes too limited or too destructive, or it is characterized by too little love and affection, or it fails to teach self-control. Close, loving relationships—necessary for providing us with the raw materials for intellectual and emotional growth—are absent. There is strong evidence that infants without close ties to others die or are psychologically harmed. There is also evidence that deprivation of affection in the early years of development has serious emotional and behavioral consequences later on. People who are not loved have a difficult time loving themselves. People who do not have close ties in childhood have a difficult time developing close ties later on and often show evidence of poor self-image. *A life without affection and support in the early years of socialization becomes an important source of misery for the actor and often for others with whom the actor interacts.*

Besides the love and close ties that early socialization must provide, *it is also important for the development of self-control.* We are supposed to become independent, to learn to deal with situations as they rise. Without learning how to effectively control his or her own actions, the individual tends to act impulsively, without understanding consequences or how he or she may affect others. Socialization is supposed to teach the ways of society, and those ways normally become ours; we internalize them, and we control ourselves accordingly. Hopefully, we develop conscience, and we are influenced to act according to the rights of others. We are able to care about others, and we are able to form relationships. Part of what we do involves thinking, problem solving, and how to learn and communicate.

For any number of reasons, some of us do not learn to develop the necessary qualities that socialization are supposed to create. Neither conscious thought nor self-control are important guides to our action. Two problems become serious. First, others become victims of our lack of self-control and we create misery for them. And second, lack of self-control may bring problems in our relationships and our own ability to achieve our own goals. A life of frustration and anger replaces a life that is goal-directed and involves successful problem solving.

### *Socialization into Directions That Bring Misery*

It is, however, not only inadequate socialization that brings people misery, but also *the directions toward which successful socialization may lead us.* Most of us are indeed socialized, often in loving and supportive homes, and we learn to control what we do. However, the socializers do not always help us in living successful and productive lives. Socialization comes to influence our view and use of illegal drugs, the value we give to our education, our choice of major and occupation, our commitment to the law or rejection of it, whether we marry, whom we marry, and how we treat our spouse. Parents, teachers, adults in our neighborhoods, and friends influence our directions in life. So do our employers and our older siblings. We observe others, listen to what they say, and watch how they react to our ideas and acts—and through it all we try directions that seem to be right for us. The content of what we are taught in socialization creates much of what we all become. In a very basic sense, those who socialize us represent society to us, and we are influenced by their rules, their values, their ideas, and their example. In *Tally's Corner* (1967), a study of a disadvantaged ghetto community in Washington, D.C., Elliot Liebow dramatically shows us that the younger men there learn how they should act by observing how the older men behave. The older men spend their time hanging out. They work at temporary, low-paying, unskilled, and often dangerous jobs. Jobs do not offer them hope; they offer only an opportunity to get through the week. With little hope for the future, these men take whatever pleasure they can get in the present. Life is bare survival, with little dignity except what one can get from others on the street corner. The younger men come to believe that these older men are what they can expect to become; in these others they see their future selves.

Of course, we are also socialized by people we never personally meet. We might be influenced to change our directions by a political leader, a basketball star, a successful singer, a wealthy businessperson, or even a ruthless criminal. We may read a book or by chance interact with someone who touches us, and it is possible for our direction to change. However, it is tempting to overemphasize the importance of distant figures. Most of the time we are socialized by people who are much closer to us and whom we interact with every day. The late Supreme Court Justice Thurgood Marshall (1979: 1) wrote about how difficult it was for him to go into a ghetto neighborhood pretending to be a role model for African Americans there: Their lives were a long way from his. Their opportunities are not the same as he had. They knew the gulf between him and themselves too well.

*Those who are socialized by people who live lives of misery are then influenced to go in directions that will bring them misery.* That is the reality of socialization. Through others close to us we become aware of what our lives will be and should be. We learn what we have a right to expect out of life: dropping out of school or getting a graduate degree from a leading university, barely surviving in poverty or living in affluence, getting a lucky break or planning for a career. For many people, role models use drugs, commit crimes, engage in destructive conflict, and treat other people with contempt, or socialization leads them to dead ends, lack of skills for good jobs, or not developing their talents or abilities. *People live in misery, in part, because that is the direction in which their socialization takes them.* To overcome the misery that one is born into, one must be socialized by realistic role models who work against that socialization and alter the direction in which one decides to go. If one is located in the midst of misery, it is almost impossible to find realistic role models to help one escape; here lies the viciousness of misery for those caught up in it.

*We should also remember that people are often socialized to hurt, oppress, and exploit others.* How we treat other people, especially those who are disadvantaged in some way, supports or creates conditions of misery. We are socialized into a society where the "bottom line"—profit, accumulation of wealth—is a dominant value. It becomes easy for those of us who direct our lives this way to forget the needs of laborers, of those who are dispossessed, those who have no resources to compete with us. We are influenced to exploit others—and believe that others are our property, our tools. We are socialized to believe everyone can make it if they try hard, I have no responsibility to anyone but myself, free competition creates the best society, and it is not the government's responsibility to help people who are somehow left out because they were not good enough. We are socialized as to how we should treat nonwhites, immigrants, women, homosexuals, the poor, children, people who we employ, people who break the law, people who are different, people who are our competitors, people who disagree with us. How we treat others often creates misery for others, and much of this is a result of our socialization.

### *Socialization into Impossible or Confusing Expectations*

The final way in which socialization produces misery for people involves the power of expectations. Most of us know people who "seem to have everything" yet are still dissatisfied with their lives. People who are beautiful think of themselves as ugly; people who are rich think of themselves as poor. People who get A's and then get one B fall apart. Some of our misery stems from the difference between objective reality (how we actually perform, what we actually have) and our expectations for ourselves. The expectations we have of ourselves arise largely from our socialization. As we are socialized by others, their expectations become important to us; their demands eventually become our own. We might rebel against overdemanding parents or friends; more commonly we never do escape their expectations, which we can never satisfy.

And often the message of others is unclear, and this too leads to problems. On the one hand, "Accept who you are, know your strengths and limits; be happy with what you have in life." On the other hand, "If you work hard enough you can be anything you want. Follow your dreams. You can always do better." We need to recognize the fact that part of personal misery is created by those whose expectations are usually unintentionally harmful. For many of us, no matter how much we achieve in our lives, we cannot be satisfied with ourselves, because others have socialized us to be impossible taskmasters over what we do, or to become confused as to how to achieve success and successfully find satisfaction with ourselves.

### *Socialization and Misery: A Summary*

Socialization is really the link between society as it exists out there and the individual. It is absolutely essential for the continuation of society and for the development of the human being. It creates order; it creates the opportunity for the individual to achieve his or her potential. It can also create disorder in society and misery for the individual.

It is important to keep in mind the many ways in which socialization enters into the problem of misery. As we see the horrible misery that serial killers, terrorists, business tycoons, youth gangs, and drug pushers bring to others, it behooves all of us to ask, Why do such people exist? If we look closely, *we can almost always identify socialization as one of the most important causes.* Perhaps we will come to understand that there are *reasons* for human problems, and recognize that those who live lives of misery are not simply "their own fault."

## The Fourth Cause of Misery: Alienation

Another source of misery, from a sociological perspective, is alienation. In its simplest sense, the word *alienation* means separation. It is a concept sociologists use to describe (1) *separation from other people* (being alone; isolation); (2) *separation from meaningful work;* and (3) *separation from ourselves as active beings.*

### *Alienation from One Another*

Alienation is a central theme in the work of Karl Marx. Capitalism, according to Marx, is an economic order based on competition rather than cooperation, exploitation of others rather than sharing, and materialism rather than love and respect. Marx (1844) describes how, in his view, people relate to one another in capitalist societies as things, as commodities to be bought and sold in the labor marketplace, as property, and as means to an end rather than as ends in themselves.

Many other sociologists acknowledge the social alienation that modern life has brought but do not lay the blame on capitalism. Max Weber (1905: 181–183), although he describes the many benefits of bureaucracy, constantly reminds us that bureaucratic society is an impersonal society, one without feeling and tradition, one that emphasizes efficiency and effectiveness in organization. We are all caught within the "iron cage of bureaucracy," planning, calculating, and solving problems as they arise, yet sacrificing friendship, close emotional commitments toward one another, and a sense of community. Charles Cooley (1909), an American sociologist who wrote early in the twentieth century, describes the importance of primary groups (face-to-face groups that entail close emotional ties) to the human being. He and other sociologists bemoan the fact that our world has increasingly become impersonal, associational, and individualistic. Intimacy and caring are increasingly replaced by social alienation.

Social alienation is probably best described in the work of Georg Simmel, a German sociologist who was a contemporary of Weber and Durkheim (all three died between 1917 and 1920). Simmel (1902–1903) sees modern life as the life of the stranger. We live in large communities in which our primary concern is with our personal needs, and our ties with others are without much depth. For many of us, urban life is a world of strangers, and the closeness that used to characterize human relationships is lost. The result for many people in modern society is loneliness and misery.

Individuality is one of the dominant themes of our century. Revolutions have been fought to free the individual from the bonds of dictatorship. Education tends to make us more individualistic, less traditional, and less communal. The city cuts our ties with the tyranny of small-town control, and affluence brings the opportunity to pull back into our homes and enjoy life without having to interact with others. Students can stay in their rooms and type away with their computers connected to the Internet, interacting with those whom they never meet face-to-face, or benefiting from not having to interact with those down the hall. Individuality has exacted a cost, however. For many of us, it has brought increasing separation from others, a decline in family and close friendships, and a concern about self without a concern for the community within which we live. Together with the impersonality and selfish exploitation that society encourages, it has contributed to our alienation from one another. *Communitarianism* is a theoretical movement in social science today. Reacting to a world that seems to worship the rights of the individual without recognizing the importance of commitment to the community, communitarianism asks us to recognize how necessary it is for the individual to be committed to the continuation of society. Individualism and freedom must be balanced with respect for the community and its culture; competition between individuals for personal goals must be balanced by cooperation and desire for the whole community to excel. There should be, according to communitarians, more than simply desiring what is best for "me." Indeed, sometimes what is really best for me in the long run is to compromise and respect the importance of the larger community. Although many sociologists wonder whether communitarians go too far in their desire for commitment to the whole, it is clear to most that the worship of the individual can often become too narcissistic, too selfish, too irresponsible. There are negative consequences for both the individual and the community: social alienation for the former, and undermining a sense of community and lack of commitment to any real sense of social order for the latter.

### *Alienation from Meaningful Work*

Social alienation in modern life is accompanied by alienation from creative work. Marx's ideas (1848) continue to be most important here. To him, the human being is a creative, hardworking, productive being. But for much of modern human history, work has meant laboring for the material benefit of owners, for wages they are willing to pay, for extrinsic rewards rather than for the intrinsic benefits found in creative work. We labor for others, and our labor amounts to contributing one small task that eventually produces a finished product that we never see. Work has lost its meaning for human beings, and this loss, too, has brought us misery.

Weber (1905) sees early capitalism as a time when people did, in fact, go out and creatively build businesses that they cared about.

Early capitalism was the period of the entrepreneur, the builder of goods and business, the creative adventurer who found real meaning in work. All that creativity has passed away in the huge, modern bureaucratic enterprises created in the name of efficiency. The actor is now a cog in a great machine, finding little meaning in work, seeking the security of position rather than the adventure of work.

Marx and Weber are the founders of the sociology of work. They ask some provocative questions, all concerned with the possibility of meaningful work, and conclude with an indictment of modern life as a place where humans are not able to find it easily. Making money has replaced meaningful work as a goal for most of us, and pursuing a satisfying life through leisure rather than through work has increasingly become the norm. Life for many is a struggle to win in a game that alienates us both from one another and from meaningful work.

As I lectured on work in an introductory class, I tried to point out how our views of work had changed in the twentieth century. I pointed out that I am part of a generation that regarded work as basic to living a productive life. We worked because we believed this is what people should do, almost as a moral responsibility. Eventually, in the 1960s many young people sought what they called "meaningful work," and this had something to do with benefiting other people or society. By the 1970s, work was becoming a means to an end, a way of achieving material success, a way to make it big in the world of business. I turned to a student in the class and asked him what his view of work was—did he regard it as something a productive life demands? Did he want to find meaningful work or was it a means of achieving material success? He looked at me and said, "I dunno." I pushed a little harder, and he replied, "Work sucks!" "It is not a way of becoming successful?" "No." "Don't you want to be wealthy?" "Sure." "Then how are you going to do it?" His final reply: "Win the lottery." Since the reply by the student, work increasingly is becoming only a necessary evil, and meaning and happiness are being sought elsewhere. It seems that excessive conspicuous consumption and retirement are used unsuccessfully to try to replace fulfilling work. If, however, work itself is really so important for what it means to live a productive, meaningful life, then the increasing alienation from work in the modern world, as predicted by Marx and Weber, is certainly a source of human misery.

### *Alienation from Our Active Selves*

To be alienated from oneself as an active being simply means that humans become passive in relation to their world. They give up. They allow government to rule them, employers to hire and fire them, neighbors to bother them, their children to demand and receive from them, and social forces to manipulate them. Their lives are not their own but are instead moved by impersonal forces that seem to be outside of their control. Passivity brings misery to many: They become victims of the whims of others, they are unable to deal effectively with problems as they arise, and, probably most important, they feel powerless in their personal life and in society.

The sociologist asks again and again, What is there about our social life—our society—that creates passivity and the feeling of powerlessness? In large part, it is part of modern society. We call ourselves a democracy, yet it is obvious that one vote matters little in a society so large, complex, and difficult to understand. We call ourselves a capitalist society, yet the market is controlled by enormous corporations and even larger and more impersonal economic forces. We call ourselves a society in which the individual matters, yet things seem to change in directions that we as individuals do not wish. Even within our own personal lives, powerlessness is encouraged by the nature of society: The efforts of our parents are frustrated by the influence of peers, television, and the general youth culture. The choice of job and neighborhood is dictated by market and interest rates. The chances for getting a decent degree from a decent university are dictated by university regulations, the assignment of an adviser, and evaluations by admissions officers, instructors, and university administrators whom we never meet. Conditions such as these breed a feeling of powerlessness, and such a feeling brings misery to many people in modern society.

## Summary and Conclusion

Human misery and its causes are difficult to understand. There are many causes, some of which are not discussed here. In truth, the psychologist, the philosopher, and the religious thinker have much to say about why misery exists in the world.

The sociologist, however, tells us all something valuable: The cause of human misery, in part, is in the nature of our society and our social life. People harm other people for social reasons that we are able to identify. People live miserable existences not because of rational free choices they make as much as because of social forces they are often not aware of or do not understand. Our social life is critical to what we all become, to whether our lives are fulfilling and productive or miserable and destructive.

It might be useful to bring together into a picture the points made in this chapter concerning the four social bases of human misery (see figure below).

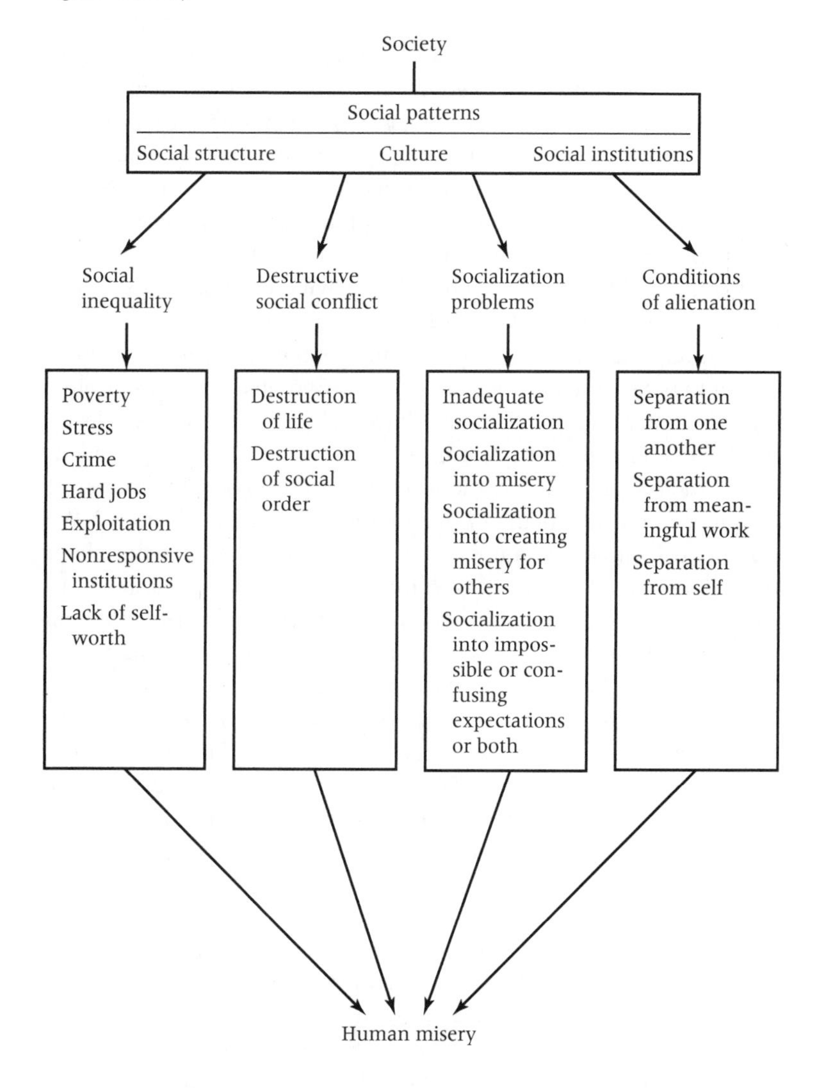

Can we alter these four broad social conditions and thus have an impact on human misery? Is human misery inevitable? Should we simply accept it? What difficult questions! All religious philosophies seek to answer them, as do all people who seek justice. Revolutions are unleashed by these questions, and even those who contribute to the misery of others will rationalize their inhumanity by declaring, "If we don't exploit these people, other people inevitably will." Human misery is probably inevitable, but it never has to be as great as it is. We can always create a society of *less misery,* or we can actually create a society of *more misery.* Poverty, for example, is more widespread today than in the 1960s; it is far less widespread than it was in the 1850s. Work is less exploitative in society than it was before the advent of labor unions and modern technology. Disease and hunger are less prevalent in the United States than in most other societies, yet some societies are far more successful in health care and in providing benefits than we are. Misery will continue to exist, but the question is always, *How much can I (or society) accept?*

It is important to recognize that much of the misery in the world is built into the nature of society itself. We cannot make progress against human misery without changing the social patterns that have developed over a long period. We cannot deal effectively with anger, alienation, and violent crime without lessening poverty, the extremes of inequality, and the linking of human dignity to material success. We cannot deal with alienation without asking important questions about the nature of work and the importance we give to individualism in this society. Of course, some of us do not want to change society (after all, we live good lives), but then we must be prepared to accept the misery of others. If we do not change the conditions that lead to misery, it will continue and may even become worse. In fact, the misery can eventually give rise to much greater change than most of us would want.

The horrible deaths of children caused by abusive parents are almost impossible for most of us to understand. Isn't the love between parent and child natural, automatic, inevitable? Increasingly, our society is uncovering a dark past of abused spouses and children too often ending up dead before they have a chance to live as productive adults. But what, we might ask, would these children have become if they had survived these conditions? Some of them might be able to come out of it remarkably well, but chances are that their socialization would produce another generation of those who prey on others. Those who bring misery to others have often themselves endured miserable conditions. Being abused encourages abuse. It is easy for those of us who are outside the actual situation to blame the abuser, as though he or she exercises free choice; but the careful student becomes the bothered student, bothered by the causes of abuse in society.

The serious student knows that ending misery may be impossible, but it is folly to believe that misery will simply go away if we complain and simply wish it away. The serious student also recognizes the folly of believing that those of us relatively free of misery will not someday be touched by those for whom it is a way of life.

## Questions to Consider

1. Is it true that misery is not dependent on social conditions but simply on subjective feeling? If this is true, then is it possible to lessen misery?
2. Which of the four social conditions that are said to lead to human misery is the most prevalent in American society today?
3. Are the rich or the poor more likely to experience misery in their lives, or are they equally likely?
4. Is social conflict ever good?
5. How does the perpetrator of destructive social conflict bring misery upon him or herself?
6. What principles should parents follow in order to lower the probability that their children will grow up into lives of misery?
7. What work might not be alienating in modern society?
8. What can be changed in society to lessen social alienation?
9. What are the most important aspects of society that make people feel they do not have power over their own lives?

# References

Most of the following works deal with the consequences of inequality. Some are good introductions to crime and to alienation. All are attempts to understand human problems in society.

**Adam, Barry B.** 1978. *The Survival of Domination: Inferiorization and Everyday Life*. New York: Elsevier.

**Akers, Ronald L.** 1994. *Criminological Theories: Introduction and Evaluation*. Los Angeles: Roxbury.

**Ancheta, Angelo N.** 1998. *Race, Rights, and the Asian American Experience*. New Brunswick, NJ: Rutgers University Press.

**Anderson, Elijah.** 1999. *Code of the Street: Decency, Violence, and the Moral Life of the Inner City*. New York: Norton.

**Apraku, Kofi.** 1996. *Outside Looking In: An African Perspective on American Pluralistic Society*. Westport, CT: Praeger.

**Aronowitz, Stanley.** 2001. *The Last Good Job in America: Work and Education in the New Global Technoculture*. Lanham, MD: Rowman & Littlefield.

———. 2003. *How Class Works: Power and Social Movement*. New Haven, CT: Yale University Press.

———. 2005. *Just around the Corner: The Paradox of the Jobless Recovery*. Philadelphia: Temple University Press.

**Aronowitz, S., and W. DiFazio**. 1994. *The Jobless Future: Sci/Tech and the Dogma of Work*. Minneapolis: University of Minnesota Press.

**Baldwin, James.** 1963. *The Fire Next Time*. New York: Dial.

**Banes, Colin, and Geof Mercer**. 2003. *Disability*. Malden, MA: Blackwell.

**Barnes, Sandra L.** 2005. *The Cost of Being Poor: A Comparative Study of Life in Poor Urban Neighborhoods in Gary, Indiana*. New York: State University of New York Press.

**Barton, Bernadette.** 2006. *Stripped: Inside the Lives of Exotic Dancers*. New York: New York University Press.

**Beeghley, Leonard.** 1983. *Living Poorly in America*. New York: Praeger.

———. 1996. *What Does Your Wife Do? Gender and the Transformation of Family Life*. Boulder, CO: Westview.

———. 2000. *The Structure of Social Stratification in the United States*. 3rd ed. Boston: Allyn & Bacon.

**Bergmann, Barbara, and Suzanne Helburn**. 2002. *America's Child Care Problem*. Oxford: Palgrave.

**Bellah, Robert N.** 1999. "The Ethics of Polarization in the United States and the World." In *The Good Citizen*, edited by David Batstone and Eduardo Mendieta. New York: Routledge/Taylor & Francis.

**Bellah, Robert N., Richard Madsen, William M. Sullivan, Ann Swidler, and Steven M. Tipton**. 1985. *Habits of the Heart: Individualism and Commitment in American Life*. Berkeley: University of California Press.

**Bensman, David, and Roberta Lynch**. 1987. *Rusted Dreams: Hard Times in a Steel Community*. New York: McGraw-Hill.

**Bergmann, Barbara, and Suzanne Helburn**. 2002. *America's Child Care Problem*. Oxford: Palgrave.

**Berrick, Jill Duerr.** 1995. *Faces of Poverty: Portraits of Women and Children on Welfare*. New York: Oxford University Press.

**Blauner, Robert.** 1964. *Alienation and Freedom: The Factory Worker and His Industry*. Chicago: University of Chicago Press.

**Blumberg, Rae Lesser.** 1991. *Gender, Family, and Economy: The Triple Overlap*. Thousand Oaks, CA: Sage.

**Boggs, Carl.** 2000. *The End of Politics: Corporate Power and the Decline of the Public Sphere*. New York: Guilford.

**Borchard, Kurt.** 2005. *The Word on the Street: Homeless Men in Las Vegas*. Reno, NV: University of Nevada Press.

**Braverman, Harry.** 1974. *Labor and Monopoly Capital: The Degradation of Work in the Twentieth Century*. New York: Monthly Review Press.

**Brown, Phillip.** 2001. *Capitalism and Social Progress: The Future of Society in a Global Economy*. New York: Palgrave.

**Carmichael, Stokely, and Charles V. Hamilton**. 1967. *Black Power*. New York: Random House.

**Chafetz, Janet Saltzman.** 1990. *Gender Equity: An Integrated Theory of Stability and Change*. Newbury Park, CA: Sage.

**Chang, Iris.** 1997. *The Rape of Nanking: The Forgotten Holocaust of World War II*. New York: Basic Books.

**Charon, Joel M.** 2002. *The Meaning of Sociology*. 7th ed. Upper Saddle River, NJ: Prentice Hall.

**Charon, Joel M.** 2009. "An Introduction to the Study of Social Problems," pp. 1–12 in *Social Problems: Readings with Four Questions*, edited by Joel Charon and Lee Vigilant. 3rd ed. Belmont, CA: Wadsworth Cengage Learning.

**Charon, Joel M., and Lee Garth Vigilant**. 2009. *Social Problems: Readings with Four Questions*, 3rd ed. Belmont, CA: Wadsworth Cengage Learning.

**Chin, Margaret M.** 2005. *Sewing Women: Immigrants and the New York Government Industry*. New York: Columbia University Press.

**Chirot, Daniel.** 1994. *Modern Tyrants: The Power and Prevalence of Evil in Our Age*. New York: Free Press.

**Clark, Kenneth B.** 1965. *Dark Ghetto*. New York: Harper & Row.

**Clarke-Stewart, Alison, and Cornelia Brentano**. 2006. *Divorce: Causes and Consequences*. New Haven, CT; Yale University Press.

**Cohen, Mark Nathan.** 1998. *Culture of Intolerance: Chauvinism, Class, and Racism in the United States*. New Haven, CT: Yale University Press.

**Cooley, Charles Horton.** [1909] 1962. *Social Organization.* New York: Schocken Books.

**Crompton, Rosemary.** 1993. *Class and Stratification: An Introduction to Current Debates.* Cambridge, MA: Polity Press.

**Crothers, Lane.** 2003. *The American Militia Movement from Ruby Ridge to Homeland Security.* New York: Rowman & Littlefield.

**Currie, Elliott.** 2005. *The Road to Whatever: Middle-Class Culture and the Crisis of Adolescence.* New York: Henry Holt and Company.

**Curtis, James, and Lorne Tepperman**, eds. 1994. *Haves and Have-Nots: An International Reader on Social Inequality.* Englewood Cliffs, NJ: Prentice Hall.

**Danzinger, Sheldon, and Ann Chih Lin**, eds. 2000. *Coping with Poverty: The Social Contexts of Neighborhood, Work, and Family in the African-American Community.* Ann Arbor: University of Michigan Press.

**Davey, Joseph Dillon.** 1995. *The New Social Contract: America's Journey from Welfare State to Police State.* Westport, CT: Praeger.

**Denzin, Norman K.** 1984. "Toward a Phenomenology of Domestic Family Violence." *American Journal of Sociology* 90: 483–513.

**Derber, Charles.** 1996. *The Wilding of America: How Greed and Violence Are Eroding Our Nation's Character.* New York: St. Martin's.

**DePrie, Jason.** 2004. *American Dream: Three Women, Ten Kids, and a Nation's Drive to End Welfare.* New York: Penguin.

**Dobratz, Betty A., Lisa K. Waldner, and Timothy Buzzell**, eds. 2001. *The Politics of Social Inequality.* New York: JAI.

**Dohan, Daniel.** 2003. *The Price of Poverty: Money, Work, and Culture in the Next Mexican American Barrio.* Berkeley: University of California Press.

**Dougherty, Charles J.** 1996. *Back to Reform: Values, Markets and the Healthcare System.* New York: Oxford University Press.

**Du Bois, William D., and R. Dean Wright**. 2001. *Applying Sociology: Making a Better World.* Boston: Allyn & Bacon.

**Durkheim, Émile**. [1893] 1964. *The Division of Labor in Society,* translated by George Simpson. New York: Free Press.

———. [1897] 1951. *Suicide,* translated and edited by John A. Spaulding and George Simpson. New York: Free Press.

**Dyer, Joel.** 1997. Harvest of Rage: *Why Oklahoma City Is Only the Beginning.* Boulder, CO: Westview.

**Ehrenreich, Barbara.** 2006. *Bait and Switch: The (Futile) Pursuit of the American Dream.* New York: Metropolitan.

**Ehrensal, Kenneth.** 2001. "Training Capitalism's Soldiers: The Hidden Curriculum of Undergraduate Business Education." In *The Hidden Curriculum.* Edited by Eric Margolis. New York: Routledge/Taylor & Francis.

**Eliade, Mircea.** 1954. *Cosmos and History.* New York: Harper & Row.

**Erikson, Kai T.** 1976. *Everything in Its Path.* New York: Simon & Schuster.

———. 1986. "On Work and Alienation." *American Sociological Review* 51: 1–8.

**Etzioni, Amitai.** 2001. *The Monochrome Society.* Princeton, NJ: Princeton University Press.

**Ewen, Stuart.** 1976. *Captains of Consciousness.* New York: McGraw-Hill.

**Ezekiel, Raphael S.** 1995. *The Racist Mind.* New York: Viking.

**Faludi, Susan.** 1999. *Stiffed: The Betrayal of the American Man.* New York: Perennial.

**Fanon, Frantz.** 1963. *The Wretched of the Earth.* New York: Grove.

**Farley, John E.** 1999. *Majority-Minority Relations.* 4th ed. Upper Saddle River, NJ: Prentice Hall.

**Faulks, Keith.** 2000. *Political Sociology: A Critical Introduction.* New York: New York University Press.

**Feagin, Joe R.** 1975. *Subordinating the Poor: Welfare and American Beliefs.* Englewood Cliffs, NJ: Prentice Hall.

**Feagin, Joe R.** 2002. *Liberation Sociology.* Boulder, CO: Westview.

**Feagin, Joe R., and Melvin P. Sikes**. 1994. *Living with Racism: The Black Middle Class Experience.* Boston: Beacon Press.

**Feagin, Joe R., and Hernan Vera**. 1995. *White Racism: The Basics.* New York: Routledge.

**Feiner, Susan F**, ed. 1994. *Race and Gender in the American Economy: Views from across the Spectrum.* Englewood Cliffs, NJ: Prentice Hall.

**Freeman, Jo.** 1999. *Waves of Protest: Social Movements since the Sixties.* Lanham, MD: Rowman & Littlefield.

**Freie, John F.** 1998. *Counterfeit Community: The Exploitation of Our Longings for Connectedness.* Lanham, MD: Rowman & Littlefield.

**Friedman, Thomas L.** 2002. *Longitudes and Attitudes: Exploring the World after September 11.* New York: Farrar, Straus & Giroux.

**Fromm, Erich.** 1962. *Beyond the Chains of Illusion.* New York: Simon & Schuster.

**Fuller, Robert W.** 2003. *Somebodies and Nobodies: Overcoming the Abuse of Rank.* Gabriola Island, BC: New Society Publishers.

**Galbraith, James K.** 1998. *Created Unequal: The Crisis in American Pay.* New York: Free Press.

**Gans, Herbert J.** 1993. *People, Plans, and Policies: Essays on Poverty, Racism, and Other National Urban Problems.* New York: Columbia University Press.

———. 1995. *The War against the Poor: The Underclass and Antipoverty Policy.* New York: Basic Books.

**Gini, A. R., and T. J. Sullivan**. 1989. *It Comes with the Territory: An Inquiry Concerning Work and the Person.* New York: Random House.

**Glassner, Barry, and Rosanna Hertz**. 2003. *Our Studies, Ourselves: Sociologists' Lives and Work.* New York: Oxford University Press.

**Goffman, Erving.** 1961. *Asylums.* Chicago: Aldine-Atherton.

———. 1963. *Stigma: Notes on the Management of Spoiled Identity.* Englewood Cliffs, NJ: Prentice Hall.

**Goldhagen, Daniel Jonah.** 1996. *Hitler's Willing Executioners: Ordinary Germans and the Holocaust.* New York: Knopf.

**Goode, Erich.** 2002. *Deviance in Everyday Life: Personal Accounts of Unconventional Lives.* Prospect Heights, IL: Waveland Press.

———, ed. 1996. *Social Deviance.* Boston: Allyn & Bacon.

**Guillaumin, Colette.** 1995. *Racism, Sexism, Power and Ideology.* New York: Routledge.

**Hall, John A., and Charles Lindholm**. 2000. *Is America Breaking Apart?* Princeton, NJ: Princeton University Press.

**Hall, Richard H.** 1994. *The Sociology of Work: Perspective, Analysis, and Issues.* Thousand Oaks, CA: Pine Forge.

**Hallinan, Maureen T.** 2005. *The Socialization of Schooling.* New York: Russell Sage Foundation.

**Hancock, Angie-Marie.** 2004. *The Politics of Disgust: The Public Identity of the Welfare Queen.* New York: New York University Press.

**Hardin, Russell.** 1995. *One for All: The Logic of Group Conflict.* Princeton, NJ: Princeton University Press.

**Hare, Bruce R**, ed. 2002. *2001 Race Odyssey: African Americans and Sociology.* Syracuse, NY: Syracuse University Press.

**Harrington, Michael.** 1963. *The Other America.* New York: Penguin.

**Hays, Sharon.** 2003. *Flat Broke with Children: Women in the Age of Welfare Reform.* New York: Oxford University Press.

**Healey, Joseph F.** 1997. *Race, Ethnicity, and Gender in the United States: Inequality, Group Conflict, and Power.* Thousand Oaks, CA: Pine Forge.

**Hedges, Chris.** 2002. *War Is a Force That Gives Us Meaning.* New York: Public Affairs.

**Henslin, James M.** 2000. *Social Problems.* 5th ed. Upper Saddle River, NJ: Prentice Hall.

**Higley, Richard.** 1995. *Privilege, Power, and Place.* Lanham, MD: Rowman & Littlefield.

**Hochschild, Arlie Russell, and Anne Machung**. 2003. *The Second Shift.* New York: Viking Penguin.

**Hochschild, Jennifer L.** 1995. *Facing Up to the American Dream: Race, Class, and the Soul of the Nation.* Princeton, NJ: Princeton University Press.

**Iceland, John.** 2003. *Poverty in America: A Handbook.* Berkeley: University of California Press.

**Ignatieff, Michael**, ed. 2004. *The Lesser Evil: Political Ethics in an Age of Terror.* Princeton, NJ: Princeton University Press.

**Israel, Joachim.** 1971. *Alienation: From Marx to Modern Sociology.* Boston: Allyn & Bacon.

**Jones, Ron.** 1981. *No Substitute for Madness.* Covelo, CA: Island.

**Katz, Michael B.** 1989. *The Deserving Poor: From the War on Poverty to the War on Welfare.* New York: Pantheon.

**Kelso, William A.** 1994. *Poverty and the Underclass: Changing Perceptions of the Poor in America.* New York: New York University Press.

**Kerbo, Harold R.** 1999. *Social Stratification and Inequality.* 4th ed. New York: McGraw-Hill.

**Koonigs, Kees, and Dirk Druijt**, eds. 1999. *The Legacy of Civil War, Violence and Terror in Latin America.* London: Zed.

**Kozol, Jonathan.** 1988. *Rachel and Her Children: Homeless Families in America.* New York: Crown.

———. 1991. *Savage Inequalities.* New York: Crown.

———. 2005. *The Shame of the Nation: The Restoration of Apartheid Schooling in America.* New York: Random House.

**Kushner, Harold S.** 1981. *When Bad Things Happen to Good People.* New York: Avon.

**Lareau, Annette.** 2003. *Unequal Childhoods: Class, Race, and Family Life.* Berkeley: University of California.

**Lee, Robert G.** 1999. *Orientals: Asian Americans in Popular Culture.* Philadelphia: Temple University Press.

**Levi, Primo.** 1959. *If This Is a Man.* New York: Orion.

**Lewellen, Ted C.** 1995. *Dependency and Development: An Introduction to the Third World.* Westport, CT: Bergin & Garvey.

**Lewis, Bernard.** 2002. *What Went Wrong? Western Impact and Middle Eastern Response.* New York: Oxford University Press.

**Liebow, Elliot.** 1967. *Tally's Corner.* Boston: Little, Brown.

———. 1993. *Tell Them Who I Am: The Lives of Homeless Women.* New York: Simon & Schuster.

**Lynch, Michael J.** 2000. *The New Primer in Radical Criminology: Critical Perspectives on Crime, Power, and Identity.* Monsey, NY: Criminal Justice Press.

**Malcolm X and Alex Haley**. 1965. *The Autobiography of Malcolm X.* New York: Grove.

**Mann, Michael.** 2004. *The Dark Side of Democracy: Explaining Ethnic Cleansing.* New York: Cambridge University Press.

**Marger, Martin.** 1999. *Race and Ethnic Relations: American and Global Perspectives.* 5th ed. Belmont, CA: Wadsworth.

**Marshall, Thurgood.** 1979. Address delivered November 18, 1978. *Barrister,* January 15, pp. 1.

**Marx, Karl.** [1844] 1964. *Economic and Political Manuscripts of 1844.* New York: International.

**Marx, Karl, and Friedrich Engels.** [1848] 1955. *The Communist Manifesto.* New York: Appleton-Century-Crofts.

**McCall, Nathan.** 1994. *Makes Me Wanna Holler.* New York: Random House.

**McCord, Joan,** ed. 1997. *Violence and Childhood in the Inner City.* New York: Cambridge University Press.

**Meissner, Steven E., and Richard Rosenfeld.** 2000. *Crime and the American Dream.* 3rd ed. Belmont, CA: Wadsworth.

**Merry, Sally E.** 2005. *Human Rights and Gender Violence: Translating International Law into Local Justice.* Chicago: University of Chicago Press.

**Merton, Robert.** 1938. "Social Structure and Anomie." *American Sociological Review* 3: 672–682.

**Mills, C. Wright.** 1959. *The Sociological Imagination.* New York: Oxford University Press.

**Mills, Nicolaus.** 1997. *The Triumph of Meanness: America's War against Its Better Self.* Boston: Houghton Mifflin/Sage Foundation.

**Moore, Joan, and Raquel Pinderhughes,** eds. 1993. *In the Barrios: Latinos and the Underclass Debate.* New York: Sage.

**New York Times.** 2005. *Class Matters.* New York: Times Books.

**Newman, Katherine.** 2004. *The Social Roots of School Shootings.* New York: Basic Books.

**Newman, Katherine S.** 1999. *No Shame in My Game: The Working Poor in the Inner City.* New York: Knopf.

**Oliner, Pearl M., and Samuel P. Oliner.** 1995. *Toward a Caring Society.* Westport, CT: Praeger.

**Ouchi, William G., with Lydia G. Segal.** 2003. *Making Schools Work.* New York: Simon & Schuster.

**Paap, Kris.** 2006. *Working Construction: Why White Working-Class Men Put Themselves—and the Labor Movement—in Harm's Way.* Ithaca, NY: ILR Press/Cornell University Press.

**Parenti, Michael.** 1998. *America Besieged.* San Francisco: City Lights.

**Parrillo, Vincent N.** 2002. *Contemporary Social Problems.* 5th ed. Boston: Allyn & Bacon.

**Penn, Michael L., and Rahel Nardos.** 2003. *Overcoming Violence against Women and Girls: The International Campaign to Eradicate a Worldwide Problem.* Landham, MD: Rowman & Littlefield.

**Perrucci, Robert, and Earl Wysong**. 2003. *The New Class Society: Goodbye American Dream?* 2nd ed. New York: Rowman & Littlefield.

**Phillips, Kevin.** 1990. *The Politics of Rich and Poor.* New York: Random House.

———. 2002. *Wealth and Democracy: A Political History of the American Rich.* New York: Broadway.

**Piven, Frances Fox, and Richard A. Cloward**. 1993. *Regulating of the Poor: The Functions of Public Welfare.* 2nd ed. New York: Vintage.

**Quadagno, Jill.** 1994. *The Color of Welfare: How Racism Undermined the War on Poverty.* New York: Oxford University Press.

**Reich, Charles A.** 1995. *Opposing the System.* New York: Crown.

**Rifkin, Jeremy.** 2004. *The End of Work: The Decline of the Global Labor Force and the Dawn of the Post-Market Era.* New York: Penguin.

**Ritzer, George, and David Walczak**. 1986. *Working: Conflict and Change.* 3rd ed. Englewood Cliffs, NJ: Prentice Hall.

**Rosoff, Stephen M., Henry N. Pontell, and Robert Tillman**. 1998. *Profit without Honor: White Collar Crime and the Looting of America.* Upper Saddle River, NJ: Prentice Hall.

**Rubington, Earl, and Martin S. Weinberg**. 2003. *The Study of Social Problems: Seven Perspectives.* 6th ed. New York: Oxford University Press.

**Russell, James W.** 2006. *Double Standard: Social Policy in Europe and the United States.* Landham, MD: Rowman & Littlefield.

**Ryan, William.** 1976. *Blaming the Victim.* Rev. ed. New York: Vintage.

**Schacht, Richard.** 1970. *Alienation.* New York: Doubleday.

———. 1994. *The Future of Alienation.* Urbana: University of Illinois Press.

**Schlosser, Eric.** 2001. *Fast Food Nation: The Dark Side of the All-American Meal.* Boston: Houghton Mifflin.

**Schwartz, Barry.** 1994. *The Costs of Living: How Market Freedom Erodes the Best Things in Life.* New York: Norton.

**Scott, Kody.** 1993. *Monster: The Autobiography of an L.A. Gang Member.* New York: Penguin.

**Seeman, Melvin.** 1975. "Alienation Studies." In *Annual Review of Sociology,* edited by Alex Inkeles, James Coleman, and Neil J. Smelser. Palo Alto, CA: Annual Reviews.

**Segrave, Kerry.** 1993. *The Sexual Harassment of Women in the Workplace, 1600 to 1993.* Jefferson, NC: McFarland.

**Sennett, Richard, and Jonathan Cobb**. 1972. *The Hidden Injuries of Class.* New York: Random House.

**Shapiro, Thomas M.** 2003. *The Hidden Costs of Being African-American.* New York: Oxford University Press.

**Silberman, Charles E.** 1980. *Criminal Violence, Criminal Justice.* New York: Vintage.

**Simmel, Georg**. [1902–1903] 1950. "Metropolis and Mental Life," pp. 00–00 in *The Sociology of Georg Simmel,* edited by Kurt Wolff. New York: Free Press.

**Skolnick, Jerome H., and Elliott Currie**. 2000. *Crisis in American Institutions.* 11th ed. Boston: Allyn & Bacon.

**Staub, Ervin.** 1989. *The Roots of Evil: The Origins of Genocide and Other Group Violence.* New York: Cambridge University Press.

**Steinberg, Stephen.** 1995. *Turning Back: The Retreat from Racial Justice in American Thought and Policy.* Boston: Beacon.

**Straus, Murray A., Richard J. Gelles, and Suzanne K. Steinmetz**. 1988. *Behind Closed Doors: Violence in the American Family.* New York: Doubleday.

**Suro, Roberto.** 1999. *Strangers among Us: Latino Lives in a Changing America.* New York: Knopf.

**Terkel, Studs.** 1972. *Working.* New York: Pantheon.

**Thrasher, Frederic.** 1927. *The Gang.* Chicago: University of Chicago Press.

**Tichenor, Veronica J.** 2005. *Earning More and Getting Less: Why Successful Wives Can't Buy Equality.* New Brunswick, NJ: Rutgers University Press.

**Tilly, Charles.** 1998. *Durable Inequality.* Berkeley: University of California Press.

**Titchkosky, Tanya.** 2003. *Disability, Self, and Society.* Toronto: University of Toronto Press.

**Tonry, Michael.** 2004. *Thinking about Crime: Sense and Sensibility in American Penal Culture.* New York: Oxford University Press.

**Travis, Jeremy, and Michelle Waul**. 2003. *Prisoners Once Removed: The Impact of Incarceration and Reentry on Children, Families, and Communities.* Washington, DC: Urban Institute.

**Trotter, Joe W., with Earl Lewish and Tera W. Hunter**, eds. 2004. *African American Urban Experience: Perspectives from the Colonial Period to the Present.* New York: Palgrave Macmillan.

**Tucker, Robert C**, ed. 1972. *The Marx-Engels Reader.* New York: Norton.

**Turow, Scott.** 2002. *Ultimate Punishment.* New York: Farrar, Straus, & Giroux.

**U.S. Bureau of the Census**. 1999. *Current Population Survey, March.* Washington, DC: Government Printing Office.

**Van den Berghe, Pierre.** 1978. *Race and Racism: A Comparative Perspective.* New York: Wiley.

**Walker, Henry A., Phyllis Moen, and Donna Dempster-McClain**, eds. 1999. *A Nation Divided: Diversity, Inequality, and Community in American Society.* Ithaca, NY: Cornell University Press.

**Wallerstein, Immanuel.** 1974. *The Modern World-System.* New York: Academic Press.

**Waters, Malcolm.** 1995. *Globalization.* New York: Routledge.

**Weber, Max**. [1905] 1958. *The Protestant Ethic and the Spirit of Capitalism,* translated by Talcott Parsons. New York: Scribner's.

**Weitzman, Lenore J.** 1985. *The Divorce Revolution: The Unexpected Social and Economic Consequences for Women and Children in America.* New York: Free Press.

**Wilson, William Julius.** 1987. *The Truly Disadvantaged: The Inner City, the Underclass, and Public Policy.* Chicago: University of Chicago Press.

———. 1999. The *Bridge over the Racial Divide.* Berkeley: University of California Press.

———, ed. 1993. *The Ghetto Underclass: Social Science Perspectives.* Newbury Park, CA: Sage.

**Witt, Griff.** 2004. "As Income Gap Widens, Uncertainty Spreads." *Washington Post,* September 20.

**Wolfe, Alan.** 2001. *Moral Freedom: The Search for Virtue in a World of Choice.* New York: Norton.

**Wolff, Edward N.** 1996. *Top Heavy: A Study of the Increasing Inequality of Wealth in America.* New York: Twentieth Century Fund.

———. 2007. "Recent Trends in Household Wealth in the United States: Rising Debt and the Middle-Class Squeeze." Working paper No. 502. The Levy Economics Institute of Bard College (http://www.levy.org).

**Wright, Erik Olin.** 1994. *Interrogating Inequality: Essays on Class Analysis, Socialism, and Marxism.* New York: Verso.

**Yankelovich, Daniel.** 1982. *New Rules: Searching for Self-Fulfillment in a World Turned Upside Down.* New York: Bantam.

**Young, Alfred A.** 2004. *The Minds of Marginalized Black Men.* New Princeton, NJ: Princeton University Press.

**Zimbardo, Philip.** 1972. "Pathology of Imprisonment." *Society* 9: 4–8.

**Zweig, Michael.** 2000. *The Working Class Majority: America's Best Kept Secret.* Ithaca, NY: Cornell University Press.

# Does the Individual Really Make a Difference?

## An Introduction to Social Change

### Concepts, Themes, and Key Individuals

- ❑ Social change
- ❑ The individual versus society
- ❑ Social influence
- ❑ Social conflict
- ❑ Social trends
- ❑ Rationalization of life

When I was an undergraduate at the University of Minnesota, I did not understand much about government and politics. I knew I lived in a democracy, but I did not seek to understand exactly what that meant. I was suspicious of communism, but I did not understand exactly what it was. I looked forward to the day I would be 21 so I could vote, although I did not know how I might do so intelligently. Looking back, I was truly a naive undergraduate, but at least I really had some interest in understanding government and politics.

In the library one day, I met another student who looked old enough to vote, so I asked him if he had voted in the last election. His answer was a simple no. I was shocked, and so I began to question him. He presented his case to me: "It doesn't matter if I vote or not. One vote will never make a difference to anything." I pointed out that if everyone thought that way, our democracy would be a farce. He replied that he was not talking about "everyone," he was talking about himself. I asked him about his influence on others around him: family, friends, other students. He replied that they did not know whether he voted; they could vote if they wanted.

So began my investigation of democracy. I have actually voted in every national and state election as well as most local ones—not because I believed it might really change the world, but because I live in a nation that strives to be democratic, and I feel responsible to contribute to it. I happen to believe that voting is the most fair and peaceful way to change leaders, and even if my vote is not super important, it is really my commitment toward the democratic society that influences me to vote. Through all my reading and discussions, however, the memory of this man from the basement of Walter Library continues to haunt me. I really want to matter to the political system, but realistically it is very difficult. Even if my candidate wins, he or she may not do what I thought he or she would do if elected. And even if he or she fights for my interests, it will be very difficult for that individual to make a real difference, given the existence of social patterns and those who oppose my position. Because we exist in large groups, communities, and societies, it is very difficult for any person to really change the patterns that organizations develop over many years.

It is fashionable in the United States, of course, to believe that individuals can do anything they set their minds to: If someone wants to make a difference to other people, society, and even the world, he or she can. It is also fashionable to believe that society changes because of the efforts of the individual. To believe these things does not make them true; to believe them is really a statement of faith taught in society. It is important for people to believe they make a difference: "I'm important." "My life matters." "I do have an effect on the lives of others." "I can shape the future of society."

This chapter will examine the power of the individual to change society. We will start small, however, and look first at how individuals can make a difference in:

- their own lives,
- the lives of those with whom they interact, and
- various organizations to which they belong.

Then, finally, we will examine society. As you will see, each topic is highly complex, and you will probably find that the actual difference the individual can make is less than you imagine.

## The Individual's Influence on His or Her Own Life

If we are free at all and exercise control over our own ideas, values, actions, and directions, then each actor makes a difference *to his or her own life.* I influence what I do. I make decisions that allow me to go one way rather than another, to believe one thing rather than another, to act a certain way, or to become a certain type of person. We examined the question of freedom in Chapter 5, and although most of the chapter emphasized that there are many social forces that operate limiting both our freedom of thought and action, we also claimed that we do have some control over our own lives, some ability to make a difference in directing what we think and do. In the context of this chapter, if we argue that we are indeed free to some extent, our lives do matter—to ourselves. We make a difference in our own lives.

But because of the many social and other factors that control us, this ability is always limited, often to a great extent. We always act in a social context. Social patterns always matter. Role, class, culture, and institutions always guide our decisions and our lives. Our problems are always linked to social problems, and our successes are linked to the state of society. The sociological view leaves some room for individual freedom, but not a great deal.

When people ask if the individual really does make a difference, however, they are usually asking about *impact on others.* Do our acts matter in terms of influencing other individuals, groups, communities, or societies? The topic really becomes one of *individual and social change:* Can my acts influence others—do they matter to anyone besides myself?

## The Individual's Influence on Other Individuals

We can begin to understand the importance of the individual in effecting change by first looking at social interaction. Do individuals make a difference in how other people live their lives? Do individuals influence their children and friends? Are they able to influence the lives of other individuals trapped by poverty or ignorance?

### *The Problem of Measuring Influence*

Let us start rather simply. In all that we do, we encounter others who follow different interests from ours, who have different views, different priorities, different problems. If we try, can we make a difference to them? The question of the influence of one actor over another is difficult to answer. Teachers, for example, often exaggerate their own influence over students: "I taught my class the locations of all the nations of Africa. They really know something." Or, "Now that I've taught my class the harmful effects of drugs, they know that if they do drugs, they'll destroy themselves." But such influence is much less than imagined.

First of all, *people almost always forget most of what they are taught because it is not useful to them.* As every teacher knows, teaching others does not guarantee that much significant learning goes on. What is taught one day is forgotten by the next; what is reviewed for an exam is forgotten a week later. What I learn from a book one week does not necessarily mean that I will remember it the next.

Second, *many are actively influencing others at the same time we are trying.* The extent of any one person's influence will depend on all of these other influences. The problems of my family may be far more important to me than what I am trying to learn about geography; the desire to be accepted by my friends may be far more important to me than a lecture on drugs by someone I barely know and who gets his or her information from a book.

Third, *my influence on another may be far more unintentional than intentional.* I may make a difference, but in a way that I may not even want. The teacher may be teaching the location of nations, but the student may be learning to hate geography; the teacher may be teaching the harmful effects of drugs, but the student may be learning that adults are hypocritical, unrealistic in their expectations, and not to be trusted. Of course, the influence can be unintentionally positive: The student in geography may become excited about visiting other societies; the student learning about drugs may be influenced to confront his or her parents about their alcohol abuse. *The influence of one individual on another is difficult to achieve in exactly the way it is intended.* We may make some difference, but it is a highly complex difference—and sometimes it is exactly what we did not intend.

Fourth, the belief that the individual matters and can influence others must take into account that *our actions will often have a harmful influence on those with whom we interact.* We should thus not assume we will always matter in a positive way. We who abuse our children will lay the foundations for misery—for our children and those with whom they interact—and that misery may last their whole lives. Those of us who teach our children that it is all right not to care about others, to hate those unlike themselves, and to harm or exploit others will also be influenced. Those of us who hurt others may cause emotional harm, and that, too, will make a difference to their lives. Indeed, as I look back at my own life I have to wonder how many of my actions may have actually harmed others; yes, as a teacher I may have made a difference to those students I failed in high school, and what happened to those students I acted too impatiently when they needed help.

Before we assume that social interaction will make it possible for the individual to make a real difference on others, it is important to recognize that *interaction—and therefore influence—is two-way.* I act, and in my action I may exert influence on your life; at the same time, however, you act, and in your action you too may have some influence on my life. To be fair and honest, if either of us claim that "Yes, I made a difference," we must realize that the other too may have made a difference. We negotiate influence; normally, no one has complete power over the other—it is two-way. I may have influenced my wife, but my wife has had much influence over my life, too. At the very least, we each made a difference to the other. My sons also have had influence on my life; sometimes it is difficult to assess our actual influence. To believe that social influence is a one-way affair exaggerates the importance of the individual in the interaction. Influence is negotiated, and before we simply see how important we are to others, we need to consider their influence on us.

It is therefore possible to make an important difference in other people's lives, but simply interacting with them does not guarantee much influence. If we really care about this issue, then we need to consider the points made here: What others learn from us is too easily forgotten; whenever we try to influence others, there are other people influencing them and those people may also be important to them; my influence may be unintentional; my influence may actually be harmful to them; and they may very well have had a greater impact on me than I had on them.

Most of us will never really know our real influence on others. We become "significant others" to them, and we might not even know it. Each of us can influence our spouse and children in a positive manner just by how we act toward them and toward other people. Teachers, friends, acquaintances, employers, rock stars, members of the clergy, and political leaders can become significant others to us, their lives and actions become examples for us, their ideas and values become our own. Sometimes an encounter with someone will change our whole direction in life, and what may at first seem to be a small influence becomes a huge one. If I examine my own history, I recognize a parent here, a band director over there, my social studies teacher, a friend, my freshman English teacher, a history lecturer, a colleague I taught with in high school, several students in my high school classes, and many others who had distinctly positive influences on me that they probably do not even recognize. They—and many others—made a big difference in my life. Even a chance encounter with a book salesman altered my life considerably; without that encounter I probably never would have tried to write a book, and my life and research might have gone in another direction.

I am writing this book primarily to influence other individuals. I am realistic. How much lasting influence can I have? For most students, not much—and we can undoubtedly list a dozen reasons. For a few, maybe I can have a small influence on their direction. For even fewer, maybe I can influence them to have a love of learning or a love of sociology or a concern for the fate of humanity that will remain important all of their lives. Unfortunately, I also fear I will supply ammunition for being intolerant, sexist, racist, anti-intellectual, or anti-sociological—unintentional influences but nonetheless real.

I respect those people who live their lives working to better the lives of others. Certain professions and situations allow individuals to truly contribute something that makes a difference. A doctor, a firefighter, a police officer can save people's lives or contribute to their quality of life. Teachers, clergy, a bartender, lawyers, activists, and simply individuals who care about others can make a positive impact on individuals. People who perform for others, write books, report the news can be very important to some people. Often, we never know the importance of what we do, but in fact we have become important in relieving pain and hardship. I will never know how I might make a difference to those around me, so perhaps I should always be kind, always be respectful of others, always pay attention to the needs of others. Certainly there are those who believe and act on this, and they make a difference to a great many people.

### *The Larger Social Context and Individual Influence*

Leave it to the sociologist to bring up the role of society in everything human. Here we go again! It is essential to recognize that individual influence always exists in a larger social context. The likelihood of influence depends on larger social trends. My likelihood of influencing you to enjoy opera as much as I do depends on whether opera is loved in society or whether it is being replaced by rock music. It depends on whether we live in Italy or the United States; it depends on whether we live in the 1990s or the 1890s. The teacher influences students who are ready and interested. The religious leader converts individuals who are seeking conversion. The political leader influences individuals who agree with his or her political philosophy or, sometimes, who are rebelling against their parents' political philosophy. Charles Manson did not appeal to everyone; he appealed to people living in a certain kind of community at a certain period of our history. A young person is influenced by a movie star who represents the culture to which that young person belongs. The individual who influences someone to try drugs is aided by a social context that regards drug taking as acceptable behavior.

Thus, the influence of one actor over another may be real, but it is often exaggerated and always facilitated or made more difficult by the social context. All of us make a difference sometimes. That difference may be in the direction of tolerance, love, caring, and growth; it may be in the direction of intolerance, hatred, and destructiveness. There is usually no way of knowing who we have affected or how strongly. For most of us most of the time interpersonal influence remains an article of faith that we believe in.

## The Individual versus Social Organization

To affect ourselves or to influence other individuals is one thing; to affect a group, formal organization, community, or society is something else again. Imagine the difficulties involved in accomplishing something important and lasting in an established social organization.

My first job after graduating from college was as a high school history teacher, and I was prepared to make a difference in the world. I remember that I wanted to contribute to all humankind. Perhaps I would teach someone who would become a great leader; perhaps I would teach ideas that would spread; perhaps I would be recognized as a model teacher for all to learn from. It wasn't clear how I would do it, but I knew I wanted to make a positive mark on society. A short time after I began, it became obvious to me that my sights were set too high. At least I could have an impact on the community of St. Paul. After teaching for a month or so, I knew that if I wanted to have an impact I would have to settle for Harding High School. My idealism changed when I realized how few students actually took my classes or knew who I was. Like most idealistic teachers I eventually came to realize that my real chance for making a difference in people's lives was to be found in the everyday interaction with 150 individuals I met five hours a week. Yet I came to see that I could not expect great things there either. I was not an important person to many of these people. Some saw me as an intruder in their lives, and some did not often understand what I was trying to teach. My actions in the classroom did, in fact, eventually influence several students, but I am afraid I had a really lasting effect on only a few and, even then, often not in the direction I had intended. I never did influence society or St. Paul, and I left Harding High School as I had found it, having had a minimal impact on it.

I really wanted to have an impact on an organization. Why couldn't I? What stood in the way? Why is it so difficult—perhaps impossible—for the individual to have a real impact on an organization?

### *The Individual Confronts Social Patterns*

We return to the existence of *social patterns.* Every organization eventually develops certain ways of doing things. That's actually what is meant by being "organized." People know what others are going to do, and they understand what they are supposed to do. Structure distributes positions in an organization, which are usually ranked and have attached to them roles, or scripts, laying out what is expected. Culture is taught to all, creating a shared set of beliefs, values, and rules that guide actors as they interact. In society there are institutions—long-established procedures that guide the individual. For example, people in American society have established marriage as an institution that the vast majority of us follow. When we marry, the general outline for what we are to do is laid out in advance: courtship, engagement, a religious ceremony, a reception with friends and family, an agreement to be loyal and to give love, money, and time to one another—and so on. In our individual marriages we establish our own social patterns over time: who does what, the degree of independence each of us has, how much money we should spend versus how much we should save. As new problems arise in our marriage, we have to discover new solutions, and they, too, become established as patterns. As we have children, change jobs, and move from one community to another, we have to alter what we do, but we normally revise the old patterns rather than simply throw them out. The revisions become new patterns we follow. The bond between us depends on our feelings, as well as on all of the patterns we have come to share. For one of us to decide to abandon these patterns or to change them radically upsets the relationship. And if one tries to change things without consent, the danger of dissolution grows.

If it is so difficult to change social patterns in a marriage, then consider how much more difficult it is in a larger social organization: a group of friends, a school, or a society. A lasting group—a class of thirty students, for example—establishes patterns early, often ones that have been developed elsewhere to guide all such classes or have been established through the demands of the teacher or even through day-to-day interaction. Any individual, including the teacher, who tries to radically change those patterns threatens the organization of the class and its success. Even the students are aware of this, and once the patterns are established, they, too, work against the rules being changed, either by the teacher or by a newcomer in the class who wants them to fit his or her needs.

A football team works the same way. Game rules and league rules govern team play. Individuals are discouraged from openly challenging them. The team itself develops special plays, procedures for play calling and substitutions, and even subtle ways of changing what individuals do as a play unfolds. Individuals who decide to go their own way threaten the team's success.

On the opposite end of the organizational spectrum stands society. Society has a very long history. It precedes every living actor, and it will be there when every living actor dies. Its patterns—social structure, culture, and institutions—have been established over many years, and these patterns confront the individual as a generally accepted reality. The individual can cry out, "I won't do what you want of me!" and can leave that society, no longer to be influenced by it. Quiet nonconformity is possible. But it is something else for the individual to try to change the dominant social patterns. To change society is to threaten the continuation of the world as it exists for most people.

Social patterns, according to Émile Durkheim, take on a life of their own. They exist "out there" someplace—invisible, real, external to us, influencing and even controlling us. When we break these patterns, we challenge their reality. See what happens when we decide to go it alone in social organizations and refuse to follow the procedures, rules, truths, and values that were established long ago. We are talking about not only laws but also many other patterns that guide us in every move we make, from greeting someone on the street to burying the dead. Actions are neither spontaneous nor random. They generally follow patterns laid out by strangers long dead.

What chance, then, does the individual actor have against social patterns? We each live in a social reality that others have become used to and generally are fearful of losing. No matter how much we might dislike our situation, there is something in most of us that cries out to keep the structure, culture, and institutions that we have. We might hate society as it is, but it is the only world we know.

### *Some Individuals Do Impact Social Patterns*

Individuals can make a difference in the successful operation of an organization—within the bounds of its social patterns. They can help the organization achieve its goals or can make a difference in the opposite way by blocking those goals. An outstanding quarterback can pull together a bad team. A good president can lead society in positive directions. An outstanding businessperson can turn around a struggling company. Such individuals can make a difference—sometimes a big difference. Normally, that kind of success does not change the social patterns in that organization (the individual is simply an outstanding actor within the established patterns). The individual will have a much more difficult time making a lasting and dramatic impact on the social patterns themselves.

In general, most people who work for change in an organization work for minor change within the existing social patterns. Such change may be critical, but it does not constitute lasting, significant change. Some individuals will make a difference, but as outstanding actors following a written script.

Certain individuals truly shape social patterns, however, and leave a great mark. In the former Soviet Union, for example, Mikhail Gorbachev rose up in a political structure that was first established in 1917. He rose because he was perceived as someone who represented and stood for the established structure, culture, and institutions. He also used these social patterns to rise to the top. Normally when one rises to the top, one becomes increasingly supportive of the system that made that rise possible. But Gorbachev was different—he made a big difference to the Soviet Union and to the world because, once in a high position, he turned around and criticized the political structure he was leading. He criticized the nature of the economy that favored people in his position. He questioned the necessity for the massive military system that formed an integral part of Soviet society, and he began to open channels of discussion on issues that were simply taken for granted in society. Here was a man who really made a difference: powerful, critical, willing to make basic changes in some of the patterns of society. Gorbachev was an unusual revolutionary in that most revolutionaries arise outside the dominant social patterns. He came from within the system itself: He had position, powerful allies, and intelligence. He calculated well, and he brought about great reforms. He made a difference, a great difference.

This was neither easy to do nor common. The dead hand of the past weighed heavily on him (as Marx in the nineteenth century had declared happens to leaders in all societies). Others who favored a system that clearly benefited them attacked him. As unproductive and unresponsive as the political, military, and economic system was, it was all that many of the people had known, so they naturally wondered if a new system would necessarily be better. Everything seemed to oppose any chance that one man would make so much difference. In fact, it is probably safe to say that Gorbachev's influence was even greater than he imagined it would be. As a result of his efforts, change took on a life of its own, and many events he did not originally intend occurred. Eventually he fell from grace, to be followed by other leaders, each trying desperately to deal with tremendously difficult problems, each trying to establish new institutions, each finding it difficult—even impossible—to change inherited institutions. The Soviet Union is no more, it will never be able to go back to its failed institutions, and Gorbachev will probably go down in history as an individual who made an important difference in the world. Without him, the world would have changed—but the way it changed, and the speed at which it changed, are a result of his efforts to a large degree.

Even revolutionaries who overthrow the scoundrels usually become scoundrels themselves without changing much of anything. We might vote out the other party only to find that the new party is not capable of making the real changes promised—not because its members are liars, but because revolutionary change is made so difficult within the social patterns already in existence. Presidents sometimes try to make a difference but end in failure. New congresses get excited about changing government, but once in a position they are faced with the great difficulty of existing social patterns. How much can one individual change social patterns that are deeply embedded, no matter what his or her intentions are? Political candidates promise change, and many really believe they can bring it about only to meet the realistic power of our inheritance as a society.

### *The Role of Social Power*

What many of us ignore when we think about the influence of an individual on an organization is the element of *social power.* The individual can change an organization only if he or she has it. Power means *the ability to achieve one's will in relation to others.* One's ability often arises from high position in an organization, but it also arises from attractiveness, large numbers of followers, wealth, weapons, intelligence, or persuasive ability. Parents have great power over children and can influence their directions, ideas, and values. As other elements compete with that power, parental influence lessens. Corporate leaders have great power, and they use it to shape policies in their interests. Sometimes it is to change society (the tax structure, the relative power of unions, the degree of governmental "interference"); more often it is to protect the social patterns as they operate (private enterprise, inheritance practices, a court system based on the ability to pay). The president of the United States has great power and thus is able to have more influence on the direction of society than the rest of us. (After all, what do we have that allows us to influence that direction? a vote? a contribution to one official? going door-to-door to get votes for our candidate?) A skillful leader in a well-armed, well-funded revolutionary group may have great influence on society, and a skillful religious leader may have an impact on a congregation of believers or even on the direction of society.

In understanding how such influence is possible, however, it is important to recognize not only that desire is necessary, but also that desire must be wedded to power. Ideas can be effectively challenged only with power. Criticism must be backed by power. New directions for an organization must result from those who have more than good ideas or good intentions, and new social patterns can arise only from people who have the power to bring them about.

Power is a complex matter. First, those with the most power usually do not want basic change in society. They benefit the most from existing social patterns. They rose through those patterns and generally approve of them. Second, one actor's social power is one part of a *social equation.* Power is exerted from both sides. Even if a person desires change and tries to bring it about, those on the side protecting the social patterns also have power—and almost always much greater power. Every organization (from the family to the society) has mechanisms for dealing with those who rise up to try to change that organization as it exists. To be able to change society is to have enough power to influence those who defend society as it exists.

Change usually occurs, therefore, not because of the efforts of one individual but because people work together, form a power base, and bring about change. A leader cannot change society alone; he or she needs a base that includes other individuals who are willing to work in the same direction for change. In 1789, French workers and the middle class united around emerging leaders and overthrew the monarchy. In the 1950s, African Americans in the South organized around Martin Luther King, Jr., and together began to bring down the system of segregation. King was important but alone, he was without power in society. *Social movements*—loose organizations of large numbers who can be effectively mobilized around leaders to march, protest, boycott, strike, and actively confront the opposition—change society. More organized protest groups—Mothers Against Drunk Driving (MADD)—change society. Various individuals within them may make a difference in the direction of change, but *it is many people working together in opposing established social patterns who have the real potential for changing society.* Each individual is a resource used to bring about change. Alone without influence, united they can make a difference.

Success is never guaranteed, no matter how much one desires impact, no matter how much power one has, no matter how noble one's cause. Our efforts bring four possibilities: (1) The social patterns may *not change;* (2) the social patterns may change but *in a direction unintended* (perhaps toward more oppression); (3) the social patterns *may actually change in the direction desired;* or (4) the social patterns *may change in exactly the way desired.* Number 1 is the most likely to occur; number 4 is the least likely.

## Social Change: A Sociological View

If sociologists recognize that the individual does, in fact, sometimes make a difference in changing a social organization, but that this difference is usually minor and often unintentional, then how else do they approach the problem of social change? If the individual is not primarily responsible for change, then what is? It is probably best to begin answering this question by listing and explaining five guiding principles that most sociologists tend to believe.

### *Principle 1: Change Is Inherent in All Social Organization*

*The first principle is that every organization always and continuously changes.* As its size changes, it changes; as it becomes older, it changes; as its environment changes, it must adjust. Each event in its history becomes part of its past and can be recalled and used to make decisions later on. Indeed, it is erroneous to see social organization as a rigid, permanent set of social patterns that represents enduring stability and order. Every action of every individual alters society a tiny bit. Every decision by our government alters society a tiny bit. Of course, some actions, individuals, and decisions are more important than others. But the point is that society never stays the same from moment to moment. Change is to be expected. Whatever one likes about society today—its music, its movies, its family patterns, its level of religious commitment—will inevitably change. Whatever exists today will be at least slightly different tomorrow.

Societies have *rates of change.* Some societies change far more rapidly than others. The rate of change is an integral part of society; it might even be understood as a social pattern. Rates of change predictably increase as society becomes modern. A number of events and individuals may impact the rate, temporarily slowing it down or speeding it up, but over time the rate of change is normally determined by larger worldwide or societal trends—such as globalization, colonization, economic depression, war, trade, industrialization. The rate of change is influenced by the level of modernization a society exists: the more a society successfully modernizes, the more rapid the rate of change. Dictators who decide to suddenly modernize *do* make a difference in the rate of change, but once modernization develops, it is difficult for any individual to determine the rate and forces of change. Change is highly interdependent. When one part of society changes—such as education, trade, or building a giant dam—much of the rest of society changes, the rate of change is altered, and the new rate of change predictably increases. By 2000 several important events and trends brought the leaders of China to decide that China must become part of the world economy. Their decision to do this has altered the rate of change in China considerably, this rate of change will undoubtedly increase itself more and more, and the leaders of China will probably find themselves unable to direct or slow down that rate, which will eventually create trends that they might not favor, but are unable to stop.

Certainly the rate of change in the United States increased dramatically during the Civil War and after. By the end of the nineteenth century industrialization, urbanization, as well as forms of transportation and communication created rates of change that transformed everything. After 1900, that change continued, and the greater the change the faster the rate. The United States in 2000 was distinctly different from the United States in 1900, and by 2000 new developments made our rate of change even faster.

Adults continuously remind the young that "the world is a lot different today than it was when I grew up." Well, look around you right now. The society you see will be noticeably different ten years from now. Whatever you like may be gone. Whatever you do not like may be improved, or it may be worse. Everything changes, but the change is generally gradual, and it is usually impossible to pin down one individual or group as responsible for the change.

### *Principle 2: Change Stems More from Social Conflict than from Individual Acts*

*The second sociological principle is that change probably results more from social conflict than from the acts of any individual.* Organization is never as peaceful and finished as it first appears. There is always disagreement and protest. It is when the authorities say *no* and others continue to say *yes* that conflict and social change arise. Rarely do such individuals have their way, but because they fight for what they believe or want, some change occurs over time, often in ways they did not even intend. The civil rights movement in American society has never achieved its goal of racial equality. However, because of the conflict that it generated—the back-and-forth struggle between the movement and those who supported the institution of segregation, the culture of racism, and the unequal racial social structure—patterns changed gradually in the direction that the movement cried out for, even though never to the point that it desired. The conflict has given rise to better educational opportunities for many nonwhites, more equal political and civil rights for most, and more economic opportunities for middle-class African Americans. However, it has not made great impact among those African Americans who have no real economic or educational opportunities; on the increasing numbers of young, single minority parents; and on serious inner-city drug abuse.

Did the civil rights movement succeed? Yes (partially): Society has become more open to middle-class African Americans, and the political and legal system is more sensitive to problems of racism. And no, equality does not prevail, racism is still widespread, and for large numbers, nothing has really changed for the better.

Karl Marx emphasizes the role of social conflict as the source for change. History is the struggle of opposing classes, he writes. Society is made of workers and owners, and over time the inevitable conflict between them alters society. For a long time the open conflict is kept in check, and then suddenly a great upheaval brings down the old and creates the new. The new society, according to Marx, is a synthesis. It is the coming together of the old and the new, those who fight to keep what they have and those who are opposed and must fight for their rights. The new arises out of social conflict. Marx sees the English and French revolutions as examples of mass conflict that created such syntheses: They were examples of societies moving from feudalism to capitalism. The revolutions were really culminations of social conflict that existed for hundreds of years; the new societies were significantly different from the old but not brand new—they were created from social and economic trends existing in the old.

Most sociologists see conflict as inevitable. As long as there are people with different ideas and different interests, there will be conflict; and as long as there is conflict, nothing will stay the same. Everything is in flux; nothing is inevitable, and everything is open to challenge and change.

Like Marx, Max Weber (1924) sees social change as arising out of conflict—between those who defend the traditional order and those who act against it. Those who rise up are revolutionary; they gather followers and sometimes overthrow the old order. But it is never completely overthrown, because some old patterns help forge the new ones. New conflict follows immediately between those who support the new patterns and those who oppose them. The new eventually loses its newness and becomes *tradition,* and new charismatic leaders eventually attract followers and fight once again. History is the struggle between tradition and revolution.

Social movements imply social conflict. Social movements protest the direction of other powerful groups, usually groups that protect the social patterns of society. We have witnessed such conflict throughout our history. Since World War II, labor unions fought for workers' rights, African Americans fought against injustice throughout society, antiwar activists fought against the Vietnam War, women fought against sexist social patterns, animal rights' activists have fought for respect of all animals, and environmentalists have fought for the preservation of forests, water, air, and the whole earth. These are only a few of the more important examples. Each one was the result of many individuals working together; each one tried to alter the dominant social patterns in society; each one was met by opposing powerful individuals and groups. In every case there was a struggle, no group actually got exactly what it wanted, yet each influenced the direction of society because it exerted itself and created social conflict. Individuals mattered, because individuals were willing to work together for change, yet it was ultimately the cooperation of many individuals that created the conflict that led to the change.

### *Principle 3: Change Is Most Likely When the Social Situation Favors It*

*A third principle is that individuals, groups, and social conflict are most likely to change an organization when the social situation favors it.* Hitler is an example of an individual who made a great difference in history: He significantly changed the social patterns in Germany to accommodate a totalitarian dictatorship; his efforts to establish German supremacy clearly led to a world war, and through his influence millions of people were killed. More than sixty years after Hitler's death, his influence is still felt the world over. Many individuals and groups are still attracted to his philosophy and hold him as a great leader. Much of the world sees him as a representative of all that is evil in human beings and all that is possible for an all-powerful individual to attain. Most of us would admit that Hitler truly made a difference: Part of it was intentional, part unintentional.

Hitler, however, was not successful simply because he wanted to change the world. He was a part of history as much as a leader of history. He was a product of German society as much as a molder of German society. Without the right social circumstances, he would not have had the impact he did.

Scholars of history remind us of some of the most important reasons that Hitler was able to come to power: the humiliating peace treaty that ended World War I; the Great Depression, which devastated the German economy; and Germany's paralyzed government, plagued by extremism from all sides. Hitler was a product of many social patterns in German culture, and he appealed to these in his rise to power: German nationalism, militarism, authoritarianism, and anti-Semitism. In large part he came to power and made a difference because he tapped into and used German cultural patterns.

Hitler also made a great impact on the world because Germany was highly bureaucratized and scientifically advanced. He was able to use bureaucratic principles to organize society, control the population, build an efficient military machine, and transport, imprison, and systematically murder millions of people. He was able to use the German scientific community to develop weapons of war that were often superior to those of his enemies.

Without this social context, Hitler would not have risen to power, and his influence on German society and the world would have been impossible. So it is with every influential leader in history. The individual makes a difference when social conditions are right: Luther, Lenin, Mao, Roosevelt, Lincoln, King, and Gorbachev are individuals who made a difference in part because society was ready for them.

Weber's analysis of charismatic authority in history underlines this point. Those who make revolution tend to exist within certain periods of history when the old world is collapsing, institutions no longer work well, and old ideas no longer seem sensible. They rise because others look to them in a world where many are dissatisfied. In short, revolutionary individuals make a difference only in a much larger social context that is ripe for their influence. Revolutionaries probably always exist; they come to make a real difference only when the times are such that others are ready for them. They make little difference when few are willing to listen.

Ideas sometimes change society, but ideas, too, exist within a social context. To create new ideas, the individual must build on what is known. Great ideas are often a synthesis of the ideas of others or the reactions to these other ideas. Newton, Galileo, Copernicus, Darwin, Freud, and Marx are all thinkers who revolutionized how people in society thought, but their ideas were built on those that had gone before. Moreover, the influence of these ideas depended on conditions in society that encouraged their promulgation. It is not only the "truth" that wins out in society (it may or may not), but those ideas that have sponsors: groups, communities, social movements, or classes willing to believe and to sell them to others. The ideas of some individuals make a difference, but there is always a social context that helps determine their acceptance or rejection.

It is tempting for the sociologist to discount the impact of the individual on society. However, certain individuals do influence other individuals and do change social patterns. Their impact is sometimes great and cannot be discounted. Yet it is vital to always see these individuals in a larger social context that (1) helped produce them and (2) made their influence significant. This same point can be made of any social organization, be it a group, formal organization, or community: They can make a difference, but they succeed, in part, because of a wider social context.

### *Principle 4: Most Lasting Change Results from Social Trends*

*A fourth principle to which most sociologists subscribe is that much of what we call social change results from impersonal social trends over which individual actors have little control.* A social trend is change that arises from the actions of many individuals who deal with their everyday situations and act in a similar direction and produce a cumulative effect on society. Few people actually intend to change society, but together their acts do, in fact, cause a change. So, for example, many people today are putting off marriage until they get older; many are deciding to divorce; many are remarrying after getting divorced. These are social trends, broadly general tendencies shared by many in society. As a result, society changes. Social trends are themselves caused by even larger trends such as industrialization and increasing individualism. They create change in spite of the fact that each actor's influence is unintentional. Important general trends in our society today might include population trends (fertility, mortality, and migration), urbanization, industrialization, increased use of technology, computerization, revolution in communication, bureaucratization, secularization, and globalization.

The United States, for example, continues to experience a revolution in computer technology. Indeed, the development of technology—the application of knowledge to solving human problems—has been a rapidly accelerating social trend for at least 300 or 400 years. Computers are altering every aspect of our lives, from education to music, from diagnosing illness to making war. Compared to the individual who makes lasting intentional change, such trends seem much more powerful. The former are less common, and the scale of their changes is much less.

Social trends are long-lasting, far-reaching, general developments that affect all the various social patterns in society. In the long run, such trends are the most important forces for social change. They set an almost irreversible direction for society. Individuals normally contribute to social change if their acts and ideas are consistent with these trends. Many in society may hate such trends and fight them, but these trends have inertia; once begun, they take on a life of their own and are difficult to turn around.

Weber (1905) maintains that a social trend he calls "the rationalization of life" is dominating Western societies. Throughout society, he writes, there is increasing reliance on "calculation," "efficiency," "problem solving," and "goal-directed behavior." This is the meaning of modern life to Weber: Instead of tradition, human beings value reaching goals—organizing themselves most efficiently, making and selling goods in the most profitable way, and calculating the most effective way of getting what they want. "That's the way we have always done it" is replaced by the ethic of "This is the smart way of doing it." Tradition is not valued; achieving our goals is. Indeed, Weber argues, neither are we any longer a people committed to value-oriented behavior; I do what I do not because of commitments to values (such as knowledge, goodness, equality, love, and freedom), but because my behavior is the most rational way of achieving my ends. Weber documents the declining importance of tradition, values, and feeling as human behavior becomes increasingly rational and calculating. Society is becoming efficient in many diverse ways. We are able to turn out millions of television sets, tons of wheat, and large numbers of college graduates. We are able to provide health care to more people than ever before in human history, and we are able to encourage more people to buy more goods without any cash. We can provide more answers than ever before as science and mathematics dominate our society, and we find our lives less and less private as computers and bureaucracies are able to monitor what we do.

This is the most important modern trend, according to Weber. Many sociologists agree. Once begun, it is difficult for any individual to turn it around. Individuals may matter in society, but compared with this trend, their influence is minimal. Like most other trends, the rationalization of life is a mixed blessing: It contributes to a better life, but it takes away something important. The rise of calculation challenges mystery and myth. The rise of bureaucracy threatens the small entrepreneur. The dominance of mind deemphasizes feeling. The desire to find the best way of doing something erases the past. This rationalization of life also has important implications for religion, not only changing what it has been in the past but also creating forces that tend to undermine its importance to society (see Chapter 9).

Other trends are identifiable, too, each one important, each one the work of many thousands of individuals going about their business in life, trying to rear children, make enough to live on, and do what they have to do. No one person has much impact alone, but together they contribute to the trends. Does the individual really make a difference? It is difficult to say yes when we look at these general trends.

It is difficult to clearly determine whether it was a certain individual who made a great difference or if it was a general trend to which many individuals contributed. The question becomes, If it had not been for that particular individual, would society have been different than it is? For example, did Elvis or John Lennon make the real changes in music that we have experienced today, or would it have been approximately the same without their influence? Or, as a colleague from the music department at my university keeps asking me, did rock music create our society or did society create our rock music? He personally believes that although the rise of rock music must be understood as rising out of some other important trends in both society and the history of music, certain individuals made a great difference in what rock music became (he is a champion of the Beatles in this regard), and rock music did indeed contribute to important changes in society, including fostering individualism, greater creativity, and the development of a powerful youth culture in society. Of course, the other side of the case is that it was changes in society (such as the large number of people who became adolescents, the successful mass marketing of music, and world events) that actually created the rock revolution in music, and it was the interaction of thousands of individuals and groups who made the real difference.

### *Principle 5: Social Patterns Persist*

*Our final principle is that dramatic change is difficult because there is a strong tendency for social patterns to hold on.* Think for a moment about what a social pattern is. People interact and over time develop routines: rules, expectations, shared values and truths, regularities in how to get goals achieved. These routines become established and an integral part of social interaction. One such pattern is that some people become more powerful than others; some have more privileges and prestige than others. A related pattern is that roles are established—that is, the expectations about how people are supposed to act in their various positions. The longer and more intense the interaction, the more important and established become the patterns. The more they are rooted in the history of an organization, the greater the likelihood that new members are socialized into these particular patterns.

Such patterns tend to hang on. This is the way we have always done something, the way we have always thought, the rules we have always believed. The past acts as a force for right. Furthermore, those in society who are relatively well-off will spend money, life, and time defending such patterns, which they honestly believe are right. In fact, most of us, no matter how critical we are of the social patterns that make up our lives, fear change, because it may threaten the existence of social organization itself. We hold on partly because we fear we will lose everything if we challenge these basic patterns.

The individual can affect other individuals or the direction of an organization if he or she works within the patterns of that organization. But basic change—change in the social patterns of an organization—is profoundly difficult to achieve, and it generally occurs for reasons other than simply the intentional acts of an individual.

## Some Implications for Living

"The truth is the truth." These were Weber's last words. It was not that he died thinking he knew the truth; far from it. More than most others, he realized that truth was extremely difficult to know. Instead, his statement reflects his commitment to seeking truth rather than security in ignorance. He understood the discomfort of many ideas.

A society develops ideas over a long period, and they become embedded. These ideas are part of what we call *culture*. It is a people's way of thinking about reality. To grow up in the United States is to confront a set of "truths" taught through our various institutions. These ideas may not be true, but they are still important to us. One of these ideas is that "individuals can do whatever they set their minds to." This notion is obviously false, but it is an important article of faith for many people. We believe not only that individuals can accomplish what they choose but also that they can have a great impact on others if they choose. Our view of social change tends to be simplistic because of our culture. It is important in our society to focus on the individual rather than on something abstract like social forces; it is important to believe that the individual is responsible for change rather than see social trends that no one individual can control.

It is comforting for me to believe I matter, that what I do will affect others in ways I want. From the point of view of social justice, it is important for me to believe my acts will make a difference. That is how the civil rights movement was able to achieve what it did: People had faith that each individual mattered. I like what that movement accomplished, and sitting here at my computer declaring that the civil rights movement was really the work of organized groups acting in the right social context seems cold and almost ruthless.

Several individuals have questioned my emphasis on the role of social power and social organization on social change rather than the individual. Some point out that I neglect the importance of individuals in social movements. They remind me that the twentieth century was a century of struggles by very committed individuals, and they point out that many sociologists do not take the position that I have written in this chapter. They are correct; some sociologists are more likely to see individuals as important. However, I have tried to be honest here; sociology does lead me to these conclusions, even though they may not be comforting. *Individuals have little impact on society, and then only when there is a power base and when social conditions favor change. In my mind, social conflict and social trends are far more likely to bring about lasting and important change.*

Does sociology necessarily lead to apathy? Is one left with no hope for much impact? Does one have to go from a life of wanting to make a difference to a life of hopeless acceptance? Not at all!

Sociology leads one to take a more *realistic* look at social change and the impact one can have on others. It helps explain who can have an impact and under what circumstances; it helps explain why such an impact is so difficult; it warns us that impacts may not be intended. It tells me I may not be able to change society's system of inequality but I can speak up in my own personal relationships against racism and injustice and realistically influence those immediately around me. Sometimes this can be my only lasting contribution. Sociology tells me that my greatest impact will be in relation to those with whom I interact the most, over whom I have the greatest power. Thus, what

I do in relation to my children may matter greatly to their future. It tells me that to have any impact on society there has to be power (which I normally do not have as an individual). So I must contribute my money wisely to movements that represent my concerns and my time and efforts to organizations that are in a position to influence policy in directions I desire. Sociology tells me to understand that change in the direction I want does not come easily and that I must balance my anger over injustice with realistic expectations. It tells me not to be fooled: Real change is in society's patterns. Simply to vote out one individual for another does not mean change. Simply to pass a law for or against something does not usually change the way in which society operates. Finally, it warns me that change is not usually in the interests of those who are successful, and that if I want it, I must fight those who benefit from the social patterns that exist. In fact, I must realize that if I want change *and* I am benefiting from the patterns that exist, I will have to make some hard choices.

Far from bringing me to my knees, sociology teaches me a realistic view of the relationships among the individual, social patterns, and social change. That view gives me more confidence in what is possible through my efforts.

## Summary and Conclusion

Social change is a difficult topic. Frankly, sociologists usually have an easier time describing order.

The individual actor exists within social forces, from those in intimate relationships to those in society as a whole. It is easiest to recognize that the individual may influence other individuals with whom he or she interacts. It is most difficult to understand how any individual can have an impact on the society and its social patterns. Some individuals, however, undoubtedly have great influence if they act within a social context that favors such influence and if they have a strong power base. Recognize that attempts to influence society are countered by the power of long-standing social patterns that are normally defended by people who have a stake in those patterns.

When sociologists examine social change, they normally go beyond the influence of the individual. Change occurs in every social organization, and it is ongoing and inevitable. It arises out of organized groups and social conflict, and it tends to be characterized by general social trends that no one really controls.

The attempts by sociologists to describe the individual's role in a changing society may not be comforting to many people, but they are realistic and useful for understanding ourselves as social beings.

## Questions to Consider

1. Does the individual really make a difference? What are the different ways one might interpret this question?
2. What is essential for someone to have a meaningful influence on another individual?
3. Who in society is in the best position to create significant change in American society?
4. What is the most important social trend in the United States today?
5. Does the sociological view of social change encourage apathy?
6. How would someone who works for the Salvation Army answer this question: Does the individual really make a difference?
7. President Obama's campaign emphasized "real" change if he was elected? Do you believe he has been successful in basic change? If so, in what ways? In not, why not? Was the social context part of the reason he was able or unable to make basic change?

## References

These works focus on social change and on the difficulties that stand in the way of the human being who is trying to make significant changes in society. Some of the works examine the massive society that confronts the individual.

**Anderson, Eric.** 2005. *In the Game: Athletes and the Cult of Masculinity.* New York: State University of New York.

**Aronowitz, Stanley.** 1998. *From the Ashes of the Old: American Labor and America's Future.* Boston: Houghton Mifflin.

———. 2001. *The Last Good Job in America: Work and Education in the New Global Technoculture.* Lanham, MD: Rowman & Littlefield.

———. 2005. *Just around the Corner: The Paradox of the Jobless Recovery.* Philadelphia: Temple University Press.

**Aronowitz, Stanley, and W. DiFazio.** 1994. *The Jobless Future: Sci/Tech and the Dogma of Work.* Minneapolis: University of Minnesota Press.

**Beeghley, Leonard.** 1996. *What Does Your Wife Do? Gender and the Transformation of Family Life.* Boulder, CO: Westview.

**Bell, Daniel.** 1973. *The Coming of Post-Industrial Society: A Venture in Social Forecasting.* New York: Basic Books.

**Bensman, David, and Roberta Lynch.** 1987. *Rusted Dreams: Hard Times in a Steel Community.* New York: McGraw-Hill.

**Berger, Peter, Brigitte Berger, and Hansfried Kellner.** 1974. *The Homeless Mind: Modernization and Consciousness.* New York: Vintage.

**Beyer, Peter.** 1994. *Religion and Globalization.* Thousand Oaks, CA: Sage.

**Blau, Peter M., and Marshall W. Meyer.** 1987. *Bureaucracy in Modern Society.* 3rd ed. New York: Random House.

**Blumberg, Paul.** 1981. *Inequality in an Age of Decline.* New York: Oxford University Press.

**Blumer, Herbert.** 2000. *Selected Works of Herbert Blumer: A Public Philosophy for Mass Society,* edited by Stanford M. Lyman and Arthur J. Vidich. Urbana: University of Illinois Press.

**Bowles, Samuel, Herbert Gintis, and Melissa Osborne-Groves.** 2005. *Unequal Chances: Family Background and Economic Success.* Brunswick, NJ: Princeton University Press.

**Bowser, Benjamin P.** 2006. *The Rise and Fall of Class in Britain.* New York: Columbia University Press.

**Brecher, Jeremy, and Tim Costello.** 1994. *Global Village or Global Pillage: Economic Reconstruction from the Bottom Up.* Cambridge, MA: South End.

**Brown, Phillip.** 2001. *Capitalism and Social Progress: The Future of Society in a Global Economy.* New York: Palgrave.

**Bruce, Steve.** 1996. *Religion in the Modern World: From Cathedrals to Cults.* Oxford: Oxford University Press.

**Buechler, Steven M.** 2000. *Social Movements in Advanced Capitalism.* New York: Oxford University Press.

**Caplow, Theodore, Howard M. Bahr, John Modell, and Bruce A. Chadwick.** 1991. *Recent Social Trends in the United States, 1960–1990.* Montreal: McGill-Queen's University Press.

**Carmichael, Stokely, and Charles V. Hamilton.** 1967. *Black Power.* New York: Random House.

**Charon, Joel M.** 2009. "An Introduction to the Study of Social Problems," pp. 1–12 in *Social Problems: Readings with Four Questions,* edited by Joel Charon and Lee Vigilant. 3rd ed. Belmont, CA: Wadsworth Cengage Learning.

**Charon, Joel M., and Lee Garth Vigilant.** 2009. *Social Problems: Readings with Four Questions* 3rd ed. Belmont, CA: Wadsworth Cengage Learning.

**Chirot, Daniel.** 1986. *Social Change in the Modern Era.* New York: Harcourt Brace Jovanovich.

——— 1994. *Modern Tyrants: The Power and Prevalence of Evil in Our Age.* New York: Free Press.

**Collins, Randall.** 1998. *A Global Theory of Intellectual Change.* Cambridge, MA: Harvard University Press.

**Coontz, Stephanie.** 2005. *Marriage, a History: From Obedience to Intimacy or How Love Conquered Marriage.* New York, NY: Viking.

**Crothers, Lane.** 2003. *The American Militia Movement from Ruby Ridge to Homeland Security.* New York: Rowman & Littlefield.

**Currie, Elliott.** 2005. *The Road to Whatever: Middle-Class Culture and the Crisis of Adolescence.* New York: Holt.

**Dahrendorf, Ralf.** 1958. "Toward a Theory of Social Conflict." *Journal of Conflict Resolution,* 2: 170–183.

———. 1959. *Class and Class Conflict in Industrial Society.* Stanford, CA: Stanford University Press.

**Davey, Joseph Dillon.** 1995. *The New Social Contract: America's Journey from Welfare State to Police State.* Westport, CT: Praeger.

**Derber, Charles.** 1996. *The Wilding of America: How Greed and Violence Are Eroding Our Nation's Character.* New York: St. Martin's.

**Della Porta, Donatella, and Mario Diani**. 1999. *Social Movements: An Introduction.* Malden, MA: Blackwell.

**Diamond, Larry.** 1999. *Developing Democracy.* Baltimore, MD: Johns Hopkins University Press.

**Dougherty, Charles J.** 1996. *Back to Reform: Values, Markets and the Healthcare System.* New York: Oxford University Press.

**Drucker, Peter F.** 1993. *Post-Capitalist Society.* New York: HarperCollins.

**Du Bois, William D., and R. Dean Wright**. 2001. *Applying Sociology: Making a Better World.* Boston: Allyn & Bacon.

**Durkheim, Émile.** 1893. *The Division of Labor in Society.* 1964 ed. Trans. George Simpson. New York: Free Press.

**Dyer, Joel.** 1997. *Harvest of Rage: Why Oklahoma City Is Only the Beginning.* Boulder, CO: Westview.

**Eliade, Mircea.** 1954. *Cosmos and History.* New York: Harper & Row.

**Emerson, Michael O., and Rodney M. Woo**. 2006. *People of the Dream: Multiracial Congregations in the United States.* Princeton, NJ: Princeton University Press.

**Ennis, Phillip H.** 1992. *The Seventh Stream: The Emergence of Rock 'n' Roll in American Popular Music.* London: Wesleyan University Press.

**Etzioni, Amitai.** 2001. *The Monochrome Society.* Princeton, NJ: Princeton University Press.

**Ewen, Stuart.** 1976. *Captains of Consciousness.* New York: McGraw-Hill.

**Faludi, Susan.** 1999. *Stiffed: The Betrayal of the American Man.* New York: Perennial.

**Farley, John E., and Gregory D. Squires.** 2005. "Fences and Neighbors: Segregation in 21st Century America." *Contexts.* 4(1): 33–39.

**Feagin, Joe R., and Hernan Vera.** 2002. *Liberation Sociology.* Boulder, CO: Westview.

**Featherstone, Mike,** ed. 1990. *Global Culture: Nationalism, Globalization, and Modernity.* London: Sage.

**Fischer, Claude S.** 2004. *The Urban Experience.* 2nd ed. Belmont, CA: Wadsworth/Thomson Learning.

**Flacks, Richard.** 1971. *Youth and Social Change.* Chicago: Markham.

**Foner, Nancy.** 2005. *In a New Land: A Comparative View of Immigration.* New York: New York University Press.

**Freeman, Jo.** 1999. *Waves of Protest: Social Movements since the Sixties.* Lanham, MD: Rowman & Littlefield.

**Freie, John F.** 1998. *Counterfeit Community: The Exploitation of Our Longings for Connectedness.* Lanham, MD: Rowman & Littlefield.

**Friedan, Betty.** 1993. *The Fountain of Age.* New York: Simon & Schuster.

**Friedman, Thomas L.** 2002. *Longitudes and Attitudes: Exploring the World after September 11.* New York: Farrar, Straus & Giroux.

———. 2005. *The World Is Flat: A Brief History of the Twenty-First Century.* New York: Farrar, Straus & Giroux.

**Fuller, Robert W.** 2003. *Somebodies and Nobodies: Overcoming the Abuse of Rank.* Gabriola Island, BC: New Society.

**Gamson, William A.** 1968. *Power and Discontent.* Homewood, IL: Dorsey.

**Garner, Robert Ash.** 1977. *Social Change.* Skokie, IL: Rand McNally.

**Giddens, Anthony.** 2000. *Runaway World: How Globalization Is Reshaping Our Lives.* New York: Routledge.

**Gitlin, Todd.** 1987. *The Sixties: Years of Hope, Days of Rage.* New York: Bantam.

**Glendinning, Tony, and Steve Bruce.** 2006. "New Ways of Believing or Belonging: Is Religion Giving Way to Spirituality." *British Journal of Sociology* 57(3): 399–414.

**Goldberg, Michelle.** 2006. *Kingdom Coming: The Rise of Christian Nationalism.* New York: Norton

**Goldstone, Jack A.** 1991. *Revolution and Rebellion in the Early Modern World.* Berkeley: University of California Press.

**Haenfler, Ross.** 2006. *Straight Edge: Hardcore Punk, Clean-Living Youth, and Social Change.* New Brunswick, NJ: Rutgers University Press.

**Hall, John A., and Charles Lindholm**. 2000. *Is America Breaking Apart?* Princeton, NJ: Princeton University Press.

**Hancock, Angie-Marie.** 2004. *The Politics of Disgust: The Public Identity of the Welfare Queen.* New York: New York University Press.

**Handel, Michael J.** 2003. *The Sociology of Organizations: Classic, Contemporary, and Critical Readings.* Thousand Oaks, CA: Sage.

**Hardin, Russell.** 1995. *One for All: The Logic of Group Conflict.* Princeton, NJ: Princeton University Press.

**Hare, Bruce R**, ed. 2002. *2001 Race Odyssey: African Americans and Sociology.* Syracuse, NY: Syracuse University Press.

**Harper, Charles, and Kevin T. Leicht**. 2002. *Exploring Change: America and the World.* 4th ed. Upper Saddle River, NJ: Prentice Hall.

**Hays, Sharon.** 1996. *Cultural Contradictions of Motherhood.* New Haven, CT: Yale University Press.

**Hedges, Chris.** 2002. *War Is a Force That Gives Us Meaning.* New York: Public Affairs.

**Henslin, James M.** 2000. *Social Problems.* 5th ed. Upper Saddle River, NJ: Prentice Hall.

**Hertz, Rosanna.** 2006. *Single by Chance, Mothers by Choice: How Women are Choosing Parenthood without Marriage and Creating the New American Family.* Oxford, UK: Oxford University Press.

**Hervieu-Leger, Daniele.** 1998. "Secularization, Tradition and New Forms of Religiosity: Some Theoretical Proposals," pp. 28–44 in *New Religions and New Religiosity,* edited by Eileen Barker and Margit Warburg. Cambridge, UK: Cambridge University Press.

**Hill, Herbert, and James E. Jones Jr.**, eds. 1993. *Race in America: The Struggle for Equality.* Madison: University of Wisconsin Press.

**Howard, Judith A., and Jocelyn A. Hollander**. 1996. *Gendered Situations, Gendered Selves: A Gender Lens on Social Psychology.* Thousand Oaks, CA: Sage.

**Jacobs, Nancy R., Mark A. Segal, and Carol D. Foster**. 1994. *Into the Third Century: A Social Profile of America.* Wylie, TX: Information Plus.

**Johnson, Heather B.** 2006. *The American Dream and the Power of Wealth: Choosing Schools and Inheriting Inequality in the Land of Opportunity.* New York: Routledge.

**Jones, Landon Y.** 1980. *Great Expectations: America and the Baby Boom Generation.* New York: Coward, McCann & Geoghegan.

**Katz, Michael B.** 1989. *The Deserving Poor: From the War on Poverty to the War on Welfare.* New York: Pantheon.

**Koonigs, Kees, and Dirk Druijt**, eds. 1999. *The Legacy of Civil War, Violence and Terror in Latin America.* London: Zed.

**Kornhauser, William.** 1959. *The Politics of Mass Society.* New York: Free Press.

**Kozol, Jonathan.** 2005. *The Shame of the Nation: The Restoration of Apartheid Schooling in America.* New York: Random House.

**Kraybill, Donald B., and Marc A. Olshan**, eds. 1994. *The Amish Struggle with Modernity.* Hanover, NH: University Press of New England.

**Larana, Enrique, Hank Johnston, and Joseph R. Gusfield**. 1994. *New Social Movements: From Ideology to Identity.* Philadelphia: Temple University Press.

**Lauer, Robert H.** 1982. *Perspectives on Social Change.* 3rd ed. Boston: Allyn & Bacon.

**Lewellen, Ted C.** 1995. *Dependency and Development: An Introduction to the Third World.* Westport, CT: Bergin & Garvey.

**Lewis, Bernard.** 2002. *What Went Wrong? Western Impact and Middle Eastern Response.* New York: Oxford University Press.

———. 2003. *The Crisis of Islam: Holy War and Unholy Terror.* New York: Random House.

**Kotlikoff, Laurence, and Scott Burns**. 2004. *The Coming General Storm: What You Need to Know about America's Economic Future.* Cambridge, MA: MIT Press.

**Lichtenstein, Nelson.** 2006. *Wal-Mart: The Face of Twenty-First Century Capitalism.* New York: New Press.

**Loe, Meika.** 2004. *The Rise of Viagra: How the Little Blue Pill Changed Sex in America.* New York: New York University Press.

**Lukes, Steven.** 2005. *Power: A Radical View.* 2nd ed. Oxford, UK: Palgrave Macmillan.

**Madsen, Richard**, ed. 2002. *Meaning and Modernity: Religion, Polity, and Self.* Berkeley: University of California Press.

**Malcolm X, and Alex Haley**. 1965. *The Autobiography of Malcolm X.* New York: Grove.

**Marx, Karl, and Friedrich Engels**. [1848] 1955. *The Communist Manifesto.* New York: Appleton-Century-Crofts.

**Masci, David, and Gregory A. Smith**. 2006. "Is Religion Giving Way to Spirituality?" *Sociology Review* 15(4): 14–16.

**Mills, C. Wright.** 1956. *The Power Elite.* New York: Oxford University Press.

———. 1959. *The Sociological Imagination.* New York: Oxford University Press.

**Nash, Kate.** 2000. *Contemporary Political Sociology: Globalization, Politics, and Power.* Malden, MA: Blackwell.

**Olsen, Marvin E.** 1978. *The Process of Social Organization.* 2nd ed. New York: Holt, Rinehart & Winston.

**Paap, Kris.** 2006. *Working Construction: Why White Working Class Men Put Themselves—and the Labor Movement—in Harm's Way.* Ithaca, NY: ILR Press/Cornell University Press.

**Parrillo, Vincent N.** 2002. *Contemporary Social Problems.* 5th ed. Boston: Allyn & Bacon.

**Perrow, Charles.** 2002. *Organizing America.* Princeton, NJ: Princeton University Press.

**Perrucci, Robert, and Earl Wysong**. 2003. *The New Class Society: Goodbye American Dream?* 2nd ed. New York: Rowman & Littlefield.

**Phillips, Kevin P.** 2002. *Wealth and Democracy: A Political History of the American Rich.* New York: Broadway.

**Piven, Frances Fox, and Richard A. Cloward**. 1993. *Regulating of the Poor: The Functions of Public Welfare.* 2nd ed. New York: Vintage.

**Postman, Neil.** 1992. *Technopoly: The Surrender of Culture to Technology.* New York: Knopf.

**Rashid, Ahmed.** 2002. *Jihad: The Rise of Militant Islam in Central Asia.* New Haven, CT: Yale University Press.

**Reich, Charles A.** 1995. *Opposing the System.* New York: Crown.

**Risman, Barbara J.** 1998. *Gender Vertigo: American Families in Transition.* New Haven, CT: Yale University Press.

**Ritzer, George.** 1993. *The McDonaldization of Society: An Investigation into the Changing Character of Contemporary Social Life.* Thousand Oaks, CA: Pine Forge.

**Rosen, Bernard Carl.** 1998. *Winners and Losers of the Information Revolution.* Westport, CT: Praeger.

**Schlosser, Eric.** 2001. *Fast Food Nation: The Dark Side of the All-American Meal.* Boston: Houghton Mifflin.

**Scott, Jacqueline L., Judith K. Treas, and Martin Richards**, eds. 2004. *The Blackwell Companion to the Sociology of Families.* Malden, MA: Blackwell.

**Scott, W. Richard.** 2001. *Institutions and Organizations.* Thousand Oaks, CA: Sage.

**Segrave, Kerry.** 1993. *The Sexual Harassment of Women in the Workplace, 1600 to 1993.* Jefferson, NC: McFarland & Co.

**Shapiro, Thomas M.** 2003. *The Hidden Costs of Being African American.* New York: Oxford University Press.

**Skocpol, Theda.** 1979. *States and Social Revolutions.* New York: Cambridge University Press.

**Skolnick, Arlene, and Jerome Skolnick**. 2000. *Family in Transition.* 11th ed. Boston: Allyn & Bacon.

**Skolnick, Arlene S. and Jerome Skolnick**. 2007. *Family in Transition.* 14th Ed. Boston: Allyn & Bacon.

**Skolnick, Jerome H., and Elliott Currie**. 2000. *Crisis in American Institutions.* 11th ed. Boston: Allyn & Bacon.

**Slater, Philip.** 1976. *The Pursuit of Loneliness.* Boston: Beacon.

**Starr, Paul.** 1982. *The Social Transformation of American Medicine.* New York: Basic Books.

**Steger, Manfred.** 2003. *Globalization: A Very Short Introduction.* New York: Oxford University Press.

**Stirk, Peter M. R.** 2000. *Critical Theory, Politics, and Society: An Introduction.* New York: Pinter.

**Suarez-Orozco, Carola.** 2001. "Immigrant Families and Their Children: Adaptation and Identity Formation," pp. 129–139 in *The Blackwell Companion to Sociology,* edited Judith Blau. Oxford, UK: Blackwell.

**Sullivan, Maureen.** 2004. *The Family of Woman: Lesbian Mothers, Their Children, and Undoing of Gender.* Berkeley: University of California Press.

**Sullivan, Oriel.** 2006. *Changing Gender Relations, Changing Families: Tracing the Pace of Change Over Time.* New York: Rowman & Littlefield.

**Suro, Roberto.** 1999. *Strangers among Us: Latino Lives in a Changing America.* New York: Knopf.

**Sztompka, Piotr.** 1993. *The Sociology of Social Change.* Oxford: Blackwell.

**Toennies, Ferdinand.** [1887] 1957. *Community and Society,* translated and edited by Charles A. Loomis. East Lansing: Michigan State University Press.

**Toffler, Alvin.** 1980. *The Third Wave.* New York: Morrow.

**Trotter, Joe W., with Earl Lewish and Tera W. Hunter**, eds. 2004. *African American Urban Experience: Perspectives from the Colonial Period to the Present.* New York: Palgrave Macmillan.

**Tucker, Robert C.**, ed. 1972. *The Marx-Engels Reader.* New York: Norton.

**Turner, Bryan S.** 1983. *Religion and Social Theory: A Materialist Perspective.* London: Heinemann.

**Turner, Ralph H.** 1968. "The Role and the Person." *American Journal of Sociology* 84: 1–23.

**Wallerstein, Immanuel.** 1999. *The End of the World as We Know It: Social Science for the Twenty-First Century.* Minneapolis: University of Minnesota Press.

**Weber, Max.** [1905] 1958. *The Protestant Ethic and the Spirit of Capitalism,* translated and edited by Talcott Parsons. New York: Scribner's.

———. [1924] 1964. *The Theory of Social and Economic Organization,* edited by A. M. Henderson and Talcott Parsons. New York: Free Press.

———. 1969. "The Social Psychology of the World Religions." *Max Weber: Essays in Sociology,* translated and edited by H. H. Gerth and C. Wright Mills. New York: Oxford University Press.

**Winant, Howard.** 1994. *Racial Conditions: Politics, Theory, Comparisons.* Minneapolis: University of Minnesota Press.

**Witt, Griff.** 2004. "As Income Gap Widens, Uncertainty Spreads." *Washington Post,* September 20.

**Wolfe, Alan.** 2001. *Moral Freedom: The Search for Virtue in a World of Choice.* New York: Norton.

**Wrong, Dennis H.** 1977. *Population and Society.* 4th ed. New York: Random House.

**Zald, Mayer N., and John D. McCarthy**, eds. 1988. *The Dynamics of Social Movements.* Cambridge, MA: Winthrop.

**Zellner, William W.** 1995. *Countercultures: A Sociological Analysis.* New York: St. Martin's.

**Zinn, Maxine Baca, and D. Stanley Eitzen**. 1999. *Diversity in Families.* Boston: Allyn & Bacon.

**Zirakzadeh, Cyrus.** 2006. *Social Movements in Politics: A Comparative Study.* New York: Palgrave McMillan.

**Zweig, Michael.** 2000. *The Working Class Majority: America's Best Kept Secret.* Ithaca, NY: Cornell University Press.

**Zweigenhaft, Richard L., and G. William Domhoff**. 1998. *Diversity in the Power Elite: Have Women and Minorities Reached the Top?* New Haven, CT: Yale University Press.

# Is Organized Religion Necessary for Society?

## Tradition, Modernization, and Secularization

### Concepts, Themes, and Key Individuals

- ❑ Religion, the sacred, the profane, and the community
- ❑ Culture, meaning, and social control
- ❑ Berger, Weber, Marx, Durkheim
- ❑ Functions of organized religion
- ❑ Tradition, modernization, and secularization
- ❑ Individual spirituality, pluralism, and fundamentalism

Religion is not an easy topic for sociologists to study and write about. For many people, religion is thought to be outside the confines of science. It is a faith one has, a matter of religious commitment or acceptance of God's word. For some people, this makes religion a sacred topic not to be examined. For others, no one but the faithful should study religion, for they know firsthand its meaning and power. For still others, religion is becoming a backward view of the universe no longer worthy of study. Yet for anyone interested in understanding society, human history, and human behavior, it is impossible, no matter what one's private beliefs are, to ignore religion as an important force in the world.

To sociologists, religion has always been seen as a central aspect of society and necessary to understand. Sociologists cannot show anyone that certain religious beliefs are correct or incorrect; sociologists assume that this is up to each believer and nonbeliever.

Is organized religion necessary for society? There are three topics discussed here that try to answer this question. Consistent with all other chapters, the discussion will be sociological. Thus, the focus will be on the importance of religion in society rather than on the importance of religion to individuals. The questions that guide the three topics are:

1. What is religion?
2. What is the role of organized religion in society?
3. Is organized religion necessary in the modern world?

Sociologists do not agree on any of these topics. However, by examining them, you should have a good background for understanding the issues that divide us and a fuller understanding of the role of religion in human society.

## Defining Religion

People may honestly disagree on what religion is, and the definition matters because it influences what we see the role of religion to be and whether it is more or less important today than in the past. Some thoughtful people refuse to even define it because they recognize that the definition itself will set the agenda of the discussion.

Some people find it easiest to define religion as a belief in God, but there are important problems with this. Buddhism does not really teach us to believe in a God, yet most would include it as one of the most important world religions. Some would claim that Scientology is a religion, yet, again, belief in God is not required. Some treat Soviet-style communism or fascism as religion, even though neither regards God to be an important part of the universe. Some people claim they are spiritual but not religious; their spirituality may or may not include God. Others may in fact believe in God, but this is not a central part of the religion they believe in. "Belief in God" might be a start, yet it tells us too little and leaves out too much. What quality is required before we call it a religion? Is religion more than simply belief, or must it include practices? Is religion an organization of people, or can each individual have his or her own religion?

## *Émile Durkheim's View of Religion*

The sociologist's search for definition usually starts with Émile Durkheim. To him, religion is a set of beliefs and practices that divide the universe into two parts: the sacred and the profane. Religious *beliefs* "express the nature of sacred things," their relationship with other *sacred* things, and their relationship to profane things (Durkheim 1915: 41). Religious *practices* determine how people are supposed to act in the presence of sacred objects.

Religion to Durkheim is really a statement that there is more to the universe than the physical, more to a meaningful life than immediate pleasure, more to life than the everyday mundane affairs of the human being. This "more to life" is the sacred; it is a part of the universe that we separate from the everyday profane. It is called "sacred" because it is special, universalistic, beyond our senses, beyond the immediate, to be honored, respected, and held in awe.

The sacred and the profane are treated differently by us; each is thought of differently; each is felt differently; each is acted toward differently. This division gives the believers a special feeling when the sacred is observed or acted upon. If a people treat things such as God, soul, morality, justice, and meaning in the universe like everything else—as simply the whim of human beings and as something physical to be used and thrown away—then all of this is simply a part of the material profane world. But if we separate some aspects of life and treat them as unexplainable simply by observation, physics, and science—then we enter into the world of the sacred. The sacred is not valuable because it can be used to achieve some goal for us; it is instead something special because it stands above utilitarian value. The sacred does not necessarily assume that the individual believes in a God; many people may regard meaning, spirit, beauty, goodness, love, and even immortality to have a special place in the universe without assuming a God.

The sacred is created by human beings in their social life. "This is what is sacred to us!" It is the human being who divides this universe; it is people who designate what is to be treated as sacred. A cemetery may be sacred, as may be individual graves; the graves of certain people we have known may become especially sacred. Such sacredness, in fact, may extend to certain wines or water, certain prayers, certain designated buildings, certain places, certain morals, certain values, certain beliefs, certain people, certain offices, and certain rituals. Some beliefs may be held sacred—there is one God, a thing of beauty is a joy forever, love thy neighbor as thyself, God is just, America means freedom, humans will be saved through faith in God. To those who hold these beliefs as sacred, these beliefs hold a special place and are to be assumed, untouched by human criticism, and held apart from other beliefs we hold. Certain life cycle events—birth, baptism, confirmation, bar and bat mitzvahs, marriage, death, funerals—are made sacred for some. Even objects that are not clearly religious—like a flag, a special photograph we cherish, a house we grew up in, an art piece, a baseball from the 1948 World Series, or a great person or novel—can become sacred if people designate them so. The twin towers of the World Trade Center in New York were not simply buildings; they clearly were sacred. They were sacred precisely because they represented the community of New York City and ultimately the community of the United States. Humans establish objects, beliefs, and actions to be sacred; sacredness does not simply establish itself in our lives.

The community and the sacred are inseparable. Violation of the sacred violates the community. Ultimately, it is the community itself that becomes sacred. "We are special." It is the society's core ideas—its values and morals and its constitution, king, dictator, or president—that are designated as representatives of the community. This is what *we believe*, this is what *we honor*, this is what *we are*. To honor it, we honor ourselves. The sacred holds the community together. And when a God represents the community, to worship God is to worship the community. To recite a prayer or to state a special belief system is to support the continuation of community and to restate what the community together believes.

The existence of the sacred reminds each individual that life is to be more than selfish individual mundane pleasure, that something more permanent and important exists and that those who attack or degrade the sacred are attacking the community and what it stands for. Religious beliefs and practices are taught to those who enter the community, cutting off people in that community from those outside it. Religion is, in truth, a universal tendency, according to Durkheim; a community can exist over time only if some form of religion establishes and reestablishes the community as sacred to those who are part of it.

### *Max Weber's View of Religion*

Max Weber, writing about the time of Durkheim, also emphasized the social aspect of religion. He emphasized religion as a central part of a people's culture, as a people's way to understand their own lives in relation to their universe. Weber treated religion as an "ethic," a cultural view, a tool a people use to understand their lives and to bring meaning to them. Religion influences what people do; it helps create truth for them. Over time, religious beliefs and practices become organized and established as a church. Their religious views influence their government, their economic world, their law, their views of people outside the community, their goals in life, their successes, and their failures. People fight wars to defend their religious beliefs. They seek peace, work hard, devote themselves to family, help or persecute their neighbors, develop democratic or capitalistic values partly because of their religion. Sometimes it is the most important cause of their actions; sometimes other causes such as material interests or pursuit of political power are more important.

Weber wrote a great deal about religion. He wrote detailed studies of Protestantism, ancient Judaism, Buddhism, and Hinduism, always trying to link them to other aspects of people's lives. His most important work was *The Protestant Ethic and the Spirit of Capitalism.* His goal was to demonstrate that human beings are moved by religious values and ideals as much as they are economic interests. Whereas those who emphasized economics as the source of change believed that religion is always shaped by economic forces, Weber tried to show that the opposite is also correct: Economic developments are also dependent on religious beliefs and practices. Specifically, he showed that the development of capitalism in Western Europe and the United States was built on a religious ethic, a Calvinistic Protestantism that taught that those who believed and acted in a certain way were good Christian people and were among the chosen or elect. Good Protestants were supposed to work hard, were successful in making a living, invested their money in their business, and contributed much of their wealth to religious and family matters rather than spending it on earthly pleasure. Throughout all of his sociological work, Weber's message was that ideas and values shape human action and society and not simply economic interests, and that religion is a highly important basis for a people's ideas and values.

Besides this, Weber's work on religion also showed the role of conflict in the history of religion, showing that the struggles between sect and established religion, between clergy and prophets, between tradition and charisma were important reasons why religion constantly changes and influences the larger society to change. He also showed that the conflict between traditional and modern society is the whole basis for the debates contained in the sociology of religion. To Weber, secularization would probably accompany the kind of world we are creating. Nothing is inevitable, but Weber wondered and sometimes worried what would happen to society if the supernatural explanations of the universe were replaced by a more rational scientific approach.

He wondered what would happen to the mystery and excitement in life when the unexplainable was dissected by science, and what would happen to people's values and feelings when religion based on tradition and feeling would be replaced by continuous change, impersonality, and efficiency.

### *Peter Berger's View of Religion*

Peter Berger, writing in the last third of the twentieth century, continues and refines both Durkheim's and Weber's definition. Berger, too, emphasizes the social essence of religion. Religion, he says, is a way that people in a community make sense out of the reality they live in. Religion is like a context—a "sacred canopy"—within which we try to make sense out of life. Where tragedy occurs, religion helps us understand that tragedy. Where chaos seems to occur, where nothing seems to make sense to us, religion helps us find order in the events. When people act in an evil way, religion helps us understand; when people do good things, religion helps us explain.

To Berger, seeing the universe from a religious perspective means rising above the scientific and profane and finding a meaningful, more permanent, and sacred universe. The idea of *meaning* is central here: Berger wants us to realize that religion helps us see our lives as important, our actions as worthwhile, and our place in the universe—although minuscule—as special and more than simply a profane physical existence. Religion helps the human being somehow transcend—or rise above—the physical universe. Religion to Berger is an answer to the question posed best by Shakespeare's Macbeth: Is life meaningful, or is it "a tale told by an idiot, full of sound and fury, signifying nothing." Many of us—maybe most, maybe even all of us—are "under an imperative to define and to live a worthy and meaningful life." This is why to Berger religion will continue to find adherents and continue to be a central aspect of human culture (Adams 1993: 9). Thus, religion to Berger is a perspective that helps people make sense out of their life events and the universe in which they exist.

### *The Definition of Religion: A Beginning*

All three sociologists—Durkheim, Weber, and Berger—believe that religion is a necessary force in society. For each, religion is a social construction. Each teaches us that religion has been necessary for every society, and because society is necessary for human life, religion becomes a central part of what humans are. Together these sociologists regard religion as a way that people in community come to define reality: It is thus a central part of a people's culture. All emphasize that religion is the recognition that something exists besides what our eyes tell us, that there is something sacred, universal, and meaningful to human existence. To Durkheim, religion is the creation of the sacred, to Weber it is a central part of a people's culture, to Berger it is a way of giving understanding and importance to human existence.

It seems that we may have come to a working definition of religion: Religion is *a view of the universe that through beliefs and practices identifies a special separate sacred world apart from our physical, mundane, profane, everyday existence. It is socially created, it is part of human culture, and it has an important impact on human action as well as the continued existence of community.* An exception to this might be those religions—such as Zen Buddhism—that create a sacred world out of almost everything in the physical universe and find meaning in the most mundane of activities. Here everything is sacred—the sacred exists but not really the profane.

"Organized religion," "religion," "religious," and "spiritual" are not the same, especially when their distinctions become important to our discussion or understanding. The large majority of people are part of an organized religion, meaning they belong to a religion that is formally structured, has its own culture, traditions, beliefs, institutions, and practices. Some people may belong to a much less formally structured religion, highly individualized, without clearly established beliefs, rituals, practices, and formal leaders. More common are those who are not part of any informal or formal religions, who contend that they "have their own religion." Because the community is so much a part of what religion is, it might be more accurate to call them "spiritual," recognizing that although they are not part of a "religion," their view of the world is not completely profane. It is important to understand that to be spiritual does not mean one necessarily believes in God: one can still believe that goodness, beauty, love, nature, humanity, truth are real and central to their lives.

Sometimes we might distinguish between a "major religion," a "denomination," a "sect," and a "cult." The terms *major religion* (referring to Catholicism, Protestantism, Judaism, and Islam, for example) and *denomination* (referring to Lutheran, Reform Judaism, and Sunni Muslim, for example) are used to describe established, traditional, organized, formal religions. The *sect* (Jehovah's Witnesses, Seventh-day Adventists, and Scientologists, for example) and the *cult* (Heaven's Gate and People's Temple, for example) are smaller, less established in society, more critical of society, and critical of the more established religions in society. The sect works within society in order to change the direction of society and to compete with or replace the established religion; the cult pulls away from society, isolates itself from those outside the cult, believes that outsiders are beyond help, and tends to be seen as illegitimate by the larger society.

This is the question that organizes this chapter: Is *organized religion* necessary for society? So far, we have examined the meaning of religion. Now we will turn to the social "functions" of religion, especially organized religion. What does religion actually do? The functions of religion will further allow us to understand what religion is and whether it is still necessary.

## The Social Functions of Religion

Why does some kind of religion seem to be universal? What is its function? Is it truly an inevitable part of the human condition? What happens if individuals or societies try to exist without religion? There seem to be three ways to approach these questions: a religious approach, an individualistic approach, and a sociological approach.

Those who explain the universality of religion from a religious approach will often contend that religion exists because the supernatural exists, and that all of us are driven to believe in that supernatural. More simply put, religion exists because God exists, because the sacred exists, because there has to be something more to existence than the profane material world, and because the spiritual world exists.

Some people move away from the purely religious view and argue that people psychologically or intellectually need religion. Religion is necessary for the individual. "The individual needs to understand the universe, and so much of it is mysterious and explainable only through a religious perspective." "Individuals need religion if they are going to be moral." "Religion gives the individual hope, a value system that makes life worth living, a meaning after death." "Individuals need certainty, answers, security, and this is the function of religion." "How can there be meaning in life without religion? Everyone seeks meaning; therefore, religion is a necessary part of existence." "When the chips are down the individual will always turn to religion; after all, there are really no atheists in foxholes." In each example, the focus is on the individual: Religion, in one way or another, organized or not, works for the individual. Thus, we must understand the needs of the individual to understand the importance of religion.

Sociologists—and others—will emphasize the social functions of religion; although individuals can be spiritual, in general, religion is almost always organized in some way and its important functions are social. Organized religion is important for society. Some sociologists—such as Durkheim—would argue that society cannot exist without organized religion in some form. Religion has been part of every known society; it is central to a people's culture. From the sociological view, the decline of organized religion in society will create a vacuum that is extraordinarily difficult—maybe impossible—to fill. To the sociologist, religion does not exist simply because it captures the truth or simply because it is necessary for the individual. In its organized form, it is also a central part of what *human society* is.

Of course, these three general explanations are not mutually exclusive. Some people will use all three to explain the functions of religion: Religion may confirm an actual truth in the universe and may be important for both the individual and society. Because the purpose of this chapter is to focus on the sociological view, society is what we shall examine. This does not, however, deny the other two views.

One more point. We are not trying here to suggest that all functions are positive for the society or for the individual. If religion is important for bringing people together in society, that does not necessarily mean that this is good for society or for the individual. It may contribute to the continuation of society, but that society might not be what you or I would call good. It may function for the individuals in society by giving them certainty in life, but such certainty may be in conflict with critical thinking and open-mindedness.

### *Social Solidarity: The First Social Function*

When I moved to Moorhead, Minnesota, over thirty-five years ago, I knew no one. Eventually, I began to meet people and made friends. However, I still did not feel part of any community. The brother of someone who became my friend died suddenly, and I went to the funeral. I did not know that brother, but through the service I slowly felt that I finally belonged in that community. A feeling of solidarity with that community became real to me. I suddenly felt close to many people I had never even met. After this experience, I became increasingly aware of the fact that religious ritual often does this to me and probably to many others. Weddings, prayers, funerals, baptisms, and namings seem to bring people closer together; the emotional experience is real and strongly felt. It is a shared experience that acts to focus attention not simply on an individual or family, but on the whole, the community, the "we." As we act together in ritual, we tie ourselves to one another. The experience reassures us that we are right and life is meaningful. Organized religion encourages solidarity of community. We belong together, and that brings us meaning in our lives. Community can be a few people, a group or organization of people, a city, a whole society, or people scattered all over the world. Each community has a special sacred belief system, objects that represent it, and rituals that bring people closer.

Of course, religion can also divide people. Opposing religious systems may actually cause friction within and between societies. Unity within one community and exclusiveness toward others can create serious conflict with outsiders. Solidification often brings the condemnation of individuals who are different in the community and intolerance of those people who do not regard the dominant religion to be theirs. Such condemnation may end up working against social solidarity in the larger society. If mutual respect among a number of religions or denominations exists in society, then a pluralistic society can be created and diverse religions can actually work to unite the whole society. This is difficult to establish, however, because the nature of religion is to claim a sacred reality belonging to that community alone.

The history of religion in the United States is unique. Throughout its history, the society has attracted people who were fleeing religious persecution, and it was assumed by immigrants that they would be able to live within their own religious community without persecution by the larger society or government. In one important sense, whatever religious communities existed were assumed to be under a much larger umbrella, a sacred "civil" religion, a commitment to principles of pluralism and diversity—a democratic ethic—rather than simply a commitment to one religion or one branch of a specific religion. A continuous tug-of-war exists between separate religious communities, each declaring a unique sacred world, and an overall "civil religion" emotionally committed to pulling these communities together into one pluralistic society with its own sacred world that is supposed to both respect individuality and freedom and expect mutual understanding among many religious communities, including those who actually do not believe in God. Civil religion is political, yet religious. It is political in its democratic belief system; it is religious in that it is meant to bring us together, create sacred objects that we feel represent us, and follow traditions that we regard central to our continuation as a society. Our civil religion interprets our history and its purpose. Occasionally, we even use God in reference to its history and its purpose. The civil religion attempts to create a consensus among the people within society by which they agree that religious differences can and should exist (Bellah 1975: 3).

Few societies in the world seek religious pluralism. In almost every nation, one religion dominates and is the official religion. The society assumes that the political representatives will be of that religion, mixes the political and educational institutions with the religious, and tends to repress minority religions that claim a different view of the sacred world.

Durkheim's insight that religion does indeed help solidify the community is important. It is especially true where one religion exists in the community. It also exists when several religious communities agree to respect one another within the larger community in which religion creates sacred beliefs and rituals.

### *Protecting Group Identity: The Second Social Function*

Religion is also an important way in which people establish and protect their group identity. Religion defends people from losing their unique place in the world. Even whole societies establish themselves through their religion. All through history and all over the world we see societies, communities, and groups fighting to hold onto their identity, and religion plays a critical part. If religion is really the recognition of the sacredness of the community, as Durkheim believed, then it is essential that each community continue to use religion as part of its claim to uniqueness. For most Mexican Americans, identity depends in part on Catholicism. Jewish people for centuries were able to survive as a community through their religious history and a unity of religious belief, language, and practices. Religion is like language; it is an integral part of culture that establishes and protects the identity of people. Religion is used by minorities in almost every society to continue to express their identity; it is used to defend against oppression by the majority and to protect the community from assimilation into the larger society or the larger world. For many minorities, to give up their religion is to lose their unique place in the world.

Every immigrant population to the United States eventually made a choice: The immigrants had to become "American" and make their former identities less important, hang onto their historical identities and quietly accept American citizenship and identity, or somehow become American while simultaneously holding onto their historical identity. In almost every case, organized religion kept the immigrants unique in the larger society for a time. As they turned away from their traditional religion, they increasingly lost their identity in that community. And as they left the religious community, they increasingly turned away from their traditional religion because organized religion, community, and identity are highly interdependent in such a diverse society as the United States. A democratic society not only encourages diversity of religion on the one hand but also brings the loss of religious identity for many individuals as they become freer to succeed in the larger society.

Who we are in the world—our identity—comes from our community. Our community is defended by our religious system; religion functions for our community's identity. Religion also functions to protect us from the larger society, which might try to assimilate or persecute us; it may also cause difficulties for the larger society unless that larger society accepts religious diversity and those in the minority are willing to accept the patterns in the dominant society as well as their own.

### *Control over the Individual: The Third Social Function*

There are many ways in which society controls the individual, but it is important to understand here that the role of religion is critical. Organized religion controls the human being through socialization into a *morality* that appears sacred. It teaches a view of *justice* that upholds that morality. It encourages *responsibility toward the community* rather than simply following self-interest.

**Control through a Moral System** Morality is fragile; it is not easily established and there are always challenges to it. Many of us question society's morality and walk a delicate path between what others tell us and what we desire to do. Many of us wonder if there is really some absolute morality, or is morality really up to the individual? If we believe there really is a true morality, then we are challenged to ask how we can recognize it and come to believe it is important to follow.

It is religion that legitimates the morality of a community by wrapping it into the sacred world apart from the profane social world in which we exist. *Religion tries to make morality seem universal.* Religion makes important rules that may be actually socially derived, human, temporary, situational, changeable, and debatable into absolute rules that are universal, sacred, and true. Society becomes a moral power; it is necessary for a "civilized humanity"; its morality is handed down from one generation to the next (Durkheim 1974: 154).

Can society exist without some general moral order, and can a general moral order exist without a religious basis? Can there be "moral freedom," a right to choose for oneself what shall be good and evil? Since the 1960s, many people in the United States have claimed moral freedom—freedom expanded to one's own moral system rather than society's. "My morality is my own; no one has a right to tell me I am wrong!" Society itself is seen to be the obstruction of all freedom. Even though many of those who believe in moral freedom may themselves be morally responsible, the sociological question is whether or not moral freedom can possibly become the standard for any *community* (Wolfe 2001).

It is important to note that religion does not always lead people to act in ways that many of us would call moral. Religion may teach love and tolerance, but at the same time it may often encourage people to justify oppression or destruction of people who are not part of the religious community. Religion may teach respect for the individual's free will, yet demand conformity to an absolute morality. Religion may teach that helping the poor is something holy, yet people's daily lives are filled with exploiting the poor. Religion may teach peace, but peace turns out to be only dominion over others, or its followers come to believe that it is only through punishment and war that peace can become real. Religion has inspired people to do things that today we regard to be highly immoral: sacrifice, slavery, expulsion or shunning from the community, silencing those who disagree. Sometimes, in making morality sacred, there is too little questioning, and people obey without value analysis, discussion, and individual decision making, and they are discouraged to truly intentionally act in a moral way.

However, for almost all societies, organized religion is an important aspect of social control. It helps to socialize the individual to accept societal morality, and it presents that morality as sacred rather than as simply social.

**Control through a View of a Just Universe** Religion also controls the individual through teaching that the universe is ultimately a just place. Religion encourages people to follow the moral life established within the community by explaining and encouraging the acceptance of a just universe. "There are always consequences to our actions." "Doing wrong will always come back to haunt the wrongdoer." "No one else may know what you have done, but you will." "Those who sin will only find unhappiness in this life and damnation to hell in eternity." "The moral life is the only meaningful life." "God rewards and punishes; in the end, justice will ultimately prevail." Such beliefs bring an order to the moral universe and show why it is important for everyone to obey the moral community. To believers, such beliefs give hope, promise, meaning, and understanding to life. Even without belief in God, a religious system that makes the community sacred tells us that personal tragedy is sometimes necessary for the continuation of the community or for building character and that personal sacrifice or failure may still contribute to a higher good. Religion teaches the believer why some will be blessed and some damned in their beliefs and actions. Creating a logic of justice linked to a religious system is an important way of controlling what people choose to do in life. It warns that those who violate the rules of the religious community will be accountable, and that those who follow the teachings and rules will be rewarded.

A view of a just universe sometimes is dysfunctional for the individual or society. When evil things happen, individuals or even a society may believe that the individual is at fault when he or she really is not. "God is punishing me." "You are getting what you deserve." Too often, events in life are not understood, but are simply explained by accepting that somehow they must be just, and that is all we need to know. Religion's view of justice also encourages people to look for scapegoats, arguing that nonbelievers are the cause of personal or social problems. This leads to persecution and oppression of innocent people and a failure to understand the real causes of any problem. Sometimes, a belief in a just universe as taught by religion leads to unnecessary punishment, horrible wars, and a defense of any tragedy that occurs in life.

However, by teaching that somehow there are always consequences for those who choose good or evil, religion reinforces the morality of society.

**Control through Commitment to Community** Religion controls individuals by teaching them to work cooperatively in the community instead of simply following selfish pursuits. We are taught that community matters, and, because it does, we must all work for other people and not just ourselves. Our actions will always have important consequences for other people and for the future of the community. Communities do not magically exist; they exist because people are willing to work and sacrifice for them. Religion reminds the individual that meaning in life is to be found in community and in unselfish pursuits.

Is commitment to community a good thing? In part, the answer depends on how individual rights are important to us. Commitment to community is essential even for individual growth, but some communities demand sacrifice and even work against our duty to our selves. Too often in the world of the twenty-first century we must be willing to blow ourselves up for the community, to go to war without reason, to sacrifice our own thoughts and our lifework for the community, or to sacrifice others for our community. Religion is important for self-control; this is necessary for community; sometimes this encourages personal growth; sometimes it calls for the unthinking destruction of oneself or others.

Although many people believe there is a conflict between individual freedom and commitment to community, for sociologists this relationship is more complex. Societal control over the individual does not necessarily mean the end of freedom and individualism. Durkheim recognized that modern society encourages the idea that individuals have dignity and worth and have a right to develop their own talents but that this occurs only within a moral order, not outside of it. In fact, Durkheim feared that freedom and individuality for some can become license and cause the dissolution of collective morals and values, and this might well end whatever freedom we have. Alexis de Tocqueville wrote that religion supports morality, morality supports the law, and the law is "the surest pledge of the duration of freedom" (quoted in Aron 1968: 255).

**Summary of Control** Organized religion controls the individual. To some extent, its purpose is to take away some choice so that the community can exist. This is a highly important function.

Perhaps some people can be moral without organized religion. Perhaps some people can believe in a just universe without organized religion. Perhaps some people can be committed to community without organized religion. However, the sociologist wonders: Can a whole society continue without organized religion in control of the vast majority of individuals? If not organized religion, what is there?

### *Defending Democracy from Tyranny: The Fourth Social Function*

Seymour Lipset, an important and insightful sociologist, makes the point that democracy never thrives in a theocracy—that is, a government controlled by religious leaders—and he is probably right (1994). It is often difficult for our religious side to always be consistent with democratic principles. One of the greatest stumbling blocks to democracy has been traditional, authoritarian, fundamentalist religion. Traditional religious beliefs and practices are often contrary to freedom of thought and speech, respect for the individual, acceptance of minority differences, human equality, and a belief that human beings should try to improve their lives in this world.

However, there is another side to religion and democracy. Sociologists and philosophers argue that organized religion can actually support both the development and growth of a democratic society. Religion can inspire people to participate in government and help ensure that democratic government does not slip into a tyranny. Alexis de Tocqueville, a prominent social thinker in the nineteenth century, visited the United States and pointed out the differences between the United States and European societies. He was impressed by how religious we were and how important organized religion was to so many people. He noted the widespread involvement of American citizens in a variety of religious and other social groups. We are joiners, he wrote, and our religious groups are among the most important groups we join. Democracy thrives in America because our involvement in groups makes us active in our community and involved in government. Through various groups, our interests are represented and we are advised as to what government is doing. We have a healthy organization that watches government and tries to limit its power. The threat to democracy, he feared, is a mass of isolated disorganized individuals unable to influence government, hopeless and helpless, and easily manipulated by the press, political leaders, businesses, and demagogues. Religious groups are therefore important for protecting us from mass society and tyranny. Of course, when religious groups use the democratic process to overturn democratic principles and the institutions of free speech, freedom of religion, freedom of the press, and respect for the individual, then they, like some political movements, can actually undermine a democratic society.

### *Understanding and Finding Meaning in the Universe: The Fifth Social Function*

Peter Berger, influenced by Max Weber, emphasizes that the most important role of religion is to help make sense of the chaos we all encounter as we look at our universe. To Berger, the purpose of religion is to build a "sacred cosmos" for people to understand and believe in.

It places the individual in space and time, and it helps give life purpose and fulfillment. This becomes a central part of a people's culture and is passed down from one generation to the next. Culture may control us, but it also provides us with a way of making sense of our world. Berger believes that all people need this, because we all seek to order our experiences in a way that makes some sense to us.

Both Berger and Weber see religion as a way in which humans find meaning. To find meaning is to *understand oneself* in relation to the universe, to find *importance* in what one is and does, and to *believe that life matters* in some way. "Religion is a set of coherent answers to the core existential questions that confront every human group" (Bell 1980: 333–334). Max Weber argued that humans are motivated to establish a cosmos that is meaningful to them. Andrew Greeley (1995: 6) writes, "The function of religion is to give meaning to life. If this needs to be valued by the human being, then there will always be religion. If not for everyone, then for most humans."

And a very big part of meaning is a fear of death, and religion almost always includes the idea that physical death is not the end of our existence. Human beings have a self—that is, they are able to look back on themselves as both subjects and objects in the universe. Ultimately, this brings questions about who we are and the nature of our own importance. It is religion that helps us find this, no matter if it is traditional and organized, a sect, a cult, an individualistic religion, or a religion without a belief in God.

The search for meaning might include artistic creation or appreciation, love, the wonder of nature, great drama, literature, or music. Someone once told me that great music was the closest thing to the supernatural for him. Of all creations in the universe, music is probably one of the most human made, not by nature so much and not by the supernatural. Some people will contend that because such activities are indeed a search for meaning above and beyond simply the physical, they are indeed "religious," or certainly "spiritual." Others would require "belief in God" to be religious, and still others would require following an organized religion. Personally, I include such activities as certainly spiritual, perhaps religious (especially when a community is involved), because they involve beliefs and practices that treat the universe as more than simply physical, they seek a higher meaning to life than can be found in simple physical pleasure, they regard some objects as sacred, and they function in all the various ways we have listed as the functions of religion. In other words, they provide social solidarity, find meaning in the universe, bring a moral sense to the individual, and, when part of community, they help maintain group identity.

Traditionally, however, it is organized religion that introduces a supernatural force within a people's culture to guide people and give them purpose and importance within the vast universe.

### *The Defense of Social Patterns: The Sixth Social Function*

Many of us are familiar with Karl Marx's view that religion is the opiate of the people. He meant that for the workers who were oppressed by their working conditions and the poverty they experienced, religion was a relief from a horrible world. Marx also believed that organized religion acts to turn people's attention away from the real source of their problems—economic conditions—and thus protect society's social patterns from criticism and reform. Marx's view of all social institutions from the religious to educational, political, and economic was that their primary function in society is to defend the economic order and those who have wealth and power in that order. By relieving the hardship of the poor and protecting the wealthy and powerful from the poor, religion, according to Marx, is instrumental for protecting society's social patterns from harm.

There is some important truth in Marx's assertion. Much of society does, in fact, function in a way that protects existing inequality; religion, as a central part of society, shares this function. It socializes people to accept their position in society and to obey authority in society; it threatens those who do not conform to society's law and morality; and, in many societies, its beliefs come to justify the accumulation of private property into the hands of a few and the resulting poverty for the many. It often teaches people that the world we live in is unimportant compared to the world we will find after death. Religions that are fatalistic teach us that there is little we can do to change the world; in fact, some religions even promise rewards after death if people are willing to accept their lot rather than question and criticize their life situation. Religious leaders wrap the political order into the sacred order, defending political authority, arguing that both the religious and political orders are linked in ways that ordinary people should not question. Religion upholds the family system, which socializes the young into society; it upholds and encourages the dominant values in a society. Churches share power with all the other institutional systems in society. Religion has defended slavery, racial inequality, gender inequality, and oppression of homosexuality. It almost always protects the successful. In this way, the arrangement of society becomes justified by religion, which wraps leaders, constitutions, laws, and morality into a sacred order, protecting established society from human criticism.

### *The Criticism of Social Patterns: The Seventh Social Function*

There is another extremely important side to religion. It not only defends society but also often becomes a critic of society. Religion is an important force for identifying problems in society, criticizing the way it works, and uniting people into social movements that make significant changes in society. In recent years, religion has played a critical part in organizing social movements that ultimately helped bring down communism in Eastern Europe. Earlier, religion was instrumental in the civil rights movement in the United States and critical in ending apartheid in South Africa. In this way, religion not only unites society but also inspires reform.

More and more social movements are organized around social issues and have been inspired by religion: for example, human rights, civil rights, racial and gender equality, and both antiabortion and abortion rights. Religion has become a vitally important force in these movements, inspiring criticism and change in the United States and the world. Religions that support the status quo are being challenged by religions that are critical of the directions that both society and the world are taking. Some religious groups oppose the loss of rural America, while some work against the excesses of capitalism. Others work hard for gender equality and gay rights. Some push for a less modern and more traditional society, and some push for a more democratic and pluralistic society. Some new religious movements exhibit great dissatisfaction with the status quo and with mainstream religions precisely because too much in society that needs to be criticized is instead ignored.

The history of religion itself becomes an ongoing dialectic—that is, a constant struggle between those who use religion to protect society and those who are prophetic and inspire new directions, between those who have become part of a conservative social structure and established culture on the one hand and those who claim that revelation from God has led them to oppose the dominant religious and political establishment. Religion is a dynamic force in societies, writes Richard K. Fenn (2001a: 14–15), whenever it provides hints of a future that is different from the present. Religion remains an important force for comprehensive social change.

### *The Social Functions of Organized Religion: A Summary*

Is organized religion necessary? It seems that it has had a central place in society. Perhaps society cannot exist without organized religion, but perhaps we also exaggerate its importance in today's world.

In this section of the chapter, we have argued that religious belief, practices, and institutions contribute to society in seven ways:

1. They help hold the community together.
2. They help retain and defend a people's identity.
3. They control the individual so that the individual acts morally and for the community.

4. In a democracy, they check the political leaders and encourage participation by the masses.

5. They are an important part of culture, giving answers to people so they can understand their meaning and importance in the universe.

6. They protect those who rule and the society as it is.

7. They also question, criticize, and challenge society as it is, making change possible.

However, organized religion also functions in less attractive ways.

It often sacrifices individualism for the community; creates persecution, oppression, and disunity within and between societies; controls individual choice; encourages intolerance of ideas and values that question its own beliefs and practices; and, in protecting and legitimating society and its institutions, often ends up protecting inequality, injustice, and ignorance. In fighting society and criticizing and questioning society's social patterns, it is not always clear if the change that takes place will make society a better or worse place.

## Is Organized Religion Still Necessary?

One of the most interesting debates in social science is the assessment of the role and importance of religion in modern society. There are two general positions concerning this issue.

First, there is a long tradition in social science and philosophy that sees the decline of religion in the world. Those who hold to this view generally describe *secularization* to be one of the most important trends in almost all societies—that is, the decreasing importance of religion in society because of important trends that accompany *modernization.* Modern life, they argue, inevitably undermines the importance of religion.

Second, there are those social scientists who declare that modern life may change religion, but it does not necessarily undermine the influence of religion, and, in many ways, may even increase its importance. Traditional organized religion may become less important, but new forms of organized religion will continue to attract people, and those people who are not attracted to organized religion will still continue to be religious as individuals. Secularization of life is not the universal trend that the first group contends.

## *Religion in Traditional and Modern Societies*

One of the most important reasons sociology arose in the nineteenth century was that many social thinkers wondered about the future of society. They realized that dramatic changes had occurred in Europe and North America, recognized the increasing speed of change, and hypothesized about where everything seemed to be going. As Europeans colonized much of the rest of the world, they encountered more traditional societies, societies that did not have the same technology, government, economic system, and culture that Europe and North America had, and these societies had slower rates of change. These social thinkers identified a rough division for all societies: One they called "traditional," and the other they called "modern." They also recognized a universal trend: As modern economic and social changes were introduced into traditional societies, all institutions would show dramatic changes. Many people thought this was a good change and that the whole world would be better off as it became modern. After all, who can argue with "progress"? Today, most thinkers see these changes differently, and much less positively, and they are more suspicious of what exactly real progress consists of. The question arises, How does one really determine whether traditional or modern society is better for people?

Modernization developed in different places at different times, and it is still absent in many parts of the world. In Western Europe, modernization can be traced at least to the creative period of the Renaissance in the fifteenth century, the breakup of one dominant church in the sixteenth century, the rise of modern science and mathematics beginning in the fifteenth and sixteenth centuries, the rise of the nation-state and eventually the fall of traditional monarchies in the seventeenth and eighteenth centuries, as well as the Industrial Revolution in the eighteenth century. Bureaucratization, the dominance of science, urbanization, and rising individualism increasingly became the norm in Western Europe. In the United States, modernization was certainly on its way at the founding of the nation, but it accelerated considerably during the Industrial Revolution in the eighteenth century, as well as with the building of the railroads and the Civil War in the nineteenth century. Much of the rest of the world was influenced to modernize in the nineteenth and twentieth centuries as societies came in contact with Western Europe and the United States. Modernization meant that rural society became increasingly urban, agricultural society became increasingly industrial, and traditional society became increasingly forward-looking and constantly changing, with traditional culture giving way to the rational and the scientific approach to problems. A society in which people formerly lived their whole lives in one place became much more mobile; and where tradition, informality, and family ties previously dominated, now individuality, progress, and efficiency became more important.

Each quality affected the nature and role of religion; thus, it is important to describe in further detail the link between modernization and religion. In general, as societies become modern everything else changes, including religion. Briefly, let us look at this important process.

1. *Traditional society is a society dominated by the past.* The past is important, including the truths people believe in from a distant past. The ways people act are laid out according to custom and rules that are anchored in the past. Religion, too, tends to be dominated by the past. Truth, morality, values, and institutions are followed because they are part of a religious community that has a long history. *Modern society, on the other hand, is focused more on the present.* We increasingly are not as easily shaped by a religious system we inherit from our ancestors. In modern society, we increasingly evaluate and alter our beliefs, values, and morality on the basis of current goals. Traditional institutions no longer have the sacred quality they once had. Increasingly, we tend to lose our memory of the past, including that of our religious past. Beliefs formed many years ago, values, institutions, and rules no longer seem to stand on their own; now they are discussed, questioned, and reinterpreted. Instead of the past being used as a guide to action, it becomes an "old-fashioned" way of thinking and acting. Whether something from the past can be applied to the present is determined by its usefulness in the present rather than its distant origins. To the extent that a religion also is anchored in the past, then it follows that it must change when society no longer looks to its past.

2. *The idea of progress is not assumed in traditional society.* For some traditional societies, there is a tendency to see time as circular rather than linear, with time like the ever-returning seasons. In other traditional societies, real progress must be

spiritual progress, not simply material progress, and there is really little hope for this in this life. Traditional society emphasizes a more fatalistic view in which this life and this world are relatively unimportant and change little—that is, life on earth is temporary and short compared to an everlasting life after death. *In modern society, progress is taken for granted.* We tend to believe that the problems we face on earth can be dealt with through understanding and technology. There is a tendency for tradition to give way to whatever needs to be done to make life better. People increasingly focus their efforts on making their own lives better, improving their social class and increasing their wealth, living longer, solving problems they encounter in this life, and making the world a better place. The focus on a future afterlife and a traditional way for seeking guidance, purpose, and wisdom are increasingly replaced by a quest to better life on this earth. The traditional society questions the whole idea of progress; modern society is driven by it.

3. *People in traditional society tend to be highly committed to community rather than to the individual.* The community is sacred; it arose from a distant past and will carry on in the future. It is characterized by continuity rather than individual fulfillment. The individual is important because he or she is a member of the community, and it is generally assumed that the individual will live his or her entire life in that community. The family's history is part of the history of that community, and usually many generations live together in such a community; after death, they are buried in that same community. People do not seek individual happiness as much as happiness in the community, which they regard as sacred, special, and eternal. Religion in traditional society emphasizes the importance of the religious community; it emphasizes the individual's contribution to community as important; it encourages meaning to the extent the individual contributes to the community.

*Modern society has brought the triumph of the individual, more interested in self than in any community, determining truth and meaning not from the authority of the community but from secular education as well as personal investigation and experience.* One's place in the universe is not part of a divine plan; one achieves his or her own place through individual effort. People in modern society become increasingly individualistic: "My life is my own." "I have a right to believe whatever I want." "I will leave any relationship if it does not benefit me as much as I want." "Morals? Whose morals? Who says? I have my own." Religion in modern society increasingly focuses on the needs of the individual. Instead of the religious community exerting itself on the individual, the individual chooses religion according to his or her needs; the individual exerts himself or herself onto the religious world. Religious belief, rules, and rituals are all right if they fit the individual's life; they are not something the individual is expected to take on simply because they have always been there. Religion in modern society may be uplifting for the individual, but it does not necessarily demand commitment to a community.

4. *In traditional societies, religion dominates how events in life are interpreted.* It is the lens through which life events are perceived. The supernatural is thought to play a major role in world events. Religion plays a central role in explaining why events occur as they do. Sacrifice, prayer, faith, and following religious law are important ways for solving problems we face in this life. *Modern society creates perspectives that compete with the religious view of the universe,* never replacing it, but certainly slowly explaining more and more without simply accepting traditional religious explanations. *Science increasingly questions the religious explanation for natural events, and the universe becomes increasingly understandable according to the discovery of natural law; the individual need not simply believe in the decisions of the supernatural.* Disease and death are traced to natural cause; the stars, earth, plants, animals, and humans are all part of a natural universe governed by natural laws. Humans can have some control over natural events as they come to understand nature; we can improve our lives, our society, and our physical universe through unlocking what traditional societies see as much more mysterious and explainable only through traditional religious ideas.

5. *In traditional society, religion has a permanent, all-encompassing, and dominant place.* The community is a united religious community, not an association of individuals pursuing individual goals. Other institutions—political, familial, health care, economic, military, educational—are influenced by and intertwined with the religious. Many leaders in the community are religious leaders; the law and punishment are heavily influenced by religious views; education reaffirms religious and community values; government and business must follow religious principles. *In modern society, on the other hand, religion tends to be far less encompassing.* This distinction is emphasized by Bryan Wilson (1982): The local small community allows religion to thrive; the large impersonal society where individuals are increasingly separated from one another is not nearly as hospitable. In the smaller traditional community, it is much easier for religion to have an overarching importance. In the larger impersonal society, there is more specialization of roles and institutions, making an overarching religion less able to control all aspects of life.

6. *In traditional society, one religion dominates the community and the individual.* The truths that people come to accept are embedded in a religious text that is believed to have been handed down by a supernatural being or a special and chosen individual. Because the truth is known, new ideas must meet the standard of past truth, and competition with that truth must be wrong. Traditional society is not open to pluralism—a competition of religions—but one accepted view of the universe that is not supposed to be challenged. Respect for different views is possible but not normally welcome, and curiosity is often blunted in one way or another. If religion is about the sacred world, then it is not up to the individual to determine what truth exists, what morality is true, or which answer we should go with today and not tomorrow. The sacred world is the true world, and we are expected to accept it. *Modern society creates an atmosphere in which people have the opportunity to choose because there are many religions that exist side by side.* Increasingly, believing in a certain religion does not necessarily exclude the rules or truths of other religions. The truth of one's own

religion becomes plausible, not absolute; there is always room for more understanding. Religion gives way to choice (Berger 1970: 45). The monopoly of one religion is undermined as society becomes modern and diverse. With diversity comes the neutral state: government that does not establish an official religion, decide which religion is right for everyone, and, eventually, determine whether people need to be part of some religious organization in order to live.

The forces of modernization have changed much of the world, and almost all of these changes have had an influence on religion. Organized religion, anchored in community and the past, and emphasizing the role of the supernatural in life and the relative unimportance of earthly existence, has been altered considerably. The importance of religion throughout societal institutions and the dominance of one religion have also become less common.

### *The Meaning of Secularization*

This division between traditional and modern society leads us to a discussion of secularization, why it occurs, and whether it is inevitable.

*The "secularization of society" refers to the decreasing importance of religious institutions in society and in the everyday life of individuals. It refers to a certain type of consciousness as people think about life with less and less emphasis on religious thought and concerns. Secularization means that people commit much less time to religious activities. Education is increasingly critical, scientific, and goal-directed. People have far more choices in life that compete with the religious, such as time and money, and the importance of religious belief and practices declines.* Steve Bruce (1996: 26) summarizes secularization as "the decline of popular involvement with the churches; the decline in scope and influence of religious institutions; and the decline in the popularity and impact of religious beliefs." Durkheim (1972: 5) summarized the secularization trend this way: "If there is one truth that history teaches us beyond doubt, it is that religion embraces a smaller and smaller portion of social life. Originally it pervades everything; everything social is religious."

**Secularization Theory** According to some social scientists, secularization is probably inevitable because of modernization. Once modern society develops, it is impossible to turn back the clock to traditional society, and religion thus loses its importance. For this group (we will call them *secularization theorists*), the following values are predominant in modern society and create secularization:

- individual over community;
- science over authority;
- the present over the past;
- progress over fatalistic acceptance;
- pursuit of the everyday "profane" rather than pursuit of the more universal "sacred" life;
- mobility of place and status rather than embeddedness in the place within which we are born; and
- competition of the secular world (family, economic, social, educational, recreational) with the secular rather than the religious world embedded in all areas of life.

All of these together mean that the religious life has become less important for increasing numbers of people.

To secularization theorists, this universal trend is epitomized in European societies in which churches exist without many people, parishioners go without ministers, schools have little concern with the religious, and many belief systems and behaviors are without a religious foundation. Those who believe this secularization thesis will point to similar trends in the United States and argue that as any society becomes modern it will become increasingly secular, too.

Of course, religion will still be part of human existence, some of its traditional functions will continue, and certain beliefs and practices might remain, but secularization will be the dominant direction of society. In fact, there will always be movements reacting against the changes brought by modernization and secularization, and many of these movements will be spearheaded by strong religious sects and cults that are critical of both the dominant religions and the trend toward modernization, which they see as leading to the loss of people's sense of community, meaning, direction, and moral values. To those who accept the secularization thesis, although reactions to modernization in the form of new organized sects and individual spiritualism will continue and even occasionally grow, these are really attacks on both modernization and secularization—neither of which we can ultimately turn back to.

**Critics of Secularization Theory** Sociologists who have written and researched religion in the United States are critical of the secularization thesis. They are not convinced that secularization is inevitable. They believe that the modernization-secularization thesis is an exaggeration. Instead, they tend to argue that *religion has changed but not declined in importance.*

They point out that much evidence shows that religion is central to the lives of many people and that in the United States, in contrast to Europe, secularization is not on the rise. They show that belief in God is as much a part of people's belief system as it ever was. They point out that data suggest that although the larger traditional denominations have declined in attracting active participants, people have turned to newer and more attractive alternatives. People still seek meaning and community, and they seek churches that give them this. More and more people are willing to sacrifice fame and fortune to commit their lives to religious pursuits. They also point to religious movements in Islamic societies, some of which have seemed to successfully reject the secularization of life, as well as active religious movements arising in formerly communist societies. They argue that it is not religion that has become less important, but that choice has become more central to religious life; and for many people, choice brings even greater commitment. Stark and Finke (2000: 42–43) consider religion to be like other areas of modern life: informed consumers weighing costs and benefits, choosing to follow a religious path because they determine this is best for them, committing themselves, and persisting in religious belief not so much because it is simply foisted on them but because they continue to recognize religion as important to them, even in a modern society.

To critics of the secularization thesis, change, pluralism, and individualism have become a continuous modern trend and have had major effects on religion. However, this does not necessarily mean change equates with the decline of religion. Instead of religion being overwhelmingly a community affair, for many people it has become increasingly individualistic, meeting individual spiritual needs. To be religious has many meanings today, and people have choices to make; no longer is religion simply a matter of the traditional religion we are born into. However, modern life has not caused a rejection of a search for a more sacred and meaningful life; the possibilities now include traditional established religion, a more liberalized established religion, individual spiritualism without church affiliation, smaller and stricter sects that reject or compete with established religions, and religious cults that reject modernization by having nothing to do with society. For large numbers of people, religion seems to be a search for individual fulfillment rather than a strong commitment to specific communities. Except for fundamentalism, there is an increasing tolerance of religious answers that are different from one's own. Many people are also finding a spiritual answer to their lives within communities that do not emphasize a supreme being: small friendship groups, therapy groups, families, social action groups, retirement communities, and in book, music, art, literature, and poetry groups. The desire to overcome the profane everyday physical present now presents a wide number of choices for people in modern society.

The modern trends of religious individualism and pluralism are epitomized by the popularity of so-called New Age religion, a loosely organized approach to the spiritual world in which the individual is the highest authority over what he or she believes and does. This religious approach to life is characterized by minimal formal organization. The divine is in the self, not in a God. Belief is what the individual wishes to accept, and there are many belief systems to choose from: herbalism, mysticism, Zen, and meditation, for example. There is little commitment to absolute truth, and the individual is encouraged to seek the good life in his or her own way. People easily move into and out of various movements without criticism by others. Science tends to be rejected in favor of subjectivity, organized religion in favor of individual spiritualism, and the mind–body division in favor of the unity of mind and body. Tradition is not simply followed, but ancient wisdom is still studied and applied when it fits. Instead of male images dominating thought and practice, there is a strong feminine image that has become important (Bruce 1999: 162–163; and Aldridge 2000: 209–210). If anything organizes this approach, it is the large quantity of books, magazines, and articles that encourage communication between believers.

Peter Berger also reminds us that empirical data support the conclusion that "more Americans than ever regularly go to religious services, support religious organization, and describe themselves as holding strong religious beliefs" (1992: 36). Two important examples of increasing religiosity in the world highlighted by Berger are the Muslim societies and evangelical Protestantism. Both are reactions to modernization, both have strong commitment, and both have a missionary spirit. Western Europe might be the epitome of secularization, but the rest of the world "is as furiously religious as ever, and possibly more so" (1992: 32).

It is true, writes Berger, that freedom from more traditional religion may bring a loss of certitude and tentative acceptance by some, but it also encourages many people to search for certitude about things that matter again and in new places. Berger (1992: 126–127) believes that the modern world does not necessarily become a secular world; and the individual can even benefit because he or she can now pursue religious truth "from a fresh start":

> I think what I and most other sociologists of religion wrote in the 1960s about secularization was a mistake. Our underlying argument was that secularization and modernity go hand in hand. With more modernization comes more secularization. It wasn't a crazy theory. There was some evidence for it. But I think it's basically wrong. Most of the world today is certainly not secular. It's very religious. (p. 974)

**Secularization Theorists Respond** Those who hold onto the secularization theory are not persuaded. Small, highly traditional religious communities are forming everywhere, although they contain a minority of people and demand commitments of energy, time, and passion that most people are no longer willing to give. These tend to be religious sects that are critical of society's trends, yet most are generally short-lived and fickle, not easily able to isolate believers from secular society, and difficult to continue over many generations. Organized religions that try hard to cater to individuals who are concerned primarily with themselves rather than any religious community will fail by cutting their ties with the past and by their willingness to sacrifice the sacredness of their truths to pluralism. It is a tough road for organized religion to prevail in modern society, because people increasingly turn their time and energies to other matters; with a predominant strong commitment to self, the community will continue to suffer in the tug-of-war over what to do with one's life. Individual choice and individual spiritualism may continue to be important. They will, however, not function for any community, and it will be difficult to hand down any sacred truth to the next generation, which may go out on its own in search of spirituality in modern society.

For those who hold to the secularization theory, certain questions are continuously posed to those who reject the theory. Can we still believe in sacred truth when we are willing to respect truths that disagree with us? Can we continue to possess a personal religious belief system without a commitment to religious community? Can we bring so much choice into religion yet retain a strong commitment to a religious community and a religious past? Can people continue to worship modernization and still accept the traditional beliefs and practices of religion? Does not choice itself undermine tradition—and is not tradition the essence of religion?

To those who believe that modernization inevitably creates an increasingly secularized society, there are inherent contradictions between religion shared in a traditional community and modern life. For them, the sociological question becomes, What kind of society can exist without a powerful religious dimension? Steve Bruce (1999: 186) maintains that although we might talk about the existence of religious community, "modernization has destroyed it." Today we choose the tradition that we follow, God is for individual purposes, and we reject control by the past. The net result is that individual beliefs which are not regularly articulated and affirmed in a group, which are not refined and nourished by shared ceremonies, which are not the object of regular and systematic elaboration, and which are not taught to the next generation or to outsiders are unlikely to exert much influence on the actions of those who hold them and are even less likely to have significant social consequences. (Bruce 1996: 58)

## The Rise of Fundamentalism

No matter what modernization brings—secularization or simply a different role for religion—there will always be many who will be attracted to a fundamentalist approach to religion.

Today, fundamentalist religion seems to be increasingly on the rise. Some sociologists would argue that this is proof that modernization may actually inspire antimodernism in the form of a greater commitment to religious life. Others would argue that one of the outcomes of an increasingly secularized world will be the rise of more committed reactionary groups who try to turn back the inevitable. The understanding of fundamentalism is linked to this discussion of tradition, modernism, and secularization.

Fundamentalism is a certain way of looking at reality that is found in certain religious communities, certain individuals, and certain social movements. It springs from an attempt by a people to preserve their distinctive identity and to fortify themselves against what they believe are their enemies. Although they are steeped in a long tradition, that tradition is updated in order to neutralize the threats that exist in their world. Fundamentalist groups are led by charismatic and authoritarian leaders, and members are disciplined according to a rigid code. Rigid boundaries are created between the group and outsiders, an enemy is identified, and converts are sought. The reconstruction of society is planned. A rigid division of labor according to gender roles is believed in and followed. Revealed truth is valued, and both science and societal norms are questioned if they violate that truth. Personal and family morality is strict in large part because this is seen to be a path to transforming society. Criticism by outsiders is common, but fundamentalists regard judgment by other human standards to be less important than God's judgment (Marty and Appleby 1993: 3–12).

Fundamentalism tends to appeal especially to people in two kinds of social situations: those who are experiencing horrible social conditions in which they seem trapped, and those who find modernization unacceptable. For the poor and dispossessed, fundamentalism makes sense out of their suffering and gives them hope. To those who reject the forces of modern life, fundamentalism is attractive because it tries to recapture the values and truths of the idealized past and gives hope for a better future. People are able to "seek fellowship and fresh meaning" in the certainty of God's word. Individualism and pluralism are not their idea of what is really important, because they bring loneliness, dissatisfaction, uncertainty, and loss of community. Instead of science and secularization, adherents to fundamentalism find answers in the "words of preachers who derive their teaching from the Bible" (William McNeill in Marty and Appleby 1993: 568). Both categories of people—the poor who seek understanding and relief from suffering as well as those who seek meaning through community, certainty, and attractive leaders—will often seek fundamentalist movements and are more open to charismatic and authoritarian leaders. The strict requirements of these movements help ensure a strong community within a world in which modernization is rapidly changing—or threatening—the more traditional religious life of society.

Fundamentalism exists within Christianity, Judaism, and Islam today. Many people wonder why, because the appeal and the experience are so foreign to their own lives, but we must remember that the modernization in which we are wrapped is something that is threatening to those who are attracted to religious tradition and strong community.

## Summary and Conclusion

The sociological question is not whether religion will continue to function for individuals because as long as meaning is important, there will always be a consideration of a spiritual religious answer, with or without an organization, and with or without a God. *The sociological question is whether organized religion will remain important for society.* If organized religion remains an important part of our community and society, then it will continue to fulfill many or all of the functions as it always has. If it loses its place in society because of modernization, then we must ask about consequences for society. Perhaps we will find that society does not need organized religion to survive; perhaps, as most early sociologists believed, society without organized religion is impossible. Richard K. Fenn (2001a: 14–15) goes back to Durkheim: "The purely personal, self-validating form of faith represents a threat to any and all forms of solidarity." It limits religious authority. The group "gives way to the dispositions of the individual." Is there a society that can exist through "responsible individualism" alone?

The first part of this chapter examined the meaning of religion. Although there is no simple and clear definition—and definition itself biases discussion of all religious issues—we tentatively defined religion as an attempt by people to carve out a special sacred space in the universe that brings them to a higher plane than their everyday profane world. Religion is a certain kind of perspective used to define the universe that is characterized by both beliefs and practices that are developed in community and that serve the community in many important ways.

The second part of the chapter looked at the various ways religion has been important to society. Seven basic functions were described: social solidarity, identity, control over the individual, involvement in democratic institutions, meaning for the individual, ideology, and instrument of social change.

Religion also has a negative side: It limits freedom of thought, speech, and actions; functions as a source of intolerance and destructive conflict; defends those who have power and wealth as legitimate and sacred; and discourages criticism and change and may ultimately defend oppression.

The third part of the chapter has focused on modernization, secularization, and the importance of religion in the modern world. According to many social scientists and philosophers, society has become far more secularized as it has moved from the traditional to the modern. *Secularization* means that religion has become less important in the lives of people and in the social institutions they follow. Thus, their answer to this chapter's question—"Is organized religion necessary for society?"—is probably "less and less so." Modernization increasingly contributes to a decline in the role of organized religion. It is unclear how organized religion can turn this trend around. It is unclear whether spiritual individualism can continue to fulfill organized religion's societal functions. It is also unclear whether such traditional functions can be taken up by other institutions. If not, then what becomes of society?

For other social scientists and philosophers, modern life has brought important changes: Societies have become religiously pluralistic, many people have become increasingly spiritual as individuals, and many people have been attracted to fundamentalism. Modern life has made religion an active choice rather than a way of thinking that people are born into. Thus, another answer to this chapter's question—"Is organized religion necessary for society?"—is, "Yes, and it is probably just as important as ever." It remains unclear, however, what role, if any, individualistic religion can play in the continuation of society.

It is difficult to predict the future. It is too early to objectively understand the role of religion in modern society. Perhaps my friend and colleague Arnold Dashefsky best describes the future:

> The challenge facing organized religion in modern society is to provide society and the individual a balance between extreme individualism and the passionate ethnocentrism that fundamentalism displays. It must bridge the divides between community and the individual, and between democratic pluralism and religious authoritarianism.

## Questions to Consider

1. Is religion necessary for the individual? Do you believe that it will always be necessary for the individual?

2. In your opinion, what is religion? Do you believe that religion assumes that a God exists? Do you contend that the definition of religion should include only organized beliefs and practices, or do you believe that individual spirituality should count as religion? Do you believe that a commitment to communism or fascism can be considered a religion? Do you believe that love of music, love of humanity, love of art, love of nature can constitute a religion?

3. Is it even important to define what religion is?

4. What is the most important contribution religion makes to society?

5. In what ways can religion be harmful to the individual?

6. In what ways can religion be harmful to society?

7. What is the future of religion?

8. What do you believe to be the most important development in modern society that affects the importance of religion?

## References

The following works deal with the issues relevant to the meaning of religion, the functions of religion, and the relationship of modernization and secularization.

**Adams, E. M.** 1993. *Religion and Cultural Freedom.* Philadelphia: Temple University Press.

**Aldridge, Alan.** 2000. *Religion in the Contemporary World: A Sociological Introduction.* Malden, MA: Blackwell.

**Ammerman, Nancy T.** 2005. *Pillars of Faith American Congregations and Their Partners.* Berkeley: University of California Press.

**Apraku, Kofi.** 1996. *Outside Looking In: An African Perspective on American Pluralistic Society.* Westport, CT: Praeger.

**Aron, Raymond.** 1968. *Main Currents in Sociological Thought,* vol. 1. Garden City, NY: Doubleday.

**Bainbridge, William Sims.** 1997. *The Sociology of Religious Movements.* New York: Routledge.

**Beckford, James A., and Thomas Luckmann**, eds. 1989. *The Changing Face of Religion.* London: Sage.

**Bell, Daniel.** 1980. *The Winding Passage.* Cambridge, MA: Abt Books.

**Bellah, Robert N.** 1967. "Civil Religion in America." *Daedalus,* 96: 1–21.

———. 1975. *The Broken Covenant.* New York: Seabury.

**Bellah, Robert N., Richard Madsen, William M. Sullivan, Ann Swidler, and Steven M. Tipton**. 1985. *Habits of the Heart: Individualism and Commitment in American Life.* Berkeley: University of California Press.

**Berger, Peter L.** 1969. *The Sacred Canopy: Elements of a Sociological Theory of Religion.* Garden City, NY: Doubleday.

———. 1970. *A Rumor of Angels: Modern Society and the Rediscovery of the Supernatural.* Garden City, NY: Doubleday.

———. 1992. *A Far Glory: A Quest for Faith in an Age of Credulity.* New York: Free Press.

**Berger, Peter L., and Thomas Luckmann**. 1966. *The Social Construction of Reality.* Garden City, NY: Doubleday.

**Beyer, Peter.** 1994. *Religion and Globalization.* Thousand Oaks, CA: Sage.

**Bruce, Steve.** 1996. *Religion in the Modern World: From Cathedrals to Cults.* Oxford: Oxford University Press.

———. 1999. *Choice and Religion: A Critique of Rational Choice Theory.* Oxford: Oxford University Press.

**Buechler, Steven M.** 2000. *Social Movements in Advanced Capitalism.* New York: Oxford University Press.

**Cipriani, Roberto.** 2000. *Sociology of Religion: An Historical Introduction,* translated by Laura Ferrarotti. New York: Aldine de Gruyter.

**Collins, Randall.** 1998. *A Global Theory of Intellectual Change.* Cambridge, MA: Harvard University Press.

**Davis, Charles.** 1994. *Religion and the Making of Society: Essays in Social Theology.* Cambridge, England: Cambridge University Press.

**Della Porta, Donatella, and Mario Diani**. 1999. *Social Movements: An Introduction.* Malden, MA: Blackwell.

**Dowdy, Thomas E., and Patrick H. McNamara**, eds. 1997. *Religion: North American Style.* 3rd ed. New Brunswick, NJ: Rutgers University Press.

**Durkheim, Émile.** [1915] 1954. *The Elementary Forms of Religious Life,* translated by Joseph Swain. New York: Free Press.

———. 1972. *Selected Writings,* edited and translated by Anthony Giddens. Cambridge, England: Cambridge University Press.

———. 1974. *Sociology and Philosophy,* translated by D. F. Peacock with an introduction by J. G. Peristiany. New York: Free Press.

**Dyer, Joel.** 1997. *Harvest of Rage: Why Oklahoma City Is Only the Beginning.* Boulder, CO: Westview.

**Emerson, Michael O., and Rodney M. Woo**. 2006. *People of the Dream: Multiracial Congregations in the United States.* Princeton, NJ: Princeton University Press.

**Etzioni, Amitai.** 2001. *The Monochrome Society.* Princeton, NJ: Princeton University Press.

**Fenn, Richard K.** 2001a. *Beyond Idols: The Shape of a Secular Society.* Oxford: Oxford University Press.

———. 2001b. "Editorial Commentary: The Sacred and the Profane." pp. 3–22 in *The Blackwell Companion to Sociology of Religion,* edited by Richard K. Fenn. Malden, MA: Blackwell.

**Fenton, Steve.** 1984. *Durkheim and Modern Sociology.* Cambridge, England: Cambridge University Press.

**Freie, John F.** 1998. *Counterfeit Community: The Exploitation of Our Longings for Connectedness.* Lanham, MD: Rowman & Littlefield.

**Friedman, Thomas L.** 2002. *Longitudes and Attitudes: Exploring the World after September 11.* New York: Farrar, Straus & Giroux.

**Giddens, Anthony.** 1979. *Émile Durkheim.* New York: Viking.

**Glendinning, Tony, and Steve Bruce**. 2006. "New Ways of Believing or Belonging: Is Religion Giving Way to Spirituality." *British Journal of Sociology* 57(3): 399–414.

**Goldberg, Michelle.** 2006. *Kingdom Coming: The Rise of Christian Nationalism.* New York: Norton.

**Greeley, Andrew**, ed. 1995. *Sociology and Religion.* New York: HarperCollins.

**Hall, John A., and Charles Lindholm**. 2000. *Is America Breaking Apart?* Princeton, NJ: Princeton University Press.

**Hamilton, Malcolm.** 1998. *Sociology and the World's Religions.* New York: St. Martin's.

**Hammond, Phillip E.** 1992. *Religion and Personal Autonomy: The Third Disestablishment in America.* Columbia: University of South Carolina Press.

———. 2000. *The Dynamics of Religious Organizations: The Extravasation of the Sacred and Other Essays.* New York: Oxford University Press.

**Heelas, Paul, Linda Woodhead, Steel Benjamin, Karin Tusting, and Baron Szerszynski**. 2004. *The Spiritual Revolution: Why Religion Is Giving Way to Spirituality.* Oxford, UK: Blackwell.

**Hervieu-Leger, Daniele.** 1998. "Secularization, Tradition and New Forms of Religiosity: Some Theoretical Proposals." pp. 28–44 in *New*

*Religions and New Religiosity*, edited by Eileen Barker and Margit Warburg. Cambridge, UK: Cambridge University Press.

**Hunt, Stephen.** 2002. *Religion in Western Society.* New York: Palgrave.

**Jacobs, Aton K.** 2006. "The New Right, Fundamentalism, and Nationalism in Postmodern America: The Marriage of Heat and Passion." *Social Compass* 53(3): 357–366.

**Jones, Robert Alun.** 1986. *Émile Durkheim: An Introduction to Four Major Works.* Beverly Hills, CA: Sage.

**Karp, David A.** 2000. *Burden of Sympathy.* New York: Oxford University Press.

**Katkin, Wendy F., Ned Landsman, and Andrea Tyree**, eds. 1998. *Beyond Pluralism: the Conception of Groups and Group Identities in America.* Urbana: University of Illinois Press.

**Kosmin, Barry A., and Seymour P. Lachman**. 1993. *One Nation under God: Religion in Contemporary American Society.* New York: Harmony.

**Levi, Primo.** 1959. *If This Is a Man.* New York: Orion.

**Lewis, Bernard.** 2002. *What Went Wrong? Western Impact and Middle Eastern Response.* New York: Oxford University Press.

———. 2003. *The Crisis of Islam: Holy War and Unholy Terror*. New York: Random House.

**Lipset, Seymour Martin.** 1994. "The Social Requisites of Democracy Revisited." *American Sociological Review* 59: 1–22.

**Lofland, John.** 1966. *Doomsday Cult.* Englewood Cliffs, NJ: Prentice Hall.

**Lukes, Steven.** 1977. *Émile Durkheim: His Life and Work: A Historical and Critical Study.* Harmondsworth, England: Penguin.

**Macrae, Donald G.** 1974. *Max Weber.* New York: Viking.

**Madsen, Richard**, ed. 2002. *Meaning and Modernity: Religion, Polity, and Self.* Berkeley: University of California Press.

**Manning, Christel J.** 1999. *God Gave Us the Right: Conservative Catholic, Evangelical Protestant, and Orthodox Jewish Women Grapple with Feminism.* New Brunswick, NJ: Rutgers University Press.

**Marty, Martin E., and R. Scott Appleby**. 1993. "Introduction: A Sacred Cosmos, Scandalous Code, Defiant Society." Part 1 in *Fundamentalisms and Society: Reclaiming the Sciences, the Family, and Education,* edited by Martin E. Marty and R. Scott Appleby. Chicago: University of Chicago Press.

**Masci, David, and Gregory A. Smith**. 2006. "Is Religion Giving Way to Spirituality?" *Sociology Review* 15(4): 14–16.

**McNeill, William H.** 1993. "Epilogue: Fundamentalism and the World of the 1990s." pp. 558–574 in *Fundamentalisms and Society:*

*Reclaiming the Sciences, the Family, and Education,* edited by Martin E. Marty and R. Scott Appleby. Chicago: University of Chicago Press.

**Mendelsohn, Everett.** 1997. "Religious Fundamentalism and the Sciences." pp. 23–41 in *Fundamentalisms and Society: Reclaiming the Sciences, the Family, and Education,* edited by Martin E. Marty and R. Scott Appleby. Chicago: University of Chicago Press.

**Miller, W. Watts.** 1996. *Durkheim, Morals and Modernity.* Montreal/ London: McGill-Queen's University Press/UCL Press.

**Mills, Nicolaus.** 1997. *The Triumph of Meanness: America's War against Its Better Self.* Boston: Houghton Mifflin/Sage Foundation.

**Morrison, Ken.** 1995. *Marx, Durkheim, Weber: Formations of Modern Social Thought.* Thousand Oaks, CA: Sage.

**Nisbet, Robert.** 1974. *The Sociology of Émile Durkheim.* New York: Oxford University Press.

**Pickering, W.S.F.**, ed. 1975. *Durkheim on Religion.* London: Routledge & Kegan.

**Postman, Neil.** 1992. *Technopoly: The Surrender of Culture to Technology.* New York: Knopf.

**Rashid, Ahmed.** 2002. *Jihad: The Rise of Militant Islam in Central Asia.* New Haven, CT: Yale University Press.

**Rouner, Leroy S., and James Langford**, eds. 1996. *Philosophy, Religion, and Contemporary Life: Essays on Perennial Problems.* Notre Dame, IN: University of Notre Dame Press.

**Scaff, Lawrence A.** 2000. "Weber on the Cultural Situation of the Modern Age." Chapter 5 in *The Cambridge Companion to Weber,* edited by Stephen Turner. New York: Cambridge University Press.

**Schoffeleleers, Matthew, and Daniel Meijers**. 1978. *Religion, Nationalism and Economic Action: Critical Questions on Durkheim and Weber.* Assen, The Netherlands: VanGorcum.

**Schwartz, Barry.** 1994. *The Costs of Living: How Market Freedom Erodes the Best Things in Life.* New York: Norton.

**Seligman, Adam.** 1995. *The Idea of Civil Society.* Princeton, NJ: Princeton University Press.

**Stark, Rodney, and Roger Finke**. 2000. *Acts of Faith: Explaining the Human Side of Religion.* Berkeley: University of California Press.

**Staub, Ervin.** 1989. *The Roots of Evil: The Origins of Genocide and Other Group Violence.* New York: Cambridge University Press.

**Thompson, Kenneth.** 1982. *Émile Durkheim.* London: Tavistock.

**Tucker, Kenneth H.** 2002. *Classical Social Theory: A Contemporary Approach.* Malden, MA: Blackwell.

**Tucker, Robert C.**, ed. 1972. *The Marx-Engels Reader.* New York: Norton.

**Turner, Bryan S.** 1983. *Religion and Social Theory: A Materialist Perspective.* London: Heinemann.

**Walker, Henry A., Phyllis Moen, and Donna Dempster-McClain**, eds. 1999. *A Nation Divided: Diversity, Inequality, and Community in American Society.* Ithaca, NY: Cornell University Press.

**Ward, Keith.** 2000. *Religion and Community.* New York: Oxford University Press.

**Weber, Max.** [1905] 1958. *The Protestant Ethic and the Spirit of Capitalism,* translated and edited by Talcott Parsons. New York: Scribner's.

———. 1969. "The Social Psychology of the World Religions," pp. 267–301 in *Max Weber: Essays in Sociology,* translated and edited by H. H. Gerth and C. Wright Mills. New York: Oxford University Press.

**Wilson, Bryan.** 1982. *Religion in Sociological Perspective.* New York: Oxford University Press.

**Wolfe, Alan.** 1998. *One Nation, After All.* New York: Penguin.

———. 2001. *Moral Freedom: The Impossible Idea That Defines the Way We Live Now.* New York: Norton.

**Wolff, Edward N.** 1996. *Top Heavy: A Study of the Increasing Inequality of Wealth in America.* New York: Twentieth Century Fund.

**Woodhead, Linda, Paul Heelas, and David Martin**, eds. 2001. *Peter Berger and the Study of Religion.* New York: Routledge.

**Wuthnow, Robert.** 1988. "Sociology of Religion," pp. 473–509 in *Handbook of Sociology,* edited by Neil J. Smelser. Newbury Park, CA: Sage.

———. 1992. *Rediscovering the Sacred: Perspectives of Religion in Contemporary Society.* Grand Rapids, MI: Eerdmans.

**Zweigenhaft, Richard L., and G. William Domhoff**. 1998. *Diversity in the Power Elite: Have Women and Minorities Reached the Top?* New Haven, CT: Yale University Press.

# Is the World Becoming One Society?

## Globalization and the Creation of a World Society

### Concepts, Themes, and Key Individuals

- ❑ Globalization
- ❑ Technology and Communication
- ❑ Capitalism and Freedom
- ❑ World Economy
- ❑ Nations and Societies
- ❑ World Society

During the so-called Middle Ages (about 400 to 1400 AD) Western Europe's economic system and the life of its people were highly localized in manors and towns. Kingdoms arose toward the end of the Middle Ages, and eventually the nation-state became the unit of territory within which people were governed, protected, and where they conducted agriculture and trade. There was always some long distance trade between other lands, but it was limited for various reasons given the technology and lack of security that existed. In the sixteenth and seventeenth centuries, the long distance trade accelerated: Slaves, rum, sugar, guns, spices, silk, and much more opened up trade beyond the European nations. In the eighteenth century, the Industrial Revolution began in Europe and the United States, and world trade became even more important.

Throughout the nineteenth century, colonies in Africa, Asia, and the Middle East became sources of raw materials and markets for goods. These places became part of the world's economic system, and with it, the lives of the people changed. The twentieth century brought world wars, political and economic empires of colonialism, and eventually the world wars and revolutions resulted in independent nations, dramatic change, and increase in world trade.

World interaction—the interconnectedness, interdependence, integration, social networks, exchanges—accelerated at the end of the twentieth century and the beginning of the twenty-first. The multinational corporations became giants, and a transportation and communication revolution occurred. The trend toward a world system became so central, altering our lives so dramatically, that several thinkers have identified this acceleration as truly dramatic. They call it *globalization*. For many, it is not simply an evolution that goes back hundreds of years, but truly a new world that has changed economically, politically, socially, and culturally.

Social interaction among people throughout the world has increased with the technology of air and ground transportation and networked communication. People meet across vast distances, face-to-face, with cell phones, with computers and the Internet, via satellites, and fiber optics. This social interaction is further intensified by trade, outsourcing, transfers of capital, expansion of markets, corporate relocations, and migration. Isolation becomes almost impossible; awareness of connections to distant peoples and places become much more common. Using Internet and television technology, we can talk to individuals who are scattered over the world. A worldwide interdependence has arisen. People's survival is tied not simply to families, communities, societies, and nations; their needs are met through individuals living and working a long way away from them. Increasingly people become knowledgeable about those who have very different cultures from them, allowing easier ways to exploit or contribute toward these others. Geography is no longer a limit to what can be done.

*Globalization is a process, a cluster of many activities, a direction toward a world system, toward an integrated interdependent world.* Social interaction among individuals, organizations, societies, and nations throughout the world create social patterns that compete with societal and national patterns created and established over many hundreds of years. To some extent, globalization is creating a global social structure, culture, and institutions. It is increasingly establishing a level of social organization that we might describe as a "world society."

## Three Views of Globalization

Globalization is controversial. The debates are highly emotional over what it is, what it will create, and whether it is a positive or negative trend. In general, there are three viewpoints:

1. *Some individuals do not believe that a revolution is occurring.* There is change taking place, but these trends have been evolving from at least the sixteenth century. It is the development of traditional societies becoming modern and modern societies becoming more global. It is the ultimate climax of capitalism, Marx would argue, while Weber would call it the ultimate development of modernization. Although there is a definite trend toward a world order, we are still primarily a world of independent nations and societies. Local and national governments are the most important political institutions in the world, nationalism and ethnocentrism are still much more important to people's feelings than commitment to a united world, world cultures are still highly diverse, and even the world economy is still divided up into nations, societies, and communities, each with its own problems, strengths, and controls.

2. *Some individuals believe that the world is truly becoming transformed in a very dramatic way. In general, this is thought to be positive.* It benefits much of the world through the development of industrialization, new technology, trade, employment, and ultimately democratic structures, democratic cultures, and democratic institutions. The nation state is losing some of its power. Boundaries between societies are breaking down. Economies are no longer national economies, and the communication and transportation revolutions, along with free trade and huge corporations, are developing a one-world society. Thomas Friedman (2005) sees the world becoming "flat" in the sense that business opportunities are opening up all over the world, people who used to have few possibilities for

becoming middle class or rich now compete with those of that status. Globalization is thought to be good for consumers because products cost less, and good for businesses because they can produce those products more efficiently and, less costly. Stockholders benefit because companies can increase profits. Ultimately, people all over the world will benefit through a higher standard of living. Globalization makes it possible to access information and a diversity of ideas through computer technology. The whole world is creating capitalistic institutions, perhaps democratic culture, and possibly more opportunities for everyone. There is hope for commerce and communication to bring peace, democracy, and affluence throughout the world.

3. *Some individuals believe that the world is truly becoming transformed in a dramatic way, but they are not nearly positive about what is taking place*. They are the critics of globalization, and they contend that owners and stockholders are benefiting greatly on the backs of all others. Globalization is really becoming a capitalist haven, with the most wealthy people in the world creating multinational corporations that are no longer regulated by national political institutions. Such multinationals can expand their investments, produce wherever they choose, transport their goods in an international environment, and achieve greater profit in what they sell. The labor pool is expanded, competition for jobs is open to people all over the world, and by shopping for the best "wage," businesses keep their labor cost low. Businesses become flexible: If one part of the world is no longer suitable for production or markets, it can be abandoned for another. Laborers have no protection from giant corporations who value profit over other values.

Detractors point out that globalization is controlled by wealthy individuals and corporations headquartered in the United States, Japan, and Western Europe. They hold that globalization creates a worldwide culture at the expense of national cultures and that more often than not it reflects the dominant tastes and interests of the United States. Monolithic world economic institutions—such as the G8—support a world economy that brings with it a loss of democracy and

> diversity. The contrarian view of globalization sees other serious problems ranging from job insecurity to massive unregulated migration, great gulfs between rich and poor, international crime, terrorism, pornography and violence, and a serious disregard of the environment.

Perhaps all three viewpoints have some important truth. Globalization does have a long history, yet what we are experiencing is also truly a dramatic change, and, like other dramatic changes, there are reasons to have hope and to be critical at the same time. And even though globalization proceeds, there is also a contrasting trend of nations working to solidify their borders.

It is difficult to know what the future holds. History shows numerous instances where people believed that change would ultimately bring peace, justice, and opportunities for all, only to find out later that old problems remained and new problems arose. History also has examples of change that solved many problems, circumvented the catastrophe predicted for it, and a better world arose.

The discussion of globalization is really a discussion of values. It is a conflict between holding on to capitalistic values and institutions versus values that seek some control over capitalism—values, that is, other than profit. For sociologists, there is the problem of objectively understanding the great changes taking place in the midst of living them. It would be nice to look back on the twenty-first century to compare globalization with its "end" results, but we do not have the longevity to look back. Joan Ferrante (2008) presents the difficulties in evaluating globalization:

> Depending on where you live and who you are, globalization plays out differently. On the one hand, it connects the economically, politically, and educationally advantaged to one another while pushing to the sidelines those who are not so advantaged. On the other hand, it connects those working at the grassroots level to protect, restore, and nurture the environment and to enhance access for the disadvantaged to the basic resources they need to live a dignified existence. (P.20)

This sociological perspective to understanding and evaluating globalization is useful. It tries to examine the global economy, but goes further in questioning the kind of world society that is being created. We will examine globalization systematically. Thus far, we have introduced it as a concept and laid out its three "schools" of thought. We continue by exploring these four themes:

- *Technology, communication, and globalization.* The creation of new technology has encouraged a revolution in communication that continues to make globalization possible.
- *Capitalism, globalization, and the global economy.* The commitment to capitalism makes globalization almost an inevitable direction.
- *Globalization and the creation of a world society.* Globalization is more than economic. In fact, the world might be truly becoming a unit of social organization, a "world society" developing its own social structure, culture, and social institutions.
- *Is globalization good for the world?* A positive view is that almost everyone benefits through economic progress, democratic development, and worldwide social movements. A critical view emphasizes worker problems and worker rights, societal dependence, economic inequality, economic volatility, international crime, less concern for people, homogenization, worldwide disease, and terrorism.

## Technology, Communication, and Globalization

Much of what we call "globalization" has to do with communication. Because of the technology created during the last part of the twentieth and the beginning of the twenty-first century, instant communication became common among individuals, nations, and societies. In the economic sector, communication has allowed a revolution in how we produce, distribute, and consume goods and services. The whole world becomes a place where corporations compete, capital is invested, markets grow, labor is found, and great worldwide wealth is accumulated. Communication also creates tremendous changes in the creation, transferring, and storing information. What we learn from others outside our society balance and supplement what our culture of origin teaches us.

Thomas Friedman in *The World Is Flat* (2005) describes the communication revolution, its economic implications, and its ability to increase the tremendous knowledge throughout the world. He points out that the destruction of the Berlin Wall in 1989 was a critical step, for it meant the possibility of communication among nations that were separated from one another during the Cold War since 1945. This event represented the possibility of spreading democracy and capitalism throughout much of the world. Increasingly, people saw that the world was becoming "a single market, a single ecosystem, and a single community" (Friedman 2005: 51). Friedman points out that totalitarianism will be impossible if there is free interaction of information across national boundaries. He describes the destruction of the Berlin Wall as the beginning of a "world that is flat," a world where economic and informational competition is open, and a world where knowledge will be accessible for all. Individuals who never had a chance before will be able to start up their own businesses.

In Friedman's flat world tremendous outcomes have arisen from technology such as the Windows operating system, the Netscape browser, Google, TiVo, and other technologies made possible or enhanced by the Internet. From these and related developments, it is much easier to communicate throughout the world. Work and projects can be outsourced inexpensively to new communities of educated and inspired white collar workers throughout Asia. These "flatteners" contribute to global business. Work has become international via instant money and informational transactions. The division of labor can be spread to different parts of the world, resulting in virtual team project collaboration in the creation of ideas and products. Factories can be in China or India and the business that owns or contracts with them headquartered in a European Union country with its stock sold on the New York Stock Exchange.

Friedman also emphasizes the noneconomic possibilities for the world. More opportunities exist for education and knowledge, entertainment, political reform, more understanding of differences in the world, opportunities for better lives for societies and individuals who have previously been left out of the world community. He takes a positive view of technology and believes that it and the other "flatteners" are making a better world. For the United States, the United Kingdom, and Western Europe, he warns that these traditional leaders of the world will now have to change vis-a-vis the societies they once colonized or dominated such as China, Russia, Latin America, and Africa. For the United States especially there is no stopping globalization and it must become a part of this new world rather than simply reject it. The competitive edge our country once had in science, math, and engineering is being lost. Increasingly we face "the numbers gap, the ambition gap, and the education gap.... [T]hese gaps are what most threaten our standard of living" (Friedman 2005: 256).

Besides the new technologies of globalization, there is the widening acceptance of capitalism throughout the world. Together technology and capitalism have created the global economy.

## Capitalism, Globalization, and the New Economy

### *Capitalism and Globalization*

Ideally, capitalism is an economic system that attempts to encourage businesses to thrive without government intervention. Private businesses, private corporations, wealth accumulation, labor issues, prices, and competition are sought to be accomplished without government. In reality, there are always several ways that government enters into the economic system, and, intentionally or unintentionally, contributes to the success and the problems of the world of business. It is thought that businesses that succeed are those that the public "votes" through buying their products, stocks, and bonds. The price of goods and services is determined by the market place—and price goes up or down depending on supply and demand. If supply is too high, price tends to go down; if demand is high, price tends to go up. Profit is the goal of capitalism, and it is thought that those who are successful should be rewarded by keeping whatever profit they accumulate. Capitalism assumes that open competition—without governmental interference—is the best way for business and for society.

For many people capitalism and democracy are thought to be intertwined, and that without capitalism there cannot be democracy. Others believe that too much capitalism actually hurts democracy. Many do not distinguish between these two concepts, others see capitalism as inevitable, and for still others, it is the only economic system that works. The issues between capitalism and democracy are highly complex and controversial. As an introduction to these issues, I will express my own understanding.

1. Capitalism is an economic system; democracy is a political system or a type of society.

2. Capitalism does not exist in any pure way. In fact, capitalism always exists "to some extent." Real economic systems are a mixture of capitalism and non capitalism. Government always plays some role in the economic system;

but each nation is different in the degree and the kind of role government plays. Government almost always works for those who are the most wealthy; it also provides important services for the general population through taxes; and it influences economic policies, such as interest rates. Government employs, and government makes contracts with private businesses to perform various projects.

3. In theory, capitalism is a system that encourages economic democracy. People have a greater opportunity to start up and create their own businesses, choose their own livelihood, and keep the profits they earn. Critics of capitalism often argue that instead of contributing to democracy, capitalism actually creates great gulfs between rich and poor. "Free enterprise" may be "free" in the marketplace, but in reality, extreme economic inequality does not nourish a democratic society. Critics point out that the outcome of capitalism is giant corporations, not free competitive businesses; and freedom and opportunity becomes a privilege for only the most powerful and wealthy.

4. Capitalism in its global form developed over many hundreds of years, first in Western Europe and then in the Americas. The nineteenth, twentieth, and twenty-first centuries increased the development of capitalism in much of the world. Always in nations the "free market" is limited through taxes for financing government programs and excesses of capitalism. Some nations—for example, Northern Europe—go further, and increase the number of universal programs paid for by more taxes, and thus less create even more limits to "free market capitalism."

### *Economic "Freedom" and Globalization*

Globalization is truly the inevitable outcome of capitalism. The more capitalistic the economic system becomes, government regulation is minimized, and business is encouraged to find new ways to lower its costs and increase its markets. When technology advances, business recognizes that the world, rather than the nation-state or society, is the economic playground. To succeed, manufacturing, labor, capital, and markets must be determined by international rather national concerns. The outcome, of course, is that corporations become worldwide giants and global competition thrives with the new technology. For example, Tricon Corporation, owner of Kentucky Fried Chicken, Pizza Hut, and Taco Bell, was formed in 1997 as a spinoff of Pepsico. "Since then it has opened more than 3,200 restaurants in over 100 countries and plans to open more than 1,000 more overseas each year for the foreseeable future" (Eitzen, 2009: 38).

If giants find problems, they restructure, selling off parts, adding new parts, being bought out, finding mergers in order to be successful.

Among those who believe in laissez faire capitalism—that is, that minimal government should be involved in the economy—globalization is simply allowing businesses to freely make rational decisions in order to succeed in a highly competitive world. Among critics, without any regulation by government, other values are put aside, and ultimately fewer and fewer benefit while a minority becomes increasingly wealthier. If people in one place do not have the money to buy, then there are other places for business to market their goods.

In the 1960s, U.S. corporations increasingly became involved in the international economy, recognizing the availability of many new markets, the possibility of producing goods overseas, and bringing them back to the U.S. market for domestic consumption. At first the unions did not consider the possible loss of jobs in the United States and initially supported "free trade." Their leadership assumed that the products their union members manufactured would open new world markets. In the 1980s, however, U.S. unions realized that free trade ultimately would cost Americans manufacturing jobs. By the 1990s, U.S. corporations received about 30 percent of their profits from outside the United States, investment of capital greatly increased overseas, and the U.S. manufacturing labor force disapproved (Sher 2000).

Corporations headquartered in the United States sell their goods and services all over the world. In 2008, among the top twenty-five U.S. corporations, ten received more than 59 percent of their revenue outside the United States. Intel is 84 percent global, Coke 74 percent, Exxon Mobil 68 percent, Hewlett Packard 67 percent, Dow Chemical 66 percent. (*Forbes* 2008: 178)

People who have capital are encouraged to invest anywhere in the world ("free capital"). Globalization becomes a constant flow of capital, businesses build wherever they choose, and are able to open businesses in multiple societies. If the government taxes the business too much, if labor demands or environmental problems are too great, the business can always go elsewhere. Pure capitalism, pursues profit, and profit means smart investing wherever manufacturing is low cost and efficiency is up. U.S. corporations invest in companies throughout the world; non-U.S. individuals and companies invest in U.S. real estate, companies, and financial enterprises. Ownership has no boundaries and commitment is to profit rather than to the people living or working in the headquartered society.

Globalization alters the playing field between owners and employees. Capitalism seeks a free market system, competition, non governmental interference, and maximum profit. Much of any business depends on low labor costs. Competition among workers keeps wages down and increases worker dependence and insecurity. This is exactly what globalization does. "There are millions of people who would have your job if you don't want it. So keep quiet and accept what we give you." According to D. Stanley Eitzen, "more than two-thirds of IBM workers, both foreign nationals and U.S. citizens, work outside the United States"(2009: 38). U.S. companies outsource call centers to India at average annual salary of $2,667 (compared to $29,000 in the U.S.) (Eitzen 2009:39). All over the world U.S. companies compete to find labor in order to manufacture goods. Work goes to those who are paid less. Temporary work, migrant labor, women and children, lack of protection by unions, and poor working conditions become the norm. Their wages are much less than those in the more industrial societies, and their jobs are much more insecure. Increasingly, women are closing men out of factories, and women are more vulnerable and less powerful, the traditional family structure is undermined, and wage discrimination is more likely. If they become pregnant, they lose their job (Ferus-Comelo 2006: 43–54). Competition for jobs is between societies, migration from one society to another increases for jobs, corporations shop, and wages decline (Robinson 2007: 101). Mexicans, Central Americans, and Asians are hired for various service jobs, including teachers, nurses, construction workers, household workers, and "sex slaves." In 1990, sixty-four percent of construction workers in California were immigrant labor, farm workers were 91 percent, and restaurant workers 69. Why? Profit only occurs where costs are low, and globalization allows worldwide competition. (Ibid.)

Globalization is truly the outcome of capitalism. As capitalism encouraged technology for labor-saving purposes and ways to create and move products, globalization became almost inevitable. Once the technology was in place, the possibility existed for a worldwide economy, and individuals, businesses, and corporations could participate. U.S. corporations, for example, saw the tremendous advantages of setting up factories in Mexico and then in China. Consumers now existed who could not have afforded products had they been made in the United States. Not only could blue collar jobs be off-shored, highly educated workers became available in India and other Asian countries who could perform white-collar jobs once only reserved for a U.S. domestic workforce.

The world economic system thrived as never before. New technology created highly flexible ways of increasing profit; and as success thrived, even newer technology was developed. People in positions of corporate power realized that the world was changing very fast and they needed to keep up or be lost.

### *The World as an Economic System*

The world has become a global assembly line. Products can be manufactured outside of the industrial West, services can be "outsourced" to Asia, nations compete, overseas workers have more income than ever, and people all over the world can buy American products that originated in America and were produced elsewhere (Ferrrante 2008: 20).

The United Sates imported goods worth $278.8 billion from China in 2006 (up from $3.9 billion in 1985). In that year China exported $91.7 billion to Japan, $44.5 billion to South Korea, $40.3 billion to Germany, $30.8 billion to the Netherlands, and $24.2 billion to Britain.... [The United States'] imbalance with China was $232.5 billion" (Eitzen, 2009: 38).

Free capital is welcomed by governments that see benefits for themselves and for their population, and corporations are pushed by stockholders and executives to find new ways to increase profit. Work changes as computers, telephones, instantaneous use of highly increased storage of information, Internet, e-mails, satellite broadcasting, global media, and fiber optics under the sea. Economic globalization includes "gigantic flows of capital," technology, trade, markets, linking national economies, "huge transnational corporations," and "economic international institutions" (Steger, 2003: 37).

Yet, the world wide economic system may not necessarily create a worldwide society where people are united, where people share culture, and where institutions are established. Perhaps we have created a world where a small wealthy minority have shared knowledge, trading, and communication, making tremendously important decisions that influences everyone and protects their own power and wealth. Perhaps the most powerful actors in the world economy might have little commitment to any particular society or to any world society. Perhaps their culture is not usually seeking democratic values, but capitalist values, a world that allows capitalists to control much of what people do and think. Barbara Ehrenreich (2000) summarizes what we might have become:

> There are 193 nations in the world, many of them ostensibly democratic, but most of them are dwarfed by the corporations that alone decide what will be produced, and where, and how much people will be paid to do the work. In effect, these multinational enterprises have become a kind of covert world government—motivated solely by profit and unaccountable to any citizenry. Only a small group of humans on the planet, roughly overlapping the world's 475-member billionaire's club, rule the global economy. And wherever globalization impinges, inequality deepens. (P. x)

Given this point of view, instead of a world community, globalization can be seen as a worldwide corporate system that seeks profits through treating the world like a huge colony. Individuals typically based in the United States, the United Kingdom, and the EU, control large corporations that have great power in the world. *Forbes* (2008: 184) lists the top twenty-five corporations and lists where they are headquartered:

- ❑ United States: 12
- ❑ United Kingdom: 3
- ❑ France: 3
- ❑ The Netherlands: 2
- ❑ Scotland, Germany, Russia, Spain, and Japan: 1each

Ferrante points out that the ten largest global corporations together represent the sixth largest economy (in revenues) in the whole world (2008:153).

Since the beginning of the Industrial Revolution, people competed within societies. With globalization comes a worldwide labor source that allows large corporations to shop the world to find the cheapest labor supply. This search is like a shopping spree, with businesses moving from one society to another to find the best bargain in labor. Friedman (2005) quotes a Chinese mayor:

> First we will have our young people employed by the foreigners, and then we will start our own companies. It is like building a building. Today, the U.S., you are the designers, the architects, and the developing countries are the bricklayers for the buildings. But one day I hope we will be the architects. (P.36)

Who benefits? Consumers often do. Those who own the giant corporations do. Those who run the giant corporations do. Stockholders do. In the short term, the workers who can get decent jobs do. In the long run, however, labor benefits least: low wages, high expectations, fewer rights, dependence, and constant insecurity. It is very important to recognize that globalization does in fact benefit many people—but not all people. It is only certain localized places where improvement occurs in China, India, Southeast Asia; all over the world the rich and the highly educated continually do well; the vast majority are left behind.

Many dictatorial elites also benefit. They exist in nations that are players in a capitalist world, but do not encourage capitalism within their societies. One very extended family owns and controls most of the oil in Saudi Arabia—oil that capitalist societies such as the United States require.

Although China has only begun to create a capitalistic economy, a government elite still directs and controls the nation's economy and determines who benefits. China itself acts like one national corporation in a capitalist world. It is becoming an important part of the world's capitalistic structure, but private businesses and corporations in China are now only starting to be private and competitive. Globalization brings all nations into a capitalistic world; even those who do not have competitive private corporations themselves.

### *Summary: Economic Globalization*

Globalization is at the center of the world's economy. It is inspired by capitalism, an economic system that tries to encourage competition, private property, and minimal government regulation. Almost all students of globalization agree that a world economy is present and dominant. Capitalism gives great latitude to how corporations work, and encourages them to measure success according to what is profitable. Corporations are multinational; labor source is worldwide. The philosophy is laissez faire capitalism, free markets, free capital, and free labor.

The question now becomes, What has happened in the world beyond the economic? Is the world becoming a single society with its own social structure, culture, and social institutions? Or are nations and societies still independent and a world society far from reality?

## Is Globalization Creating a "World Society"?

### *Society, the Nation, and the World*

Have we created a world society that is competing with and even replacing what we now call separate and independent societies?

A *society* is a form of social organization and the largest social organization that people identify with. Inside of a society, other social organizations exist. Organizations are characterized by five qualities: social interaction, social structure, culture, social institutions, and emotional commitment.

*Nations are political units with physical boundaries, laws, and armies.*

Often societies become nations; over time nations may become societies. Sometimes a nation, such as the former Soviet Union, governs several societies. The Kurds, the Gypsies, and Basques are societies divided across one or more nations. South Africa may have ten or more societies. Africa consists of nation-states whose borders in many cases are no different from the colonies from which they originated after being granted their freedom by their former European masters or after winning it from them in revolutions.

Since 1648 nation-states as we now know them have been established as sovereign independent entities with the right to govern themselves, lay claim to territory, and be recognized by other nations. After World War I, the principle of "self-determination" became widespread in Europe, which, simply stated, adheres to the principle that each society should have its own independent nation. After World War II, self-determination spread to Europe's colonies in Africa and Asia. National self-determination is still an important principle to this day as the breakup of the Soviet Union and its East European satellite states and Yugoslavia have revealed. Then there is the question of Israel and Palestine in the Middle East and of nation-building in Iraq. During the recent 2008 Summer Olympics in Beijing, the question of Tibetan independence became an issue. Many other countries have minority societies that desire self-determination.

What this shows, however, is that existence of nations is fluid, and sometimes fragmentation rather than integration occurs. Nations become smaller; societies become divided. What is the future of societies and nations if the world becomes one society?

### *Is Social Interaction Worldwide?*

A society begins as individuals continuously interact with one another over time. Gradually there is unification and interaction among the individuals. They learn about each other. Patterns and a common history develop. Sometimes there is conflict in this interaction; usually there is cooperation and unity. Occasionally, the conflicts destroy the interaction.

If a world society is to be established, people need to move out from traditional neighborhoods, groups, formal organizations, communities, and societies in order to establish contacts, communication, and cooperation with individuals far away. Boundaries, isolation, distinct networks of people must give way to globalization. World leaders must interact face to face and through phones and video communication. Individuals and families must interact on cell phones, Internet, and occasional long trips. Students must study and make contacts and friendships within a worldwide student community. People must travel, vacation, work, communicate, trade, buy, and sell with individuals and businesses scattered all over the world. They must share knowledge, blog, write, and read each other's ideas, poetry, music, and jokes.

Lots of evidence of worldwide interaction exists. Television, movies, essays, and books get discussed and recommended. We write letters—very often e-mail—to our heroes, political leaders, teachers, professors, friends, news stations, corporate presidents, and medical professionals. We join communities that only exist on the Internet, interacting with neighbors, workers, friends, and family that we never even meet face to face. We can create social movements, protest groups, boycotts, and contribute money to political candidates through the Internet. Planners, lawyers, producers, distributors, salespeople, consumers, advertisers, government regulators are able to interact in new ways because of technology. Inventions, new ideas, and creative products are able to question and spread knowledge of what is produced.

In 2005, over 97 million trucks and personal vehicles crossed the border from Mexico to the United States (Eitzen 2009: 38). Migration due to natural disaster brings large numbers of families across borders. The desire for a decent job brings workers from Africa to Europe and Mexicans to the United States. Some workers leave their home society only temporarily, and send their earnings back. Eichen (2009: 39) predicts that approximately $20 billion is sent back to Mexico by workers in the United States. Some of the migrant workers eventually go back to their homelands. Some may only work seasonally while others seek permanent residence. Both groups, those who settle in the host society and those who simply earn their wages in it often encourage others to come. There is constant active interaction.

Some workers end up as modern slaves. These individuals are encouraged to migrate and are promised a good job that will help them get ahead. They may pay thousands of dollars for transportation and employment—but then are dumped into horrible situations where they work for almost nothing.

Large populations are on the move, leaving established, rural, traditional societies for modern urban societies. The international economy as well as transportation and communication technologies make such movement possible. In 2007 over 42 million foreign visitors entered the United States (excluding day-trippers); over 25 million U.S. travelers go to foreign countries each year (U.S. Department of Commerce, Office of Travel and Tourism Industries 2007).

Language barriers are also breaking down. This is especially true as the world economy brings salespeople, CEOs, managers, scientists, and workers together. They *interact.* They use cell phones, blackberries, and I-phones, and speak in English. Indeed, fewer and fewer languages now exist. And in many of the cities where these people work, there a increasing numbers of foreign born. 59 percent of Miami's residents are foreign born. Toronto has 44 percent, Los Angeles 41 percent, Vancouver is 37 percent, and New York City is 36 percent (Ferrante 2008: 243).

Isolation is almost impossible for most of us. Societies and groups that used to be highly local have extensive worldwide social interaction. Through migrating, finding work, communicating, interacting, selling, buying, sharing, and arguing, people can know and befriend anyone in the world. Increasingly, the word "foreigner" becomes much less meaningful; citizens of nations become citizens of the world.

### *Is There a Worldwide Social Structure?*

Out of social interaction, social patterns develop.

Over time worldwide interaction creates a world social structure. Individuals, nations, businesses, classes, corporations, societies, and communities, are placed in positions related to other positions. All are locked in, each having a role in the world, an identity, a perspective, a rank based on power, prestige, and privilege. The world's structure brings individuals, groups, formal organizations, communities, and what used to be societies into a huge interdependent set of positions. The structure of the world separates societies into specialized roles; interdependence is created through a division of labor.

Social class becomes international. Billionaires control large corporations, the factory workers in the United States lose their place as the world's most productive and best paid, and some become part of the worldwide unemployed class, moving to another town, city, society, or continent to find work. Some owners and stock holders become very wealthy and powerful. Some create new businesses, some find places in corporations, and some become professionals. Ferrante (2008: 204), using data from the Internal Revenue Service 2007, points out that the 9,656 tax returns over $500,000 account for twenty percent of all reported taxable income in the United States (the top .001 percent earn 4,000 times more of the poorest twenty percent of all the tax returns, and that does not include those who report no taxable income). Middle class workers in the United States become less secure in the world. Emerging middle classes in Mexico, India, and Southeast Asia are hired for white collar jobs, buy homes and cars, only to lose their jobs when corporations find cheaper workers in another society. Noncitizens become a source of cheap labor. As a social class, they are more dependent and exploited, lacking "citizenship and civil, political, and labor rights . . . controllable" (Robinson, 2007:102). A class structure throughout the world becomes very important.

Besides a class structure, there is also a structure of nations. Here the social structure means that nations exist in relation to other nations. Nations have positions in the world, roles, identities, power, prestige, and privilege. The United States continues to dominate the world's economy. Capital and corporate headquarters remain there. The European Union, China, Russia, India, and Japan are quickly becoming powerful. Throughout the world are nations that we might call developing, industrial, nuclear, sources of natural resources, financial, leaders, colonies, for example. The world's social structure among nations includes economic, military, political, geographic, and historical aspects. Division of labor, cooperation, specialization, interdependence, alliances, security and environmental aspects become elements of the international structure of nations.

Emmanuel Wallerstein (1995) researched the rise of an international structure of nations beginning in the sixteenth century. He called it a "world system." Some nations, he argued, become "core" nations—the most industrialized, most wealthy, highly diversified, and having a strong stable government. Today they would be the G8 nations, for example: United States, France, Great Britain, Germany, Italy, Canada, Japan, and Russia. Some nations are "peripheral," having such characteristics as a very large population and underdeveloped and unstable economies that are also dependent on a very few commodities or single mineral resources. As a result, they are dependent on the core nations. Much of Africa, several nations in Latin America, Burma, Cambodia, North Korea are examples of peripheral nations. Between the core and the peripheral are "semi-peripheral" nations. These have some wealth and diversification, but in contrast to core nations, they have much greater social inequality and less economic power in relation other nations. India, China, Mexico, Indonesia, and Brazil are examples. Given this structure of core, peripheral, and semi-peripheral, nations are arranged as specialized, interdependent, some being losers and some winners, some highly dependent, and others in the position of helping or exploiting. Each has a role in the world, each is ranked in power, privilege, and prestige, and each becomes an identity within the world.

Durkheim emphasized that modern society would eventually bring people together, not by a common culture as much as a structure, an interdependent division of labor. It seems that a world structure is being built; whether or not the world can be called a society is debatable. It is no longer useful to simply understand structures within what were called "societies." They must now be seen in a global context, as part of a global social structure.

### *Is There a Worldwide Culture?*

Culture is the second social pattern that I would like to focus on. Culture is developed through social interaction and is necessary for society. When many people think of culture, they look at the material aspects such as the food people eat, their tools, their pottery, the goods that they produce and consume, their garbage, their buildings, their artwork, and so on. In this book, we have defined culture as a *shared perspective* in society, a set of *beliefs, values, and norms.* Material culture is important too, but the material culture is really the product of a people's shared perspective. For this reason we have defined culture as a perspective, a people's view of their world.

There is no one simple world culture. Besides the various cultures of individual societies all over the world, we might argue that there are cultural themes—or trends—in the world, and individual societies differ greatly in the relative importance of these themes. These might be called the "modern-global-economy" culture, the "traditional-religious" culture, and the "human rights-democracy" culture. Every society is a mixture of these themes, every society is different in their emphases, and every society responds to these according their own historical experiences. These three are very different directions in culture, and they are often in conflict with one another.

The *modern—global—economy* cultural theme has been described thus far in this chapter. In addition to other Western nations, the United States has taught the world that capitalism, globalization, and modernization are the inevitable future. Competition, material success, consumerism, progress in this world, belief in individualism, computerization, faith in worldwide bureaucratic corporations, business, free trade, free labor, free capital, technology, science, and rationality are part of this theme, and they exist in almost every society to some extent. We believe in material products and in accumulation of wealth, and we read, talk, and listen about spending and saving. Successful business figures become ideal heroes. Famous and rich entertainers, sports figures, music stars, poker champions, computer and Internet wizards are respected. Those who have the expensive cars, boats, houses, playthings, and rich tastes are models many would like to emulate.

There is a second cultural trend. Let's call it "traditional-religious." Just as modernization is a reaction to tradition, people who are traditional have become a reaction to capitalism, globalization, and modernization. Beliefs are often integrated with religion. Values are more spiritual; tradition and community rather than individualism are more important. Science and secular society is suspect, material progress in this life is not important, history of community is known and respected. Although this cultural theme emphasizes values such as family, commitment to others, and a strong sense of right and wrong, traditional culture also tends to become ethnocentric, isolationist, sometimes intolerant of differences, usually certain of what they believe. Those who are to be respected are those who are in agreement with the culture. Rituals, rules, morals, traditions are accepted, and become guides in their lives. Material culture might include church, religious symbols, religious schools, cemeteries. Priests, ministers, rabbis, prophets, televangelists, and village elders are to be respected and heard. Tradition brings direction, meaning, truth, and feeling of community. Much of the nonindustrial world has this traditional cultural thrust, but in the industrial world there are also many who are critical and often reject the cultural thrust of capitalism, economic globalization, and modernization, and turn to tradition.

Here are two cultural themes. They are often in conflict. The values, norms, and beliefs are very different. Indeed, each develops through criticism of the other. Both are important, and they will continue to clash for a long time. The mixture of the two in each society will continue to encourage distinct individual cultures.

There is a third cultural theme in the world. It is probably less noticeable than the first two, but its presence is increasing. I call it a "human rights-democracy" perspective. There is a recognition that humanity is in danger, that unless we change what we are doing we will not have much left of a decent world. Individuals who exemplify this believe that the environment and the animal kingdom are being exploited, spoiled, and destroyed. Others act on behalf of political prisoners all over the world. Others work to censure and punish tyrannical leaders. There are individuals and groups who have a different agenda from tradition or global capitalism. Some are religious people who actively contribute to the future of humanity. Some are political liberals; some are conservative. They are activists and volunteers helping other people and improving the planet. Some are doctors, lawyers, teachers, mechanics, social workers, researchers, nurses, business leaders, who help abused women and children, who care for young people, who defend those who are imprisoned, who inspire private businesses, and who work to help addicts, homeless, sick, and those who are poor. This is a cultural theme that believes that all humans have rights, and it is everyone's responsibility to see that these rights are protected and nurtured.

This human rights–democracy cultural theme is emerging beside the modern-global-economy culture and tradition–religion themes. It has certainly penetrated into what is an ever changing United States culture. Along with human rights, there is a democratic aspect that emphasizes freedom; equality of opportunity; limited power over government, military, business, and religion; encouragement for diversity; and respect for minorities. Although we sometimes compromise and violate these principles in the United States, we also debate endlessly about what they mean, and many of us recognize that this is also a central part of American culture.

Which of these directions will ultimately be the prevailing factor in a world culture? That culture that stems from economic globalization and capitalism, that culture which stems from tradition and religion, or this newer awareness, this culture of human rights and democracy? Their presence now continues to bring diversity to the world. Almost all societies are mixtures of the three cultural themes, and as long as the traditional-religious theme thrives, distinct societal cultures will continue.

### *Are There Worldwide Social Institutions?*

When a society is established, it needs institutions as well as a social structure and culture. Social institutions are usually created (and defended) by those individuals who dominate the social structure. Society exists only if institutions are established and work effectively. By definition, it takes time for institutions to be created. They are the ways a people are supposed to follow. They are taken for granted and defended. They are patterns for organization used to deal with political, economic, religious, legal, military, media, health care, entertainment issues.

Various countries have banded together to form economic institutions that have a global reach. Examples include free trade, outsourcing, market system pricing, multinational corporations, international banking, overseas manufacturing, and private property. The World Bank and the International Monetary Fund are controlled by five wealthy nations: the United States, the United Kingdom, France, Germany, and Japan. These institutions loan money, advise on economic issues, aid developing nations, and the like. They are also "motivated by an extraordinary devotion to the free-market model" (Scher et al. 2000).

Other examples exist. In 1994 the World Trade Organization (WTO) was created. It functions like a court in conflict resolutions among national laws that violate free trade. The WTO, for example, will attempt to change the policy of a nation that promotes trade barriers with another nation (Scher et al. 2000). The European Union (EU) and the North American Free Trade Agreement (NAFTA) are part of global trend that may form economic unions that are broader in scope.

On the labor front, the International Labor Organization (ILO), now part of the United Nations, was formed to penalize nations for substandard labor conditions. Unlike the global organizations that foster free trade, however, the ILO is yet to become effective nor is its authority recognized by many countries, the United States being a conspicuous example.

The United Nations has become an important international institution, helping people in crisis and poverty, contributing peacekeeping and humanitarian assistance, protecting victims of ethnic cleansing, in preventing civil wars, and working hard to improve education and health care for its member nations. Its headquarters in New York City is where nations can discuss, argue, cooperate, and express their concerns.

The United Nations has even gone to war in such places as Korea and Bosnia. It has attempted to keep the peace in the Middle East and elsewhere. The UN has been very active in human rights, beginning with the adoption of the "Universal Declaration of Human Rights" in 1948. The UN has long promoted human rights in the world. It has developed international law and, through the International Court of Justice, it has attempted to bring actions on issues such as environment, refugees, organized crime, drug trafficking, and AIDS.

Critics of the United Nations contend that it is not an effective institution, that it is unable to deal with the major issues facing the world, that voting is too often "political" and not just, and that financial contributions are not fair. Some governments and many individuals are suspicious of an international government. Few national governments wish to give up their sovereignty to a world government—unless, of course, it is the "national interest" at the expense of a rival nation.

Other institutions exist like the UN that are global in scope. The International Court of Justice, located in The Hague, Netherlands, and affiliated with the United Nations, prosecute and judge an increasing array of international crimes, including crimes by the leaders of rogue states who have violated human rights. These courts also have faced some resistance from the United States and other countries over matters of national sovereignty.

NATO—the North Atlantic Treaty Organization—began as a Cold War–era mutual defense organization now has expanded its role from Western Europe to Eastern Europe and further. The Geneva Convention goes back to 1925. Its purpose is to regulate the way war is conducted, prevent war crimes, and to guarantee the humane treatment of prisoners of war.

Religious institutions exist in almost every society. In most societies there is a dominant religion, in some there are two or three dominant and competing religions, and occasionally there is a number of highly diverse religions in the society. Each religious community has its own institutions—separation of church and state or state religion, special holidays, church/synagogues/mosques, rituals, special days, sacred texts, type of prayers, role of clergy, gender roles, life-cycle events, for example. These are the pillars, the grooves that make each religious community successful and unique. Each directs the practices of the members. The result of religious institutions do not bring about a world society, but a large number of distinct religious communities, competing, often in conflict, and each arguing they are true and sacred. *Christianity* (Catholicism, Eastern Christianity, the multitude of Protestant denominations, Mormonism, and others), *Buddhism* (Japanese, Vietnamese, Indian, Cambodian, and other denominations or sects), *Islam* (Sunni and Shiite), *Hinduism* (at least four major denominations), and *Judaism* (Orthodox, Conservative, Reform)—is only the beginning of distinct religious communities, each with their own institutions. The existence of highly diverse religions, continuously splintering, each holding onto sacred institutions and beliefs, does not work to unite the world.

The other social institutions, such as the artistic, entertainment, educational, scientific, journalistic, media, have become worldwide largely because of the communication revolution and the leadership of the United States. CNN, The New York Times, The Economist, BBC, Google, Internet, cell phones, blogging are some of the institutions that have become international. The world of music—the symphony orchestra, the television, satellite radio, worldwide music tours, I-Phones, DVDs, rock, country, jazz, and electronic music—are examples. Casino gambling, football, golf, baseball, tennis, and soccer are international contests. The Olympics is an important worldwide social institution, where almost all nations participate, and people all over the world become involved through television. Indeed, a televised Super Bowl contest can involve the world in a single game. Nobel prizes for Peace, Chemistry, Economics, Physics, and Medicine are important ways people are honored. Movies, books, newspapers, radio, television, and computers communicate news, information, history, analysis of issues, and different views of life. Isolation becomes much less possible. World educational institutions are becoming alike and interrelated. Students are able to transfer anywhere in the world, subject areas are similar, academicians and scientists continuously exchange their research and writings.

There are important world institutions and they contribute to world unity. The most definite and powerful institutions are economic. Political and religious institutions are much less international. Other institutions continue to be national or societal, but over time, they have moved toward common institutions. Conflict, fear, and self interests assure the impossibility of worldwide military institutions.

### *Is There Commitment to a World Society?*

Actors in a society must feel like citizens. They need to have a degree of loyalty, an allegiance, a mechanism in which their identities are embedded in their society. Society exists more than simply the force of government. Interaction and social patterns normally encourage this. That said, is there a growing commitment for a world society, to all human beings throughout the world? Is there a view of oneself as responsible to the whole world, a rising tide of world citizens?

Those who are actively involved in the world economy increasingly may come to see themselves as citizens of the world. There is a loyalty to making wealth that can override national ties. On the other hand, those who flee from famine, war, and/or unemployment, must leave their geographic anchors and plant their community elsewhere, find a new identity where they can be secure, or become lost without community. Becoming citizens of the world, where they feel part of a world community, is not likely.

Worldwide social interaction brings many people who see one another as persons, rather than simply Americans, English, Chinese, or Russian. For many there is a commitment to a world society. Their efforts have turned to the future and the necessity for changing their ways. Global warming, environmental destruction, disrespect for human rights are real issues because people feel a commitment to a world community.

For the vast majority, however, traditional societies, local communities, and nations are still the most important loyalties, and commitment to a world society is not much of a reality. It is not easy for people to give up nation-states or traditional societies for an abstract humanity; it seems too distant, too strange, perhaps less secure in a vast world society.

Only structural interdependence, cultural agreement, and established institutions can bring commitment to a world society over time. It is not simply communication, wealth, trade, and struggle to seek work. Without commitment to a world society there may be interaction, but not really a society.

### *Summary: Is There a World Society?*

Alexander the Great sought a world society. The Persians did. Rome did. The Christians and Mongols did. Islam did. The Catholic Church did. England did. Nazi Germany did. The Soviet Union did. Fundamentalists today seek it, and sometimes it seems that the United States seeks it. Capitalists seek it in the economic arena.

But any unity of a world order is not likely to arise through empire, but through social interaction, modern technology, economic interdependence, and widespread communication. Any unity that is felt is not from the top down, but from the bottom up. It is in fact the ongoing interaction that brings us all together. It is only the ongoing development of a social structure, culture, and social institutions that allow us to organize a world society. It is the feeling of becoming a world citizen that ties us to a world society. We are probably far from a world society. What still limits a world society is a continuous fear of change: belief in tradition, local communities, societies, nations, and social movements; criticism of laissez faire capitalism; and fear of international government.

## Is Globalization Good for the World?

### *A Positive Picture*

Defenders of globalization argue that everyone, including the disadvantaged societies that do not now partake of the wealth generated by capitalism, will gain through industrialization and capitalism, and ultimately democratic cultures and governments, will come to dominate all the world. Throughout the nineteenth and twentieth centuries, many Western societies exhibited a system of beliefs called "modernization," a view of economic and social progress through industry, individualism, and science. Modernization, capitalism, and globalization influence one another. Technology, a higher standard of living, an enlarged educated—and democratic—middle class will be good for all societies. Globalization, it is believed, will bring opportunity for everyone; it will help prevent disease, poverty, intolerance, and tyrannical rulers. Over time the world will become democratic, not simply by voting, but by respecting the individual, encouraging freedom, limiting power by government, and bringing opportunity.

Defenders of globalization do not simply believe that only the rich benefit. Instead, almost everyone benefits. Those who own or produce goods have a larger market to sell their goods. Those who buy the goods are able to improve in their standard of living. Those who have opportunities to work are able to have a more optimistic view of their future. Education, better health care, decent housing, and increased choices become more available. Human rights can be extended, and fear can be lessened. Traditions that have oppressed many can be limited or eradicated through being publicized, censured, and changed. Mass poverty, population explosion, energy misuse, global warming, and the destruction of the environment can be dealt with in a more rational way since the policies can become cooperative and international.

Hopefully, making access to a diversity of cultures, individuals will become more knowledgeable, will be able to understand people who are different from them, and become more tolerant and respectful of those outside their own culture. Giddens argues that this is why globalization is dangerous to fundamentalists for it "lies behind the expansion of democracy" (2000: 23).

The Internet allows for millions of people to oppose injustice and organize information and movements to change serious problems. If certain societies violate labor standards, there can be an organized response. If corporations continue to destroy the environment, a worldwide response can do something opposing those who are responsible. Many international social movements and volunteer groups such as Greenpeace, Habitat for Humanity, Doctors without Borders, Amnesty International, The Salvation Army, the Red Cross, and hundreds of private and public groups change people's ideas, do good deeds, and influence change.

Educators, scientists, clergy, social workers, psychologists, and business people are moved to volunteer to share their skills to those who can benefit. Globalization informs people about the serious problems in the world and inspires many to do something. Global travel and media bring this awareness; and global interaction and travel bring organization and power.

This view of the future is, in part, *ideology,* a rationalization for exploitation by Western and industrial societies, and protection of wealthy individuals. In part, however, it is also a *theory* that scholars and optimistic political leaders honestly believe in and hope for a better world.

> What we now need is not just an alliance *against* evil, but an alliance *for* something positive—a global alliance for reducing poverty and for creating a better environment, an alliance for creating a global society with more social justice (Stiglitz 2000: A21).

## A Critical View

The benefits of modernization, capitalism, and globalization are probably much more complex and optimistic than their defenders argue. Tradition, community, and values other than economic ones still have an important purpose. The rise of a modern and global world society is not necessarily the progress that defenders promise.

*Globalization means change, the stakes are very high, and mistakes become giant ones.* Everyone is influenced. Huge bureaucratic impersonal corporations that exist primarily for profit make decisions that affect everyone. The problem, of course, is that fewer and fewer individuals are making more and more decisions that affect more and more people. Corporations grow, merge, or go out of business, affecting workers and other companies all over the world, moving offices or jobs to other societies in order to compete effectively. Then when a new nation offers cheaper and better labor supplies, these corporations move again, leaving the second nation's workers unemployed. These are big changes because of the numbers of people who win or lose, and because those who win or lose are without any control over their own future, and any attempts to create democratic institutions become hollow.

*Globalization is especially difficult for workers.* Insecurity and helplessness become more prevalent as people must alter their whole lives on decisions that are made by a few people in another part of the world. Certainly this has been part of capitalism from the beginning, but globalization magnifies the problems. The future of families, communities, societies, and nations depend on competitive wages, competitive skills, and luck. Even those who work in prosperous corporations can find themselves out of a job simply because someone else in the world does it better or cheaper. Friedman warns that job security is not a high value in the global economy: "If you are an American, you better be good at the touchy-feely service stuff, because anything that can be digitized can be outsourced to either the smartest or cheapest producer, or both" (2005: 15). By outsourcing and creating factories in other countries, globalization has already "cost" many blue collar jobs in the United States. Service jobs are being outsourced as well, being replaced by computers and less-educated professionals. Globalization creates lots of problems for *workers* who end up pawns in a giant world. Robinson (2007) summarizes what workers have become, both migrant workers, and citizen workers:

> Global elites and dominant groups around the world have imposed new capital-labor relations on all workers based on oppressive new systems of labor control and cheapening of labor. This involves . . . devalued labor, including subcontracted, outsourced, and flexibilized work, deunionization, casualization, informalization, part-time, temp, and contract work replacing steady full-time jobs, the loss of benefits, the erosion of wages, longer hours. (P. 104)

Robinson continues, describing workers as "commodities" that can be positioned like capital goods "throughout North America and utterly dehumanized in the process" (2007: 104).

*Globalization increases specialization among nations of the world.* Oil in the Middle East, natural resources in Africa, factories in China and India, banks in the United States and Europe threaten situations where-by nations are not able to independently be successful. This is an important trend. Even U.S. defense goods are contracted overseas; we use airplane parts produced overseas, outside of American society.

*Although no one really knows exactly how the social structure of the world will become, undoubtedly, there will be great poverty and great wealth.* Pessimists see a loss of net decent jobs, increasing exploitation of the poor, a small minority of billionaires increasing their wealth, a class of educated knowledgeable intelligent meritocracy, and a great number of people who have great expectations without realistic opportunities. Capital—the investment of creating wealth—becomes the weapons of war in the global economy. With an international class system, there might be an opportunity to aid the poor societies, but it is much more likely that the international upper class will be free to control their income and investments without being taxed or committed to a society or nation. We might call the world capitalistic or democratic or free but the reality simply may be increasingly characterized by the survival of the fittest (or luckiest). *The problem is that profit trumps healthy environment, profit trumps workers, and profit trumps consumers, and as long as there are so many ways for corporations to move around the world, they do not have to care about people in communities or societies.* Using taxes for human services will undoubtedly become less.

*Critics argue that "free capital" and specialization inevitability create economic volatility.* Great success also creates great failure. Investment is in the hands of a few. Investment is welcomed; investment leaves. Instability and recession in one part of the world influences the rest. Globalization increases the possibility of worldwide recession. Worldwide financial crises, unemployment, new technology, population movements, changing markets, and increasing open competition will mean big stakes and continuous ups and downs in the world economy.

*As the economy becomes global, national governments are not in the position to regulate corporations and businesses.* Globalization gives the corporation many ways to get around environmental regulation, finding cheaper workers, and selling unsafe or inferior products. From the perspective of corporations, regulation is nondemocratic, too inefficient, and too unprofitable. In addition, crime becomes more global, and it is more difficult for national governments to control it. New kinds of crime arise from globalization. According to the Central Intelligence Agency (CIA), about 18,000 to 20,000 people each year are trafficked to sex slaves, and nearly "30,000 women are trafficked to the United States to work in sweatshops or as domestic servants" (Eitzen, 2009:37). Crime exists online, as well, with "too many dark dangerous corners and too little law and order" in the form of pirated DVDs, gambling, porn sites, drug dealers, scam artists, and terrorists (Sager et al. 2002: 740).

*Political globalization may mean even more crises, may increase conflict and problems rather than create a peaceful democratic world.* The decline of the sovereignty of nations does not necessarily mean greater democracy for the world. More political coalitions, threats, fear, one or few powerful nations, powerful corporations, armies, ruthless individuals or groups may be much worse than we face today. Sometimes, on a highly abstract level, it would be nice if we all could come together and unite in peace; but there is too much evidence that a real-world government would not be any better, and perhaps worse. Ending the diversity all over the world, ending commitment to our present societies, ending our different traditions and our own interests, do not mean a democratic world government.

*Many critics of globalization focus on the "homogenization" of the world.* Instead of rich diversity, we are arriving at a time where everything is becoming the same. It does not matter where you travel, a corporate sameness exists on almost every corner. McDonalds, Starbucks, GAP, Dominos—all brands that have spread across the world. Diversity is reduced as worldwide businesses replace local businesses. The color and character of a city becomes simply gray. Popular culture, influenced greatly by the United States, has spread all over the world. Someness thrives in music, dance, toys, food, drinks, clothes, heroes, slang, film, art, literature, gadgets, and television.

*Globalization means worldwide epidemics and other dangers.* Planes, trains, ships, buses, cars—every means of transportation that puts people together and allows them to move from one country to another—increases, the probability of catastrophic worldwide communicative disease.

> HIV-AIDs has infected some 60 million people worldwide since crossing over from chimpanzees in the 1960s. Some thirty new diseases have cropped up since the mid-1970s, causing tens of millions of deaths, including SARS (severe acute respiratory syndrome), West Nile virus, and Ebola hemorrhagic fever (Eitzen 2009:40).

In addition to disease, globalization brings with it unsafe foods, toys, and other goods created in one place and spread throughout the world.

Free movement in the world also encourages *terrorism.* The technology that we rave about also makes it much easier for a small number of people to murder thousands even millions. A small group or sleeper cells do not need armies to terrorize the population. Terrorism becomes easier to succeed when specialization exists, when migration of peoples are made easier. Indeed, to some extent, globalization, modernization, and capitalism actually become targets for terrorists . Those who seek traditional societies and those who are critical of what modern society becomes are encouraged to tear down what they see as threatening.

## Conclusion: Globalization and a World Society

A trend toward a world society may be inevitable, especially in regard to social interaction, communication, and economic activities. Yet, a trend does not necessarily mean that someday a full-blown world society will be reality. There are forces that make globalization less inevitable, and not as dramatic.

*Nationalism* still competes with globalization; people are still committed to belonging to a particular nation or society. Many societies still seek nationhood. Independence, local communities, a desire for diversity, are limiting the growth of globalization. *Ethnocentrism* limits globalization. One's identity is often more important than economic interests. Giving up or criticizing one's own ethnicity or national origin is not easy for many.

*Fear of change itself* will bring reaction against globalization and perhaps retard its acceleration. People become used to certain societal economies, armies, governments, laws, institutions, cultures; it is often difficult to alter whatever one already has, even if he or she has little. Social patterns tend to hold on.

*Tradition* also limits globalization. For many people, scientific discoveries, materialism, rapid change, and efficient organization are not very attractive, and a people's history pulls them away from what the world is becoming. They see that globalization is not right, and that it is not the way they have done it before. It is not the way their parents, their people, their God, their family, their community have always lived. They are tied to village life, rural life, familiar occupations they have had in their family, the crafts they have made with their own hands rather than through modern technology.

*Human agency and social movements* also limit globalization. It is globalization that makes it possible for individuals and groups to welcome *or oppose* globalization. Many right-wing militias fear and act against globalization because they fear national independence; left wing groups act against globalization because they fear capitalism is shadowing all other values. Labor, religious groups, ethnic groups, feminist groups, small business organizations, and the middle, working, and disadvantaged classes who fear for their future become active critics of globalization.

The attempt to make conclusions about globalization is very difficult. Tentatively, however, I conclude the following:

- Capitalism and technological breakthroughs is the basis of economic globalization.
- Economic globalization is the most advanced type of globalization.
- The ongoing social interaction throughout the world is a very definite trend toward a world society.
- A world-level social structure is highly developed.
- A world culture is not inevitable. There is no simple world culture. Three cultural themes exist and are in conflict. Each society has a distinct mixture of these trends, and as long as a tradition-religion trend exists, world culture is not inevitable.
- Although some global institutions have been created, especially in the economic world, societal and national institutions continue to thrive. Religion, tradition, history, ethnocentrism, and insecurity limit development of global institutions.
- Probably most important limits to a global society is a lack of world commitment; we still are largely committed to communities, societies, and nations.
- Both defenders and critics of globalization are adamant, and it is is very important to understand both sides objectively.

## Questions to Consider

1. What is your view? Is the world really becoming a single society?
2. Do you believe that someday there will be a world government? Is it something we should fear or welcome?
3. Is the United Nations a worthwhile international institution?
4. Do you believe that the United States will benefit or be hurt by economic globalization?

5. What is capitalism? Is globalization an inevitable result of a capitalistic world?
6. Is globalization ultimately a step toward environmental responsibility?
7. Is popular culture becoming homogenized or diverse because of globalization?

## References

**Altheide, David.** 2006. *Terrorism and the Politics of Fear*. Landham, MD: Rowman & Littlefield.

**Brecher, Jeremy, and Tim Costello.** 1994. *Global Village or Global Pillage: Economic Reconstruction from the Bottom Up*. Cambridge, MA: South End.

**Dorman, Peter.** 2000. "The ABCs of the Global Economy," *Dollars & Sense: The Magazine of Economic Justice,* March/April (http://www.dollarsandsense.org/archives/2000/0300collect.html).

**Ehrenreich, Barbara.** 2000. "Forward." Pp. ix–x in *Field Guide to the Global Economy,* edited by Sarah Anderson and John Cavanagh. New York: New Press.

**Eitzen, D. Stanley.** 2009. "Dimensions of Globalization," pp. 37–41 in *Globalization: The Transformation of Social Worlds,* edited by D. Stanley Eitzen and Maxine Baca Zinn. 2nd ed. Belmont, CA: Wadsworth Cengage Learning.

**Eitzen, D. Stanley and Maxine Baca Zinn**, eds. 2009 *Globalization: The Transformation of Social Worlds.* $2^{nd}$ ed. Belmont, CA: Wadsworth Cengage Learning.

**Faux, Jeff.** 2004. "Nafta at 10: Where Do We Go From Here?" *The Nation,* February 2, 2004, pp. 11–14.

**Ferrante, Joan.** 2008. Sociology: A Global Perspective. 7th ed. Belmont, CA.: Wadsworth Cengage Learning.

**Ferus-Comelo, Anibel.** 2006. "Double Jeopardy: Gender and Migration in Electronics Manufacturing," Pp. 43–54 in *Challenging the Chip,* edited by Ted Smith, David A Sonnenfeld, and David Naguib. Philadelphia: Temple University Press.

**Foner, Nancy.** 2005. *In a New Land: A Comparative View of Immigration*. New York: New York University Press.

**Forbes Magazine.** 2008 "The Biggest Companies in the World." *Forbes,* April 21, pp. 148–208.

**Friedman, Thomas L.** 2005. *The World Is Flat A Brief History of the Twenty-First Century*. New York: Farrar, Straus & Giroux.

**Giddens, Anthony.** 2000. Runaway World: *How Globalization Is Reshaping Our Lives*. New York: Routledge.

**Goldberg, Michelle.** 2006. *Kingdom Coming: The Rise of Christian Nationalism*. New York: Norton.

**Harper, Charles, and Kevin T. Leicht.** 2002. *Exploring Change: America and the World*. 4th ed. Upper Saddle River, NJ: Prentice Hall.

**Huntington, Samuel P.** 2000 "Culture, Power, and Democracy." Pp. 3–13 in *Globalization, Power, and Democracy*, edited by Marc F. Plattner and Aleksander Smolar. Baltimore: Johns Hopkins University Press.

**Kotlikoff, Laurence, and Scott Burns.** 2004 *The Coming General Storm: What You Need to Know about America's Economic Future*. Cambridge, MA: MIT Press.

**Lichtenstein, Nelson.** 2006. *Wal-Mart: The Face of Twenty-First Century Capitalism*. New York: New Press.

**Merry, Sally E.** 2005. *Human Rights and Gender Violence: Translating International Law into Local Justice*. Chicago, IL: University of Chicago Press.

**Penn, Michael L., and Rahel Nardos.** 2003. *Overcoming Violence against Women and Girls: The International Campaign to Eradicate a Worldwide Problem*. Landham, MD: Rowman & Littlefield.

**Plattner, Marc F. and Aleksander Smolar.** 2000. "Introduction." Pp. ix–xx in *Globalization, Power, and Democracy*, edited by Marc F. Plattner and Aleksander Smolar, Baltimore: Johns Hopkins University Press.

**Robinson, Wiliam.** [2007] 2009. "Globalization and the Struggle for Immigrant Rights in the United States." Keynote Presentation for "El Gran Paro Americano II Immigrant Rights Conference, Feb. 3–4, 2007, Los Angeles. Pp. 99–105 in *Globalization: The Transformation of Social Worlds*, edited by D. Stanley Eitzen and Maxine Baca Zinn. 2nd ed. Belmont, CA.: Wadsworth Cengage Learning.

**Sager, Ira, Ben Elgin, Peter Elstom, Faith Kennan, and Pallavi Gogoi.** 2002. "The Underground Web: Drugs, Gambling, Terrorism, Child Pornography: How the Internet Makes Any Activity More Accessible than Ever." *BusinessWeek*, September 2, pp. 67–74.

**Scher, Abby**, and the *Dollars and Sense* collective. 2000. "The ABCs of the Global Economy." *Dollars & Sense: The Magazine of Economic Justice*, March/April (http://www.dollarsandsense.org/archives/2000/0300collect.html).

**Steger, Manfred B.** 2003. *Globalization: A Very Short Introduction*. New York: Oxford University Press.

**Stiglitz, Joseph E.** 2002. "Globalism"s Discontents. *American Prospect* 13(1): A16–A21.

**U.S. Department of Commerce**, Office of Travel and Tourism Industries. 2007. "International Visitation in the United States" (http://www.tinet.ita.doc.gov/outreachpages/inbound.general_information.inbound_overview.html).

**Wallerstein, Emmanuel**. 1995 *Historical Capitalism with Capitalist Civilization*. London: Verso.

**Rifkin, Jeremy.** 2004. *The End of Work: The Decline of the Global Labor Force and the Dawn of the Post-Market Era.* New York: Penguin.

**Suarez-Orozco, Carola.** 2001. "Immigrant Families and Their Children: Adaptation and Identity Formation. Pp. 129–139 in *The Blackwell Companion to Sociology,* edited by Judith Blau. Oxford, UK: Blackwell.

**Waldinger, Roger David, and Michael I. Lichter.** 2003. *How the Other Half Works: Immigration and the Social Organization of Labor*. Berkeley, CA: University of California Press.

**Witt, Griff.** 2004. "As Income Gap Widens, Uncertainty Spreads." *Washington Post,* September 20, pp. 00–00.

# Why Study Sociology?

## Understanding, Questioning, and Caring

### Concepts, Themes, and Key Individuals

- ❑ Liberal arts education
- ❑ Sociology and democracy
- ❑ Sociology and social order, social change, and social problems
- ❑ Sociology and social relationships, social power, and social class
- ❑ Sociology and caring about society and the individual

In the final analysis, it may be true that ignorance is bliss. It may be true that people should be left alone with the myths they pick up from interacting among themselves. It may be true that a liberal arts education that does not have immediate practical value is worthless.

I do not believe any of the ideas above, but I wonder about them a lot.

I believe that the university should be a place where one prepares for a career, but the university must also aim toward providing an education from which students can better understand themselves, their society, the world, and the universe. If a university education is ultimately an attempt to encourage this broader understanding and to motivate students to wonder, investigate, and carefully examine their own lives in relation to their society, then sociology is one of the most important disciplines.

The whole purpose of sociology is to deal with the questions posed in this book. It is to encourage students to carefully and systematically study an aspect of their lives that most people only casually and occasionally think about in a critical fashion. Sociology lets people understand what culture is and recognize that what they believe is largely a result of their culture. It is to get them to see that they are born into a society that has a long history, that they are ranked and given roles in that society, and that, ultimately, they are told who they are, what to think, and how to act. It is to get them to see that the institutions they follow and normally accept are not the only ways in which society can function—that there are always alternatives. It is to get them to realize that those whom they regard as sick, evil, or criminal are often simply different. It is to get them to see that those they hate are often a product of social circumstances that should be understood more carefully and objectively.

In short, the purpose of sociology is to get people to examine their lives and their society objectively. This process is uncomfortable and sometimes unpleasant. As I teach the insights of sociology, I keep asking myself, "Why not just leave those students alone?" Frankly, I usually have no answer for this question. We are socialized into society. Shouldn't we simply accept that which we are socialized to believe? Isn't it better for society if people believe myth? Isn't it better for people's happiness to let them be?

I usually come back to what many people profess to be one primary purpose of a university education: liberal arts. To me, the liberal arts should be "liberating." A university education should be liberating; it should help individuals escape the bonds of their imprisonment through bringing an understanding of that prison. We should read literature, understand art, and study biology and sociology in order to break through what those who defend society want us to know to reach a plane where we are able to see reality in a more careful and unbiased way. In the end, sociology probably has the greatest potential for liberation in the academic world: At its best, it causes individuals to confront their ideas, actions, and being. We are never the same once we bring sociology into our lives. Life is scrutinized. Truth becomes far more tentative.

## Sociology and Democracy

### *The Meaning of Democracy*

Liberation, as you probably realize, has something to do with democracy. Although democracy is clearly an ideal that Americans claim for themselves, it is not usually clearly defined or deeply explored.

Sociology, however, explores democracy, and it asks rarely examined questions about the possibility for democracy in this—or any—society. For many people, democracy simply means "majority rule," and we too often superficially claim that if people go to a voting booth, then democracy has been established and the majority does, in fact, rule. Democracy, however, is far more than majority rule, and majority rule is far more than the existence of voting booths.

Democracy is difficult to achieve. No society can become perfectly democratic;, and few societies really make much progress in that direction. Alexis de Tocqueville, a great French social scientist who wrote *Democracy in America* (1840) after traveling throughout much of the United States in 1831, believed that here was a thriving democracy, one with great future potential. Tocqueville pointed out many of our shortcomings—most important, the existence of slavery—but he believed that we probably had a more democratic future than any other society in the world. What Tocqueville did was examine the nature of our society—our structure, culture, and institutions—and then show what qualities of our society encouraged the development of democracy. For example, he identified our willingness to join voluntary associations that would impact government, strong local ties, and the little need we had for a central government. Although much has changed since Tocqueville, his lasting importance was to remind us that *democracy is a difficult society to achieve, that certain social conditions make it possible, and certain patterns support its continued existence.* It is also, he wrote, easy to lose.

Democracy is also difficult to define. When I try, I usually end up listing four qualities. These describe a whole society, not just the government in that society. Although not everyone will agree that these are the basic qualities of a democracy, I think they offer a good place to begin.

1. *A democratic society is one in which the individual is free in both thinking and action.* People are in control of their own lives. To the extent that a society encourages freedom, we can call it a democratic society.

2. *A democratic society is one in which political and economic power are limited by people throughout society.* Those who have positions in government neither do whatever they choose, nor are they controlled by a minority of people who have

great wealth or organizational power. Instead, voting, law, a multiple number of organizations of people (representing the interests of everyone), and constitutions effectively limit their power. To the extent that government and large economic and social organizations are effectively limited in these and other ways, we call it a democratic society.

3. *A democratic society is one in which human differences are respected and protected.* There is a general agreement that no matter what the majority favors, certain rights are reserved for the individual and for minorities who are different from the majority. Diversity is respected and even encouraged. To the extent that both individuality and diversity are respected and protected, we call it a democratic society.

4. *A democratic society is one in which all people have equal opportunities to live decent lives.* Privilege is not inherited; people have equality before the law—in educational opportunity, in opportunity for material success, and in whatever is deemed to be important in society. To the extent that real equality of opportunity exists, we call it a democratic society.

These four qualities that make up the definition of democracy described here must be tentative descriptions, and people should debate their relative significance. Some will regard other qualities to be more important, and some will regard only some of these to be necessary. I am only trying here to list qualities that make good sense to me and that guide my own estimate of whether the United States and other societies are democratic.

Democracy, however, should be more than an isolated act such as allowing people to vote. For democracy to succeed, people living in society need a *culture of democracy,* a set of beliefs and rules they are willing to live by. Democracy to me is a philosophy that has developed out of the U.S. Declaration of Independence, the U.S. Constitution, the French Declaration of Rights of Man, and Lincoln's Gettysburg Address, followed by many laws, amendments, speeches, essays, policies, books, court decisions, and treaties. In my mind, no society can be perfectly democratic, but for me society's structure, culture, and social institutions need to be directed toward these principles. This is, of course, extremely difficult to achieve.

If you look back through this book, you will see that democracy is a dominant theme. Because sociology focuses on social organization, social structure, culture, institutions, social order, social class, social power, social conflict, socialization, social change, and religion, *sociology must continually examine issues that are relevant to understanding a democratic society.* In addition, because *sociology critically examines people and their society, it encourages the kind of thinking that is necessary for people living in and working for a democratic society.* If we look once more at the questions and thinking in this book, the theme of democracy stands out. One might, in truth, argue that *the study of sociology is the study of issues relevant to understanding and living in a democratic society.* It is certainly more than this, as I will point out in the last half of this chapter, but everything written in this book implies something about the understanding of democratic society, its development, its possibilities, its limits, and its future.

### *Sociology: An Approach to Understanding Democratic Society*

This book is an attempt to introduce sociology by asking questions and then reacting to them as sociologists would. By trying to deal with these questions, I realized that each one of them is relevant to understanding democratic society. Here are my thoughts.

Previously, we dealt with the nature of the human being and the formative role of socialization and culture. To ask questions about human nature is to ask questions about the possibility for democratic society, a society built on qualities that often are not widespread in society: respect for individual differences, compromise, and concern over inequality and lack of freedom. The sociological approach to the human being makes no assumption of fixed qualities; but it has a strong tendency to see human beings as living within social conditions that are responsible for forming many of their most important qualities. A society tends to produce certain types of people and certain social conditions, encouraging one value or another, one set of morals or another, one way of doing things or another. Conformity, control of the human being, tyranny, and pursuit of purely selfish interests can be encouraged; but so, too, can freedom, respect for people's rights, limited government, and equality. *The possibilities for and the limits to a human being who can live democratically are part of what sociology investigates through its questions concerning culture, socialization, and human nature.*

Those who think about society must inevitably consider the central problem of social order: How much freedom and how much individuality can we allow and still maintain society? Those who favor greater freedom will occasionally wonder, How can there really be meaningful freedom in any society? As long as society exists, how much freedom can we encourage without destroying the underlying order? Are there limits? If so, how can we discover them? What are the costs, if any, of having a democratic society? Those who fear disorder and the collapse of society might ask, How much does the individual owe to society? Such questions are extremely difficult to answer, but they are investigated throughout the discipline of sociology, and they push the serious student to search for a delicate balance between order and freedom. Too often, people are willing to sell out freedom in the name of order; too often, people claim so much freedom that they do not seem to care about the continuation of society. The sociologist studies these problems and causes the student to reflect again and again on this dilemma inherent in all societies, especially those that claim to be part of the democratic tradition. There can be no freedom without society, Émile Durkheim reminds us, for a basic agreement about rules must precede the exercise of freedom. *But how many rules, how much freedom? There is no more basic question for those who favor democracy, and there is no question more central to the discipline of sociology.*

The question of social order also leads us to two questions discussed in Chapter 3: What constitutes a nation? What constitutes a society? These issues may not seem to have much relevance to democracy, but they do. It is easy for those who profess democracy to favor majority rule. It is much more difficult for any nation to develop institutions that respect the rights of all societies within its borders. A nation is a political state that rules over one—or more—societies. If it is democratic, then the nation does not simply rule these societies but responds to their needs and rights, from true political representation to a decent standard of living. If it is democratic, the nation faces not the question of whether we can mold that society to be like the dominant society, but rather how we can create an order in which many societies can exist. If it is democratic, then the nation must balance the needs of each society's push for independence with the need for maintaining social order. *The whole meaning of what it is to be a society, as well as the associated problems of order and independence, are central sociological—and democratic—concerns.*

It is the question of control by social forces over the human being that places sociology squarely within the concerns of democracy. Much of sociology questions the possibility for substantial freedom. Democracy teaches that human beings can and should think for themselves. Much of the purpose of sociology, however, is to show us that our thinking and action are created by our social life and that, although we may claim our ideas and actions as our own, they result from our cultures, our positions in social structure, our social institutions, our socialization, and social controls. Even to claim "We are a democracy!" can simply be part of an ideology, an exaggeration we accept because we are victims of various social forces. Sociology seems to make democracy an almost impossible dream and to some extent the more sociology one knows, the more difficult democracy seems. Yet it is important to remember that sociology also tries to show how some freedom is possible and necessary. Sociology links freedom to understanding: It is really impossible to think for oneself or to act according to free choice unless one understands the various ways in which we are controlled. For example, it is only when I see how my conception of being a "man" has been formed by society that I can think and act independently of that society. For example, when I understand that powerful advertising has developed my personal tastes as well as personal values, I can then step back and direct my own life. It makes freedom relative: There is no such thing as a perfectly free society, free actor, or perfectly free act—but there are degrees of freedom. *Freedom, in the sociological perspective,* is *made far more complex, difficult, and limited, which, in turn, makes democracy itself more complex, difficult, and limited.*

The study of social inequality is probably the central concern of sociology and the primary issue in understanding the possibility for a democratic society. It seems to be the nature of society to be unequal. Many forces create and perpetuate inequality. Indeed, even in our groups and our formal organizations, great inequalities are the rule. Why? Why does it happen? And what are its implications for democracy? If society is characterized by great inequalities of wealth and power, then how can free thought and free action prevail among the population? If a society—in name, a democracy—has a small elite that dominates the decision making, then what difference does going to the polls make? If large numbers of people must expend all of their energy to barely survive because of their poverty, where is their freedom, their opportunity to influence the direction of society, their right to improve their lives? If society is characterized by racist and sexist institutions, then how is democracy possible for those who are victims? *More than any other perspective, sociology makes us aware that many problems stand in the way of a democratic society, not the least of which are social, economic, and political inequalities.*

This focus on social inequality causes many individuals to look beyond the political arena to understand democracy. A democratic society requires not only limited government but also a limited military, a limited upper class, limited corporations, and limited interest groups. Limited government may bring freedom to the individual, but it also may simply create more unlimited power for economic elites in society, which often produces an even more ruthless tyranny over individual freedom. *Sociology, because its subject is society, broadens our concerns, investigates the individual not only in relation to political institutions, but also in relation to many other sources of power that can and do limit real democracy and control much of what we think and do.*

The democratic spirit cares about the welfare of all people. It respects life, values individual rights, encourages quality of life, and seeks justice for all. Sociology studies social problems. It tackles many problems, but in this book we have focused on the problems associated with human misery. Many people live lives of misery, characterized by poverty, crime, bad jobs, exploitation, lack of self-worth, stress, repressive institutions, destructive conflict, inadequate socialization, and alienations of various kinds. These are more than problems caused by human biology or human genes; these are more than problems caused by the free choices of individual actors. Something social has generally caused misery to occur. *Although it is impossible for sociology—or a democratic society—to rid the world of such problems, it is part of the spirit of both to understand them, to suggest and carry out ways to deal with them. Democracy is shallow and cold if large numbers of people continue to live lives of misery.*

What does ethnocentrism have to do with democracy? Is this central concern in sociology relevant to understanding and living in a democratic society? We return to the issue of respect for minorities mentioned earlier. Ethnocentrism, although perhaps inevitable and even necessary to some extent, is a way of looking at one's own culture and others in a manner antagonistic to a basic principle of democracy: respect for human diversity and individuality. To claim that our culture is superior to others is to treat other cultures without respect, to reject them for what they are, to believe that everyone must be like us. Such ideas encourage violent conflict and war and justify discrimination, segregation, and exploitation. Sociology challenges us to be careful with ethnocentrism. We must understand what it is, what its causes are, and how it functions. An understanding of ethnocentrism will challenge us to ask: "When are my judgments of others simply cultural and when are they based on some more defensible standards (such as democratic standards)?" "When are my judgments narrow and intolerant; when are they more careful and thought out?" Even then, an understanding of ethnocentrism will not allow us to judge people who are different without seriously questioning our judgments. *Sociology and democracy are perspectives that push us to understand human differences and to be careful in condemning those differences.*

Previously, we examined social change and the power of the individual. This discussion, too, challenges many of our taken-for-granted "truths" concerning democracy. The sociologist's faith in the individual as an agent of change is not great. Democracy is truly an illusion if it means that the individual has an important say in the direction of society. But if sociology teaches us anything about change that has relevance for democracy, it is that intentionally created change is possible only through a power base. If a democracy is going to be more than a description in a book, people who desire change in society—ideally, toward more freedom, limited government, equality of opportunity, and respect for individual rights—must work together and act from a power base, recognizing that the existing political institutions are usually fixed against them. And before we go off armed with certainty, we should remember that our certainty was probably also socially produced and that through our efforts we may bring change we never intended and may even lose whatever democracy we now have. Social change is complex, depends on social power, and is difficult to bring about in a way we would like. *The sociologist will examine the possibility for intentional social change in a democratic society and will be motivated to isolate the many barriers each society establishes to real social change.*

Religion and society also have much to do with democracy. Is religion consistent or inconsistent with democratic values? As religion has changed because of modernization, have we become so individualistic that religious community becomes impossible? And if religion is embedded on truths located in the past, then how can there be much freedom of thought and action? And if some more traditional societies are dominated by one powerful religion, can democracy flourish? Indeed, these and similar questions ask us to reflect on the prerequisites of democracy and how these prerequisites have implications for religion. *The sociologist will examine the complex relationships among tradition, modernization, religion, and democracy, emphasizing that democracy does not just happen but comes into existence only through important social conditions.*

Globalization is already influencing democracy's future. What is the relationship between capitalism and democracy? What trends lead us to a world society? Will it be one of more opportunity for more people, respect for human differences, less violence and more cooperation, and democratic political institutions? Or will it be one of greater inequality, loss of individuality, homogenization of culture, and worldwide tyranny arising from global political institutions, giant corporations, and the super wealthy? *Labor rights, the future of work, poverty, diversity, immigration, government, freedom, violence, cooperation, human values, and social power—these are the many issues that globalization highlights, that involve democracy, and that sociology studies and teaches.*

To evaluate and critique what we have been learned via culture is part of the essence of democracy. A democratic society is a place where people engage in rational conflict over what is to be believed. It is through careful investigation that citizens must pursue truth, generalizing, categorizing, understanding without stereotyping and without being motivated by their prejudices. Freedom of thought and speech—one of the basic pillars of a democratic society—is possible only if people live in a society where ideas clash, evidence is evaluated, and "truths" are always open to questioning. *The principles of science and democracy are similar. There is no greater test of this than the discipline of sociology: an attempt to apply scientific and rational principles to understand that for which we are all taught to feel a special reverence.*

## Sociology Is More than the Study of Democracy

Perhaps democracy is emphasized too much in this chapter. Isn't sociology more than the study of democracy? I have tried to show that the chapters of this book can be brought together around the study of democracy. This is central to me. However, there are other themes that should be discussed, for democracy is probably too narrow a topic for many sociologists.

Anyone who is interested in understanding order, law, morality, crime, conformity, and nonconformity is well served by the sociological perspective. Sociologists want to understand the meaning of crime, the cause of crime, the consequences of crime, the ways to treat crime. Law is a social creation, as is society, and thus understanding the origins and functions of various laws as well as law itself is an important part of what sociology does. The study of police, courts, prisons, probation, and parole from a sociological perspective is extremely important to those who wish to intelligently understand these matters. An orderly society is not automatic. As many people learned in the past ten years, societies can become threatened, chaotic, lawless, and murderous. Social order—the attempt to achieve social order, the threats to disorder, the acts of terrorism by government and by various individuals and groups, civil war and war among nations—is a dominant theme in sociology. *Many sociologists would argue, therefore, that ultimately sociology is the study of social order.*

Every one of us exists in a social world where traditions are very important, yet as we get older, our traditions are challenged by the next generation and what we have accepted as a certainty becomes a choice that people now make. In my own lifetime, I had to force myself to understand changing male and female roles, changes in sexual traditions, music, art, jobs, nation-states, religion, education, and the role of government. I had to come to terms with widespread changes in higher education, from using PowerPoint rather than blackboards to recognizing the role of computers and the Internet college. We all live in a world of change; people everywhere are affected by change. Change is difficult for almost anyone. What we have achieved we defend against change. Social movements that wish to direct change become the source of justice to many of us and the source of evil to others. For every individual, every group, every community, every society, change has always been an issue that people worry about, talk about, welcome, work for, or try to prevent. *For many, sociology is ultimately the understanding of social change.*

Almost all of us would like to know that what we believe is true. However, education and experience question our views, and what we thought was true is no longer true. Capturing reality, recognizing the difficulty in the pursuit of truth, as well as our tendency to easily accept untruth, are all important issues we confront now and then. And some of us who value the discovery of new ideas confront such issues almost habitually. Who is presenting "truth" to us? Who benefits from these "truths"? What is the perspective of science? Does it uncover "truths"? What is the role of culture in the "truths" we hold? Sociology does not deny the existence of truth; rather, it simply shows us how important our social lives, our social interaction, our social organizations, and our societies are in shaping what we come to believe is "truth." For me, this has been a guide to my whole approach to understanding the ideas I have encountered, thought about, discussed, and sometimes accepted. For many of us, *the real underlying theme in sociology is the critical and analytical attempt to understand ourselves and our society.* It is an attempt to understand and question those people who have socialized us.

We live in a society that emphasizes the role of the physical brain, human nature, DNA, and biological predispositions. It is too easy today to overlook society, yet any balanced view of cause must deal with the forces society and socialization present to us. Society is very difficult to study because it is not nearly as concrete as other entities that are more physical. However, not understanding the role of society in our lives is to leave out perhaps the source of our most important qualities as humans. No matter how much we understand about the biological side of the human being, even that factor does not make them identical as it does among other animals, even if they could be cloned like, say, dogs or sheep . Humans may be cloned to a point, but they will still remain different from one another. Every human being is born with influences arising from their biology, but immediately they interact with others, they are born into a particular community and society, they will have their own unique social history, and they will learn to use language, self, and mind to determine to a great extent their own values, goals, thoughts, and directions. Whatever intelligence we might be born with will be used in many different ways. Whatever talents we might be born with will be expressed in many possible ways. Whatever intelligence or talents we are born with will be nurtured in society, encouraged by society, and ignored or thwarted. It is to our own peril, according to sociologists, to forget society, social problems, socialization, social institutions, culture, and social structure. Anyone who wants to understand the human being must understand society, and this is what, for many sociologists, sociology focuses on. *For many, therefore, sociology is the study of the social essence of the human being, recognizing that society itself is the essence of the human being and a major cause of what human beings become.*

Each generation in society has its own crises. This will be true for future generations, but it is difficult to identify and predict them. Sociology tries hard to identify causes and consequences for coming crises, and suggests ways that individuals can adjust productively, and how society might be able to alleviate these so that people's lives are made better. Three crises that the United States faces, from my point of view, are the widespread decline of decent work, the decline of decent education for the vast majority, and the increasing destruction of large numbers of people by individuals who have a mission and are not afraid to use terror to achieve their mission. Of course, everyone sees a different crisis in our future. Some see modernization itself, the end of authority, the continuation of secular life. Others see the crises as the continued influence of religion in the political arena, loss of privacy, racial inequality, increasing poverty, sexual exploitation, or increasing gender inequality. Still others see the crises in our declining institutions in the family, government, religion, and the economy. However, almost every crisis we can identify is a crisis related to the operation of society. Understanding, describing, explaining, predicting, and alleviating social problems must be the essence of sociology to many who study and use it. Of course, political science, economics, psychology, and cultural anthropology, as well as good fiction, philosophy, art, and journalism, can tell us a great deal. Yet, *the core of sociology for many sociologists remains the understanding, describing, explaining, predicting, and alleviating social problems.* This remains a dominant reason that students find sociology interesting and valuable.

All human beings spend their lives in social relationships, have social histories, and exist in many groups and organizations. Our lives are social through and through. How to build relationships, communicate, cooperate, share, love, interact, negotiate, understand one another, play and work together, build families, neighborhoods, and businesses, create friendships, engage in conflict and compromise—all this—is the heart of what society and social life are made of. This is how most of us experience society every day: through social interaction with other people in school, neighborhoods, and businesses, on the street, on computers, and in traveling to other communities throughout the world. *As an academic discipline that studies and teaches the various ways people act around one another, for some sociologists the theme that dominates all sociology is really the study of social interaction and human relationships.*

Every time I enter a new group or organization, my question is always, "Who really has power here?" My own specialization in graduate school was political sociology, and I often taught classes in social power. Through all this I became increasingly aware that "social power" is a concept most people think they know but cannot define. And I came to recognize that few people have an appreciation of its universal role in every social relationship. I was influenced by Bertrand Russell's statement that social science in the end is really the study of social power.

I was inspired by C. Wright Mills's *The Power Elite* and much of the research throughout the 1960s and 1970s to study the corporation as the seat of power in our society, influencing—able to exert power successfully over—employees, government, communities, and the world economy. Mills recognized three centers of power throughout our history, and each epoch sees one of these arising to the top: sometimes it is the political power of the office of the president, sometimes it is the economic power of those who control the major corporations, and sometimes it is the military power of those who control the armed forces. I was impressed by Suzanne Keller's book on the role of elites in modern society, identifying seven elites who compete for control in the United States, and I studied Alexis de Tocqueville, Seymour Lipset, and Arnold Rose's descriptions of the possibilities of pluralistic democracy in the United States, and G. William Domhoff's description of a powerful traditional social class that is at the very top of society. My dissertation discussed the role of authority in society, and I became increasingly interested in ways that people are able to exert power successfully. For most of my life as a sociologist, it was power that fed my interest. I recognized that any analysis of society without understanding social power is losing a very important aspect of what all social life contains. Family life, corporate life, political life, economic life, social classes, and the creation and maintenance of social institutions all have a strong element of social power. Karl Marx described society as a system for and by the rich and powerful, and Max Weber considered legitimate power to be the essence of social order and predicted that bureaucracy in the twentieth century would create far more powerful governments and armies than ever before in the history of humankind. Of course, Robert Michels described how powerful elites are inevitable in every organization. I realized that among those who call the United States a democracy are those who would systematically ignore the issues related to social power, refuse to treat poverty as powerlessness, or describe family life without addressing its power structure. *A strong case can be made for sociology as the study of power in society; certainly if we gloss over the universal importance of power in human organization the student will never understand what organized life is.*

As I go through my library and look at articles that I have saved, books that I have read, individual chapters that I find important, it occurs to me that sociology really comes down to class inequality. Studies and theories of class weave themselves through almost every topic. To the sociologist, all the trends in the United States and the world have implications for class structure. The reason outsourcing, decline of the power of labor unions, computerization, and the changing patterns in higher education are all important for American society is that these forces impact class, and changes in class, in turn, impact what our society is going to become. Increasingly, sociologists are arguing that there is definitely a decline in the numbers and power of the middle class. One excellent book by Robert Perrucci and Earl Wysong, *The New Class Society* (2003), argues that we have become a two-class society, with a "privileged class" (twenty percent of the population) and a "new working class" (eighty percent). The two classes are broken down into five subclasses with a "superclass" at the top made up of one to two percent of the population, and the "excluded class" at the bottom made up of ten to fifteen percent. The whole book describes this class structure, emphasizing the decline of the middle class. The point of the book is to examine what this new class structure means for every aspect of society. Two other excellent books, one by Erik Olin Wright titled *Class Counts* (1997) and another by Stanley Aronowitz titled *How Class Works* (2003), highlight the power relationships among the various classes. These three books alert me that underneath it all, *sociology is the study of social class, its origins, its consequences in both life chances and lifestyle, its relationship to government, religion, education, crime, and health, its future, its history, its functions, and its role in social change.*

I also believe that sociology encourages people to care, to care about people who are disadvantaged, oppressed, and exploited, those who are victims of crime and those who commit crime. It matters to us why evil things are caused; it matters when justice is ignored. In its study of the individual actor, sociology always analyzes the societal context, and rarely do we simply blame the individual who is the victim. An underlying belief is that some people are more fortunate than others rather than the belief that society is a just system that rewards the good and condemns the bad. There is, within the sociological perspective, a faith that we can do better if we understand the forces that make a particular society what it is. I used to declare to students that sociology is the realization that "there but for the grace of social interaction go I." One cannot easily understand the sociological perspective and ignore the role of society in creating conditions that no one deserves. For many of us, caring more than just understanding becomes part of our sociological perspective. *Therefore, many of us believe that sociology is really understanding and caring about those who are left out of the many benefits of society.*

## Summary and Conclusion

Each sociologist makes a different argument as to what sociology is and why it is so important. Each has a different slant. In fact, I really have a difficult time isolating two or three slants I agree with. That is because all are legitimate. My discussion of sociology as the study of democracy seemed to be a great answer in earlier editions of this book, until one critic correctly let me know he believes that sociology is much more than the study of democracy. I do still believe that democracy is a dominant theme, a focus, a way of summing up what sociology is all about. However, in listing these various other themes or focuses I realize all of these make good sense, and I now recognize that all of these must be put side by side if we are going to grasp what sociology is and what it can achieve.

1. Sociology is the study of democracy.
2. Sociology is the study of social order.
3. Sociology is the study of social change.
4. Sociology is the analytical and critical study of ourselves and our society.
5. Sociology is the study of the social essence of the human being.
6. Sociology is an attempt to understand, describe, explain, predict, and alleviate social problems.
7. Sociology is the study of social interaction and human relationships.

8. Sociology is the study of social power.
9. Sociology is the study of social class.
10. Sociology is the study of understanding and caring about those who are left out of the many benefits of society.

Of course, in a book titled *Ten Questions* I would ultimately list ten ways of describing what sociology is and what it can achieve. Please believe me: it just happened this way. I ran out, and I leave it to you the reader and instructor to add or delete what is written here. Sociology is a broad, interesting, and many-sided, academic perspective. Those people who come to understand and use it become committed to one or more of these themes. To describe it only one way falls short of what it means to so many people who study sociology. And to go back to the standard definition, "sociology is the study of society," misses the wonder and excitement that many sociologists feel for their work.

## Questions to Consider

1. What is the most important idea to be learned from sociology?
2. Should sociology be a required class in the university?
3. Is democracy a good society to work toward?
4. What is democracy?
5. Are sociology's ideas harmful to the individual or society?
6. How might a university president answer this question: Is sociology important?

## References

Almost any sociological work might be included in the following list. However, I have tried to identify works that focus directly on democracy or the meaning of sociology—or both.

**Adams, E. M.** 1993. *Religion and Cultural Freedom*. Philadelphia: Temple University Press.

**Apraku, Kofi.** 1996. *Outside Looking In: An African Perspective on American Pluralistic Society*. Westport, CT: Praeger.

**Arendt, Hannah.** 1958. *The Origins of Totalitarianism*. Cleveland, OH: Meridian.

**Aronowitz, Stanley.** 2003. *How Class Works: Power and Social Movements*. New Haven, CT: Yale University Press.

**Bellah, Robert N.** 1999. "The Ethics of Polarization in the United States and the World." Pp. 00–00 in *The Good Citizen*, edited by David Batstone and Eduardo Mendieta. New York: Routledge/Taylor & Francis.

**Bellah, Robert N., Richard Madsen, William M. Sullivan, Ann Swidler, and Steven M. Tipton**. 1985. *Habits of the Heart: Individualism and Commitment in American Life*. Berkeley: University of California Press.

**Berger, Bennett**. 1990. *Authors of Their Own Lives: Intellectual Autobiographies of Twenty American Sociologists*. Berkeley: University of California Press.

**Berger, Peter**. 1963. *Invitation to Sociology*. Garden City, NY: Doubleday.

**Berger, Peter, Brigitte Berger, and Hansfried Kellner**. 1974. *The Homeless Mind: Modernization and Consciousness*. New York: Vintage.

**Berger, Peter, and Hansfried Kellner**. 1981. *Sociology Reinterpreted*. New York: Doubleday.

**Best, Joel**. 2004. *More Damned Lies and Statistics: How Numbers Confuse Public Issues*. Berkeley: University of California Press.

**Blumer, Herbert**. 2000. *Selected Works of Herbert Blumer: A Public Philosophy for Mass Society*, edited by Stanford M. Lyman and Arthur J. Vidich. Urbana: University of Illinois Press.

**Charon, Joel M.** 2002a. *The Meaning of Sociology*. 7th ed. Upper Saddle River, NJ: Prentice Hall.

———. 2002b. *The Meaning of Sociology: A Reader*. 7th ed. Upper Saddle River, NJ: Prentice Hall.

———. 2009 "An Introduction to the Study of Social Problems." Pp. 1–12 in *Social Problems: Readings with Four Questions*, edited by Joel Charon and Lee Vigilant. 3rd ed. Belmont, CA: Wadsworth Cengage Learning.

**Charon, Joel M, and Lee Garth Vigilant**. 2009. *Social Problems: Readings with Four Questions*, 3rd ed. Belmont, CA: Wadsworth Cengage Learning.

**Cohen, Mark Nathan**. 1998. *Culture of Intolerance: Chauvinism, Class, and Racism in the United States*. New Haven, CT: Yale University Press.

**Collins, Randall**. 1998. *A Global Theory of Intellectual Change*. Cambridge, MA: Harvard University Press.

**Collins, Randall, and Michael Makowsky**. 2004. *The Discovery of Society*. 7th ed. New York: McGraw-Hill.

**Cuzzort, R. P., and E. W. King**. 1994. *Twentieth Century Social Thought*. 5th ed. New York: Harcourt College.

**Davey, Joseph Dillon.** 1995. *The New Social Contract: America's Journey from Welfare State to Police State*. Westport, CT: Praeger.

**Diamond, Larry**. 1999. *Developing Democracy*. Baltimore, MD: Johns Hopkins University Press.

**Domhoff, G. William**. 1967. *Who Rules America?* Upper Saddle River, NJ: Prentice Hall.

**Du Bois, William D., and R. Dean Wright**. 2001. *Applying Sociology: Making a Better World*. Boston: Allyn & Bacon.

**Ehrensal, Kenneth**. 2001. "Training Capitalism's Soldiers: The Hidden Curriculum of Undergraduate Business Education." Pp. 00–00 in *The Hidden Curriculum*. Ed. Eric Margolis. New York: Routledge/Taylor & Francis Books.

**Farley, John E., and Gregory D. Squires**. 2005. "Fences and Neighbors: Segregation in 21ST-Century America" *Contexts* 4(1): 33–39.

**Faulks, Keith**. 2000. *Political Sociology: A Critical Introduction*. New York: New York University Press.

**Feagin, Joe R., and Hernan Vera**. 2002. *Liberation Sociology*. Boulder, CO: Westview.

**Fernandez, Ronald**. 2003. *Mappers of Society: The Lives, Times and Legacies of the Great Sociologists*. Westport, CT: Greenwood.

**Friedman, Thomas L.** 2005. *The World Is Flat: A Brief History of the Twenty-First Century*. New York: Farrar, Straus & Giroux.

**Fromm, Erich**. 1941. *Escape from Freedom*. New York: Holt, Rinehart & Winston.

**Fuller, Robert W.** 2003. *Somebodies and Nobodies: Overcoming the Abuse of Rank*. Gabriola Island, BC: New Society.

**Gans, Herbert J.** 1999. *Making Sense of America*. Lanham, MD: Rowman & Littlefield.

**Hallinan, Maureen T.** 2005. *The Socialization of Schooling*. New York: Russell Sage Foundation.

**Hare, Bruce R.** (Ed.). 2002. *2001 Race Odyssey: African Americans and Sociology*. Syracuse, NY: Syracuse University Press.

**Hochschild, Jennifer L.** 1995. *Facing Up to the American Dream: Race, Class, and the Soul of the Nation*. Princeton, NJ: Princeton University Press.

**Iceland, John**. 2003. *Poverty in America: A Handbook*. Berkeley, CA: University of California Press.

**Ignatieff, Michael** (Ed.). 2004. *The Lesser Evil: Political Ethics in an Age of Terror*. Princeton, NJ: Princeton University Press.

**Jones, Ron.** 1981. *No Substitute for Madness*. Covelo, CA: Island.

**Keller, Suzanne Infeld**. 1963. *Beyond the Ruling Class: Strategic Elites in Modern Society*. New York: Random House.

**Kennedy, Robert E.**, Jr. 1989. *Life Choices*. 2nd ed. New York: Holt, Rinehart & Winston.

**Killian, Lewis M.** 1994. *Black and White: Reflections of a White Southern Sociologist*. Dix Hills, NY: General Hall.

**Koonigs, Kees, and Dirk Druijt**, eds. 1999. *The Legacy of Civil War, Violence and Terror in Latin America*. London: Zed.

**Kozol, Jonathan**. 2005. *The Shame of the Nation: The Restoration of Apartheid Schooling in America*. New York: Random House.

**Lambert, Stephen E.** 2003. *Great Jobs for Sociology Majors*. 2nd ed. Chicago: VGM Career Books.

**Levi, Primo.** 1959. *If This Is a Man*. New York: Orion.

**Lewis, Bernard**. 2002. *What Went Wrong? Western Impact and Middle Eastern Response*. New York: Oxford University Press.

———. 2003. *The Crisis of Islam: Holy War and Unholy Terror*. New York: Random House.

**Liebow, Elliot**. 1967. *Tally's Corner*. Boston: Little, Brown.

**Lipset, Seymour Martin, Martin Trow, and James Coleman**. 1956. *Union Democracy: The Inside Politics of the International Typographical Union*. New York: Free Press.

**Marx, Karl, and Friedrich Engels**. [1848] 1955. *The Communist Manifesto*. New York: Appleton-Century-Crofts.

**Michels, Robert**. 1915. *Political Parties: A Sociological Study of the Oligarchical Tendencies of Modern Democracy*, translated by Eden and Cedar Paul. New York: Hearst's International Library.

**Mills, C. Wright**. 1956. *The Power Elite*. New York: Oxford University Press.

———. 1959. *The Sociological Imagination*. New York: Oxford University Press.

**Moore, Barrington**. 1966. *Social Origins of Dictatorship and Democracy*. Boston: Beacon.

**Murphy, John W., and Dennis L. Peck**. 1993. *Open Institutions: The Hope for Democracy*. Westport, CT: Praeger.

**Myrdal, Gunnar**. 1944. *An American Dilemma*. New York: Harper & Row.

**Nisbet, Robert A.** 2002. *Sociology as an Art Form*. New Brunswick, NJ: Transaction.

**Oliner, Pearl M., and Samuel P. Oliner**. 1995. *Toward a Caring Society*. Westport, CT: Praeger.

**Perrucci, Robert, and Earl Wysong**. 2003. *The New Class Society: Goodbye American Dream?* 2nd ed. New York: Rowman & Littlefield.

**Phillips, Kevin P.** 2002. *Wealth and Democracy: A Political History of the American Rich*. New York: Broadway.

**Ravitch, Diane, and Joseph P. Viteritti**. 2001. *Making Good Citizens: Education and Civil Society*. New Haven, CT: Yale University Press.

**Reich, Charles A.** 1995. *Opposing the System*. New York: Crown.

**Rose, Arnold Marshall**. 1967. *The Power Structure: Political Process in American Society*. New York: Oxford University Press.

**Rousseau, Nathan**. 2002. *Self, Symbols, and Society: Classic Readings in Social Psychology*. New York: Rowman & Littlefield.

**Schwartz, Barry**. 1994. *The Costs of Living: How Market Freedom Erodes the Best Things in Life*. New York: Norton.

**Seligman, Adam**. 1995. *The Idea of Civil Society*. Princeton, NJ: Princeton University Press.

**Skolnick, Arlene S. and Jerome Skolnick**. 2007. *Family in Transition*. 14th ed. Boston: Allyn & Bacon.

**Staub, Ervin**. 1989. *The Roots of Evil: The Origins of Genocide and Other Group Violence*. New York: Cambridge University Press.

**Stephens, W. Richard**. 1995. *Careers in Sociology*. Boston: Allyn & Bacon.

**Tocqueville, Alexis de.** [1840] 1969. *Democracy in America*. 1969 ed. New York: Doubleday.

**Tonry, Michael**. 2004. *Thinking About Crime: Sense and Sensibility in American Penal Culture*. New York: Oxford University Press.

**Touraine, Alain**. 1997. *What Is Democracy?*, translated by David Macey. Boulder, CO: Westview.

**Turner, Stephen Park, and Jonathan H. Turner**. 1990. *The Impossible Science: An Institutional Analysis of American Sociology*. Newbury Park, CA: Sage.

**Walker, Henry A., Phyllis Moen, and Donna Dempster-McClain**, eds. 1999. *A Nation Divided: Diversity, Inequality, and Community in American Society*. Ithaca, NY: Cornell University Press.

**Weber, Max**. [1924] 1964. *The Theory of Social and Economic Organization*, edited by A. M. Henderson and Talcott Parsons. New York: Free Press.

**Western, Bruce** 2006. *Punishment and Inequality*. New York: Russell Sage Foundation.

**Wilson, Everett K., and Hanan Selvin**. 1980. *Why Study Sociology? A Note to Undergraduates*. Belmont, CA: Wadsworth.

**Winant, Howard**. 1994. *Racial Conditions: Politics, Theory, Comparisons*. Minneapolis: University of Minnesota Press.

**Witt, Griff**. 2004. "As Income Gap Widens, Uncertainty Spreads." *Washington Post*, September 20.

**Wolfe, Alan**. 1989. *Whose Keeper? Social Science and Moral Obligation*. Berkeley: University of California Press.

———. 2001. *Moral Freedom: The Impossible Idea That Defines the Way We Live Now*. New York: Norton.

**Wright, Erik Olin**. 1994. *Interrogating Inequality: Essays on Class Analysis, Socialism, and Marxism*. New York: Verso.

———. 1997. *Class Counts: Comparative Studies in Class Analysis*. Cambridge: Cambridge University Press.

**Zweigenhaft, Richard L., and G. William Domhoff**. 1998. *Diversity in the Power Elite: Have Women and Minorities Reached the Top?* New Haven, CT: Yale University Press.

# Should We Generalize about People?

## Generalizing, Categorizing, Stereotyping, and the Importance of Social Science

### Concepts, Themes, and Key Individuals

- Generalization
- Categorization
- Stereotyping
- Evidence
- Social science
- Sociology

This chapter is about the human search for truth and the difficulties humans have when they seek it. It is about objectivity, good and bad thinking, and understanding and judging people. It gets to the heart of what social science is, and in some ways, it brings us back to the very first chapter on science. It is an important question to ponder and debate, it is essential to deal with it, and it was the most difficult chapter for me to write.

In fact, it is a topic that lies behind almost every discussion about human beings. It is implicit every time we try to understand people: all people, some people, or a given individual. It comes up whenever we try to label others or are labeled by others. It is part of every discussion of prejudice, the nature of American life, Russians, men, women, young people, liberals—in discussions that involve any attempt to categorize people. The question highlights a conflict that characterizes almost all of us: We categorize other people to understand them; yet we want to cry out whenever others try to categorize us. "I wanna be me. I'm like no one else who ever lived! Treat me for what I am. Do not assume that I am like any category of people!"

Science categorizes nature and makes generalizations about objects in nature. Social science does the same for human beings. Is this good for understanding? Does it contribute to stereotyping and inhumane treatment of those unlike ourselves? Here is the issue we are looking at here. The tenth question is:

***Should we generalize about people?***

## Categories and Generalizations

### *The Importance of Categories and Generalizations to Human Beings*

Sociology is a social science; therefore, it makes generalizations about people and their social life. "The top positions in the economic and political structures are far more likely to be filled by men than by women." "The wealthier the individual, the more likely he or she will vote Republican." "In the United States, the likelihood of living in poverty is greater among the African-American population than among whites." "American society is segregated." "Like other industrial societies, American society has a class system in which more than three-fourths of the population end up in approximately the same social class as they were at birth."

But such generalizations often give me a lot of trouble. I know that the sociologist must learn about people and generalize about them, but I ask myself, "Are such generalizations worthwhile? Shouldn't we simply study and treat people as individuals?" An English professor at my university was noted for explaining to his class that "you should not generalize about people—that's the same as stereotyping and everyone knows that educated people are not supposed to stereotype. Everyone is an individual." (Ironically, this is *itself* a generalization about people.)

However, the more I examine the situation, the more I realize that all human beings categorize and generalize. They do it every day in almost every situation they enter, and they almost always do it when it comes to other people. In fact, we have no choice in the matter. "Glass breaks and can be dangerous." We have learned what "glass" is, what "danger" means, and what "breaking" is. These are all categories we apply to the situations we enter so that we can understand how to act. We generalize from our past. "Human beings who have a cold are contagious, and, unless we want to catch a cold, we should not get close to them, and we should be careful shaking hands with them." We are here generalizing about "those with colds," "how people catch colds," and "how we should act around those with colds." In fact, every word we use is a generalization that acts as a guide for us. The reality is that we are unable to escape generalizing about our environment.

That is one aspect of our essence and strength as human beings. This is what language does for us. Sometimes our generalizations are fairly accurate; sometimes they are unfounded. However, we do, in fact, generalize almost all of the time! The question that introduces the chapter is really a foolish one. "Should we generalize about people?" is not a useful question, simply because we have no choice. A much better question is:

### *How can we develop accurate generalizations about people?*

The whole purpose of social science is to achieve accurate categorizations and generalizations about human beings. Indeed, the purpose of almost all academic pursuits involves learning, understanding, and developing accurate categories and generalizations.

For a moment, let us consider other animals. Most are prepared by instinct or simple conditioning to respond in certain ways to certain stimuli in their environments. So, for example, when a minnow swims in the presence of a hungry fish, that particular minnow is immediately responded to and eaten. The fish is able to distinguish that type of stimulus from other stimuli, and so whenever something identical to or close to it appears, the fish responds. The minnow is a concrete object that can be immediately sensed (seen, smelled, heard, touched), so within a certain range the fish is able to easily include objects that look like minnows and to exclude those that do not. Of course, occasionally a lure with a hook is purposely used to fool the fish, and a slight mistake in perception ends the fish's life.

Human beings are different from the fish and other animals because we have *words for objects and events* in the environment, and this allows us to *understand* that environment and not just respond to it. With words we are able to make many more distinctions, and we are able to apply knowledge from one situation to the next far more easily. We are far less dependent on immediate physical stimuli. So, for example, we come to learn what fish, turtles, and whales are, as well as what minnows, worms, lures, and boats are. We read and learn what qualities all fish have, how fish differ from whales, and what differences fish have from one another. We learn how to catch fish, and we are able to apply what we learn to some fish but not others. We begin to understand the actions of all fish—walleyes, big walleyes, big female walleyes. Some of us decide to study pain, and we try to determine if all fish feel pain, if some do, or if all do not. Humans do not then simply respond to the environment, but they label that environment, study and understand that environment, develop categories and subcategories for objects in that environment, and constantly try to generalize from what they learn in specific situations about those categories. Through understanding a category we are able to see important and subtle similarities and distinctions that are not available to animals that do not categorize and generalize with words.

Generalizing allows us to walk into situations and apply knowledge learned elsewhere to understanding objects there. When we enter a classroom, we know what a teacher is, and we label the person at the front of the room as a teacher. We know from past experience that teachers give grades, usually know more than we do about things we are about to learn in that classroom, have more formal education than we do, and usually resort to testing us to see if we learned something they regard as important. We might have also learned that teachers are usually kind (or mean), sensitive (or not sensitive), authoritarian (or democratic); or we might have had so many diverse experiences with teachers that whether a specific teacher is any of these things will depend on that specific individual.

If we do finally decide that a given teacher is, in fact, authoritarian, then we will now see an "authoritarian teacher," and we will now apply what we know about such teachers from our past.

This is a remarkable ability. We are able to figure out how to act in situations we enter because we understand many of the objects we encounter there by applying relevant knowledge about them that we learned in the past. This allows us to act intelligently in a wide diversity of situations, some of which are not even close to what we have already experienced. If we are open-minded and reflective, we can even evaluate how good or how poor our generalizations are, and we can alter what we know as we move from situation to situation.

The problem for almost all of us, however, is that many of our generalizations are not carefully arrived at or accurate, and it is sometimes difficult for us to recognize this and change them. Too often our generalizations actually stand in the way of our understanding, especially when we generalize about human beings.

To better understand what human beings do and how that some-times gets us into trouble, let us look more closely at what "categories" and "generalizations" are.

## *The Meaning of Categorization*

Human beings categorize their environment; that is, *we isolate a chunk of our environment, distinguish that chunk from all other parts of the environment, give it a name, and associate certain ideas with it.* Our chunks—or categories—arise in interaction; they are socially created. We discuss our environment, and we categorize it with the words we take on in our social life: *living things, animals, reptiles, snakes, poisonous snakes, rattlers.* Categories are created, and once we understand them, we are able to compare objects in situations we encounter. The number of distinctions we are able to make in our environment increases manifold. It is not only nouns that represent categories (*men, boys*) but pronouns (*he, she, her, they*), articles (*and, the, or*), verbs (*run, walk, fall*), adverbs (*slow, fast*), and adjectives (*weak, strong, intelligent, married*). In fact, the purpose of almost every word is to apply its meaning to a whole set of instances, to generalize its meaning. Much of our learning is simply aimed at understanding what various categories mean, and this involves understanding the qualities that make up each of these categories and the ideas associated with them.

Through learning about people (a category) we come to recognize that "all people" possess certain qualities, some of which they share with other animals (cells, brains, reproductive organs), and some of which seem unique to them (some of our DNA, dependence on language, stereoscopic vision, conscience). We understand that people can be divided into young and old, white and black, men and women, single and married. Most of us have a pretty good idea of what a male and a female are. If asked, we could explain who belongs to the categories of homosexual and heterosexual. We do not simply recognize objects that do or do not belong, we *understand* the category by being able to describe the qualities we believe belong to objects that fit and objects that do not. We might say that a male has a penis, an old person is anyone older than 70 (or 65, or 60, or . . . ), a teacher is someone who transmits knowledge, a human being is an animal who has a soul.

We argue over these definitions, and the more we understand, the more complex these definitions become. But categories and definitions are a necessary part of all of our lives. Armed with these, we go out and are able to cut up our environment in complex and sophisticated ways. We see an object and determine what it is (that is, what category it belongs in), and because we know something about that category, we are able to apply what we already know to that object. This allows us to act appropriately in many different situations.

It is necessary for all human beings to categorize, define, and understand their environment. (This statement is itself a generalization about all human beings.) If we are honest with ourselves, then we recognize that each of us has created or learned thousands—even tens of thousands—of categories that we use as we look at what happens around us. The purpose of a biology class is to create useful categories of living things so that we can better understand what these things are—how they are similar, how they differ from nonliving things, and how they differ from one another. Musicians, artists, baseball players, political leaders, students, parents, scientists, con artists, and police—all of us live our lives assuming certain things about our environment based on the categories we have learned in interaction with others.

A *role* is a category we use to understand the situations we encounter. It is a set of expectations that people have of an actor in a position within a social situation. If you are a ticket taker, I expect you to ask me for my ticket. If you are an employee in the theater working behind a candy counter, I expect you to ask me if you can help me. If you are someone sitting in a theater watching a movie, I expect you to keep quiet. If you are a receptionist in a doctor's office, I expect you to tell me when I may see the doctor; if you are a nurse, I expect you to ask me a series of questions; and if you are a physician, I trust you to respect my body. Every set of role expectations I have for others and for myself is an attempt to categorize people. It helps them know what to do; it helps me know what they will do, and what I should do in relation to them. Such expectations are an inevitable part of our lives.

## The Meaning of Generalization

*A category is an isolated part of our environment that we notice and identify.* As we come to understand the category through specific instances of objects we associate qualities in the category. We make general beliefs, aspects, distinctions, causes, effects, and all kinds of information we can apply to other objects that fit in the category (as well as objects that do not fit.) We watch birds build a nest, and we assume that all birds build nests out of sticks (including birds other than the robins and sparrows we observed). We continue to observe and note instances where birds use materials other than sticks, and then we learn that some birds do not build nests but dig them out, discover holes that can serve their needs, or reuse the nests of other birds. More often, our generalizations are a mixture of observation and learning from others: We learn that a wealthy person often drives a Mercedes and that police officers usually carry guns. On the basis of generalizing about a category, we are able to predict future events where that category comes into play. When we see a wealthy person, we expect to see a Mercedes (or something that we learn is comparable); and when we see a police officer, we expect to see a gun. That is what a generalization is.

*A generalization describes the category. It is a statement that characterizes objects included in the category and defines similarities and differences with other categories.* "This is what an educated person is!" (in contrast to an uneducated person). "This is what wealthy people do to help ensure that privilege is passed down to their children." "This is what U.S. presidents have in common." "This is what Catholic people believe in."

As we shall see shortly, a generalization sometimes goes beyond just describing the category. It also explains why a particular quality develops. In other words, *a generalization about a category will often be a statement of cause.* "Jewish people are liberal on social issues because of their minority position in Western societies." "U.S. presidents are male because ..." "Wealthy people send their children to private schools because ..."

Human beings, therefore, categorize their environment by using words. On the basis of observation and learning, they come to develop ideas concerning what qualities are associated with those categories. They also develop ideas as to why those qualities develop. *Ideas that describe the qualities that belong to a category and ideas that explain why those qualities exist are what we mean by generalizations.*

## The Stereotype

When it comes to people, generalization is difficult to do well. The principal reason for this is that we are judgmental, and too often it is much easier for us to generalize for the purpose of evaluating (condemning or praising) others than for the purpose of understanding them. When we do this we fall into the practice of stereotyping.

*A stereotype is a certain kind of categorization.* It is a category and a set of generalizations characterized by the following qualities:

1. *A stereotype is judgmental.* It is not characterized by an attempt to understand, but *by an attempt to condemn or praise the category.* It makes a value judgment, and it has a strong emotional flavor. Instead of a simple description of differences, there is a *moral evaluation* of those differences. People are judged good or bad because of the category. Examples: "The poor are lazy and no good." "Students are a bunch of cheaters nowadays." *Stupid people, crazies, heathens,* and *pigs* are some names we give to people we would have a difficult time understanding, given the emotional names we attribute.

2. *A stereotype tends to be an absolute category.* In other words, there is a *sharp distinction made between those inside and those outside the category.* There is little recognition that the category is merely a guide to understanding and that, in reality, there will be many individuals—even a majority—within a category who are exceptions to any generalization. Examples: "Men are oppressive." "Women are compassionate." "Politicians are all dishonest." "All moral people are Christian." "African Americans are poor."

3. *The stereotype tends to be a category that overshadows all others in the mind of the observer.* All other categories to which the individual belongs tend to be ignored. A stereotype treats the human being as *simple and unidimensional,* belonging to only one category of consequence. In fact, we are all part of a large number of categories. There is an assumption that if someone belongs to that particular category, *that is all one needs to know about the person.* Examples: "He is a homosexual. Therefore, he lives a gay lifestyle." "She is a woman. Therefore, she must be sexually attracted to that man." "He is a churchgoer. Therefore, he can't be guilty of theft." No matter how accurate one category is, it is critical to remember that all of us are complex mixtures of many categories. Who, after all, is the African-American divorced intelligent poet, who never finished his freshman year of college, who is a Baptist, bisexual, father of

three, and grandfather of four? Which category matters? When we stereotype, it is the category that emotionally matters to us but not necessarily to the actor.

4. *A stereotype does not change with new evidence.* When one accepts a stereotype, the category and the ideas associated with it are *rigidly accepted,* and the individual who holds it *is unwilling to alter it.* The stereotype, once accepted, becomes a filter through which evidence is accepted or rejected. Examples: "Students don't care about college anymore—I really don't care what your study shows."

5. *The stereotype is not created carefully in the first place.* It is either learned culturally and simply accepted by the individual or created through *uncritical acceptance of a few concrete personal experiences.* Examples: "Politicians are bureaucrats who only care about keeping their jobs." "Obese people simply have no willpower. My sister was obese and she just couldn't stop eating."

6. *The stereotype does not encourage a search for understanding about why human beings are different from one another.* Instead of seeking to understand the cause as to why a certain quality is more in evidence in a particular category of people, a stereotype aims at exaggerating and judging differences. There is often an underlying assumption that *this is the way these people are,* it is part of their *essence,* and there seems to be little reason to try to understand the cause of differences any further than this. Examples: "Jewish people are just that way." "Poor people are just lazy." "Women don't know how to drive. That's the way they are."

Stereotypes are *highly oversimplified, exaggerated views of reality. They are especially attractive to people who are judgmental of others and who are quick to condemn people who are different from themselves.* They have been used to justify ethnic discrimination, war, and systematic murder of whole categories of people. Far from arising out of careful and systematic analysis, stereotypes *arise out of limited experience, hearsay, and culture.* Instead of aiding our understanding of the human being, *they always stand in the way of accurate understanding.*

It is not always easy to distinguish a stereotype from an accurate category. It is probably best to consider stereotypes and their opposites as extremes on a continuum. In actual fact, most categories will be neither perfectly accurate nor perfect examples of stereotypes. There are, therefore, *degrees of stereotyping* that we should recognize:

| Stereotype | Accurate Categorization |
|---|---|
| Judgmental | Descriptive |
| No exceptions | Exceptions |
| All-powerful category | One of many |
| Rejects new evidence | Changes with evidence |
| Not carefully created | Carefully created |
| Not interested in cause | Interested in cause |

One final point: Stereotypes, we have emphasized, are judggmental. They are meant to simplify people so that we know which categories of people are good and which are to be avoided or condemned. This is the link between stereotypes and prejudice. *Prejudice* is an attitude toward a category of people that leads the actor to discriminate against individuals who are placed in that category. Always the category is a stereotype (judgmental, absolute, central, rigid, cultural, and uninterested in cause). When a prejudiced actor identifies an individual in the category, a lot is assumed to be true and disliked about that individual, and a negative response results. And once he or she acts in a negative way toward that individual, there is a ready-made justification: the stereotype ("I discriminate *because* this is the way they are!"). Stereotypes are oversimplifications of reality, and they act as both necessary elements of prejudice and rationalizations of it. Unfortunately, the stereotype also acts as a set of role expectations for those in the category, and too often people who are judged negatively are influenced to judge themselves accordingly.

## Social Science: A Reaction to Stereotypes

Creating categories about people and generalizing intelligently are quite difficult to do unless we work hard at it. A big part of a university education is to uncover and critically evaluate stereotypes in order to obtain a better understanding of reality. Each discipline in its own way attempts to teach the student to be more careful about categorizing and generalizing.

Because this book focuses on the perspective of sociology and social science, I would like to show how social science tries to rid us of stereotypes through the careful development of accurate categories and generalizations about human beings. *Social science is a highly disciplined process of investigation whose purpose is to question many of our uncritically accepted stereotypes and generalizations.* Social science does not always succeed. There are many instances of inaccuracies and even stereotyping that have resulted from poor science or from scientists simply not being sensitive to their own biases. It is important, however, to recognize that even though scientists make mistakes in their attempts to describe reality accurately, the whole thrust and spirit of social science is to control personal bias, to uncover unfounded assumptions about people, and to understand reality as objectively as possible. Here are some of the ways in which social science (as it is supposed to work) aims to create accurate categories and generalizations about human beings:

1. *Social science tries hard not to be judgmental about categories of people.* We recognize that generalizations and categories must not condemn or praise but must simply be guides to understanding. To stereotype is to emphasize qualities in others that we dislike or to emphasize qualities in others that are similar to our own that we like. To say that a category of people is lazy is to stereotype; to say that a group has a higher unemployment rate than another is to generalize carefully. To say that a group is filthy rich or trashy is to stereotype; to say that a group has a higher average income than other groups is to generalize carefully. It is sometimes difficult to draw the line between a stereotype and a generalization about people, but, in general, the purpose of each is different: To generalize simply is to try to understand other people; to stereotype is to put understanding aside in order to take a stand about other people, usually a negative one. Stereotyping prevents the individual from understanding reality.

    I am not claiming that people should stay away from making value judgments about categories of people. We all have values we believe in, and we must include these when we act around others. I try to stay away from violent individuals; I try to change racists and sexists. I make judgments about students who plagiarize and employers who do not treat employees with respect. However, such judgments should be made carefully and explicitly (on the table), and only after categories and generalizations have been developed out of a process whose purpose is to *understand.* Good social science tries to separate making value judgments about people from understanding people, because if we do both simultaneously, stereotyping will inevitably be the result. It may be all right to condemn certain qualities in others, but it should be based on objective categorization, not stereotyping. Perhaps one goal for students should be to work toward developing informed value judgments.

2. *Categories and generalizations in social science are rarely—if ever—absolute.* Social scientists begin with the assumption that it is difficult to generalize about people and that every time we do exceptions—and often a large number of them—are likely. By definition, all atheists do not believe in God, but there is absolutely nothing else we can say about all atheists. However, we might contend that

atheists tend to be more educated (but there are many exceptions to this), male rather than female (but there are many exceptions to this), and raised by atheist parents (but there are many exceptions to this). We can tease out generalizations about atheists from carefully studying them, but we will never find a quality that all of them have other than their belief that there is no God. This goes for every category of people we try to understand: those who commit suicide, those who abuse drugs, those who commit violent acts against children, serial killers, and students who do not finish college. We can generalize, but we must be careful, and we must assume exceptions within every category we create.

The scientific generalization is treated as a probability rather than as an absolute. So we might say that, among young adults, less than 10 percent regularly use illegal drugs. Stereotyping, on the other hand, admits to few exceptions and involves rigid and absolute divisions between people, assuming that each individual within the category has the quality identified. ("Young people today are a bunch of drug addicts.") To claim that "Jewish people are rich" is to stereotype. To declare that Jewish people in the United States have the second highest average income per capita of any religious group is to generalize carefully. When averages are used to compare people, we must be probabilistic: We will recognize that there is a whole range of incomes among Jewish people, but that their average will simply be higher than most, lower than some. Whether a certain individual who is Jewish will be found to be wealthy is not easily predictable: Most Jewish people are not wealthy, and many are poor ("wealthy" too must be carefully defined).

Good social science even attempts to identify exactly how many exceptions within a category actually exist. We try to make the exceptions precise: In 1993, for example, 11 percent of the white population was poor (89 percent was not), as were 33 percent of the African-American population and 29 percent of the Latino population. We are unwilling to say that "everyone is getting divorced nowadays." Instead, we might say that "if the divorce rate remains as high as it is now, in the current generation of those who marry approximately one in two marriages will end in divorce" (there is, approximately, a 50 percent exception rate to "everyone getting divorced").

3. *Categories in social science are not assumed to be all-important for understanding the individual.* A stereotype is itself an assumption that a certain category necessarily dominates an individual's life. We might meet a young African-American single male artist. The role of each of these categories may or may not be important to the

individual. For some individuals, being male or single or an artist will be most influential; for others it will be being African American. For those of us who stereotype by race, it will almost always be African American.

People who are gay live in the world of the working or middle class, too; the business, professional, or artistic worlds; the city or the rural community; the religious or the nonreligious community. Human beings should never be pigeonholed into one category if we are to be accurate.

4. *Social science tries to create categories and generalizations through carefully gathered evidence.* Stereotypes tend to be cultural; that is, they are taught by people around us who have generalized based on what they have simply accepted from others or what they have learned through personal experience (which is usually extremely limited in scope, unsystematic, subject to personal and social biases, and uncritically observed). Science tries hard to encourage accurate generalizations by explicitly describing how generalizations must be arrived at. In fact, those who stereotype rarely know exactly where their categorization has come from, and normally when they are pushed,they will admit that it is something they picked up or it is based on limited experience. Scientists, on the other hand, are supposed to know exactly where their generalizations have come from. They normally can point to evidence that has been derived from studies that have been reported and analyzed over and over again. Scientists—as well as most intellectuals—put their faith in process (how ideas are arrived at); most of the rest of us (who too often stereotype) rarely question the process by which we have arrived at our generalization; instead, we simply accept it.

5. *Generalizations in social science are tentative and subject to change because new evidence is constantly being examined.* Stereotypes, on the other hand, are unconditionally held. Once accepted, a stereotype causes the individual to select only that evidence that reaffirms that stereotype. A stereotype resists change. When we believe that whites have superior abilities to nonwhites, we tend to notice only those individual cases that support our stereotype. If we believe that politicians are selfish bureaucrats, then we tend to ignore all those political leaders who are unselfish and who get things done. (Note: The category "politician" gives away that one is stereotyping rather than simply generalizing, because "politician" has come to mean someone who cannot be worthy of our respect.) Because the purpose of a stereotype is to condemn or praise a category of people, it becomes difficult to evaluate evidence. The stereotype is embedded in

the mind of the observer, it takes on an emotional flavor, and evidence that might contradict it is almost impossible to accept.

A generalization in social science about a category of people is subject to change as soon as new evidence is discovered. The final truth about people is never assumed to have been found. The generalization is always taken as a tentative guide to understanding rather than a quality that is etched in stone.

6. *Scientists do not categorize as an end in itself.* Instead, scientists categorize because they seek a certain kind of generalization: They seek to understand cause. In social science, that means we seek to know why a category of people tends to have a certain quality. We seek to understand the cause of schizophrenia, but we can do this, only after we understand what characterizes those people who are schizophrenic. How is this category of people different from others? We seek to know why poverty exists in this society. First, we must understand what poverty is (that is, we try to describe the lives of the people whom we label *poor*). What do these people have in common, if anything? How are their lives different from other people's? Then we ask, what has led them to poverty? For example, how many come from families who were also poor? How many are single parents? How many are children? How many are products of corporate restructuring? How many have job skills that are no longer needed? Can we, through our studies, determine some of the social conditions that led to poverty among most or even a significant minority of these people? That is what social science tries to understand.

Those who stereotype are not normally interested in cause. The category stands by itself as important. Often, according to Roger Brown (1965: 181–189), it is good enough simply to believe a certain quality is "part of their essence" and to ignore its cause.

*Why* is there increasing individualism among Americans? *Why* do some people graduate college and not others? *Why* are women almost absent from the top political and economic positions in American life? *Why* do some women make it to the top? *Why* is there an increase in the number of people who are experiencing downward social mobility in the United States? *Why* is there a rising suicide rate among young people? In every one of these cases, we find a category, we describe those who make up that category, and we attempt to generalize as to why a certain quality exists in that category. To judge? No. True of everyone in the category? No. The only category of importance? No. A fixed category that clearly and absolutely distinguishes between one group and another? No. A generalization that we can regard as true without reservation? No.

## Summary and Conclusion

Social science is sometimes misinterpreted by the public. Remember that those of us who stereotype seek evidence to support our stereotypes. We lie in the weeds—so to speak—waiting to pounce on any evidence that supports us (ignoring all evidence that does not). As careful as social science might be, what scientists find can be stretched a long way and misused. There is evidence, for example, that African Americans do less well on standardized intelligence tests than whites. For the social scientist, this is a tentative generalization; it is puzzling and needs more understanding. The social scientist wants to know why and will look at any inherent bias in the test as well as the social conditions that might lead to this discrepancy. There is no sweeping, absolute generalization here. After all, we are talking about averages. There is no attempt to condemn or defend ethnic groups, to justify or rationalize racism. To the racist, however, this might become more evidence that whites are superior people. It might be used to reinforce a stereotype. This is why, unfortunately, people who try hard to understand categories of people carefully and objectively (as social scientists are supposed to do) become frustrated by those who exaggerate and twist what is found to fit their stereotypes.

Before we forget where we began this discussion, let me remind you once more:

- Human beings categorize.
- Human beings generalize.
- We must categorize and generalize.
- It is important to generalize carefully, and, when it comes to people, we should try to keep away from stereotyping if we want to understand them.
- Our generalizations about people must attempt to understand; our generalizations must be considered only tendencies among certain people; they must be accepted as open, tentative generalizations; and we must become aware of how we have arrived at our generalizations, always keeping in mind the importance of good evidence.

Generalizations must also respect the complexity of the individual, and we should seek to understand why people differ and be suspicious of those who simply categorize in order to condemn.

Those of us who are *victims* of stereotyping know full well the dangers of sloppy generalization. It is one thing to stereotype plants or rocks or stars; it is quite another to stereotype people. When we stereotype people, our carelessness normally has a negative effect on individuals who are part of that particular category. We unfairly place them at a disadvantage, not giving them a chance as individuals, making judgments about them based on inaccuracies and on our own unwillingness to evaluate our generalization critically.

Even those of us who are not victims will occasionally cry out, "I am an individual! Do not categorize me." We *are* individuals. *No one is exactly like us.* Yet, if we are honest, we must also recognize that those who do not know us will be forced to categorize us, and those who honestly want to understand humans better will have to. It is not a problem for us if the category is carefully created; and it is not a problem for us if the category is a positive one. If we apply for a job, we want the employer to categorize us as dependable, hardworking, knowledgeable, intelligent, and so on. Actually, we will even try to control how we present ourselves in situations so that we are able to influence others to place us in favorable categories: I'm cool, intelligent, sensitive, athletically talented, educated. When I write a letter of recommendation for students, I place the individual into several categories so that the reader will be able to apply what he or she knows about that category to the individual. The doctor may tell people "I am a physician" so that they will think highly of him or her as an individual. The person who announces himself as a boxer is telling us that he is tough; the rock musician is telling us that she is talented; the minister that he or she is caring—in many such cases it does not seem so bad if we are being categorized. For almost all of us, however, it is the *negative* categorization that we wish to avoid. And this makes good sense: No one wants to be put into a category and negatively judged without having a chance to prove himself or herself as an individual.

But no matter how we might feel about others categorizing us and applying what they know to understanding us as a member of that category, the fact is that, except for those we know well, human beings can be understood only if we categorize and generalize. If we do this carefully, we can understand much about them, but if we are sloppy, we sacrifice understanding and end up making irrationally based value judgments about people before we have an opportunity to know them as individuals.

We should not throw careful generalizations out the window in the name of treating all people as individuals. As much as every individual might deserve being understood and treated as an individual, knowledge about anything—including human beings—is possible only through generalization. HIV is spread through the transmission of bodily fluids through sexual contact, blood transfusion, or drug use— that is a generalization that can cause death if ignored. The history of African Americans has been one of active and subtle discrimination by the white community—that is a key to understanding many of the problems that are important in American society today. The upper class in American society has more privilege than any other class in the political, educational, and legal systems—that is an important generalization that sensitizes the individual to the limitations of our democracy. None of these generalizations is absolute, unbending, or certain, and none is meant to condemn or defend any category of people. They are not stereotypes.

Social science—and sociology as a social science—is an attempt to categorize and generalize about human beings and society, but always in a careful manner. Its purpose is to reject stereotyping. It is a recognition that generalizing about people is necessary and inevitable, but stereotyping is not.

If we have to generalize, let's try to be careful. Stereotyping does not serve our own interests well because it blocks understanding; nor does it help those we stereotype.

## Questions to Consider

1. What exactly does it mean "to understand" something?
2. What principles should be followed if one is going to generalize intelligently about people? Or is generalizing about people not an intelligent thing to do?
3. Can one judge a group without stereotyping?
4. What dangers exist as a result of overgeneralizations made by social scientists?
5. How would a poet answer the question, Should we generalize about people? How would a psychologist answer the question?

## References

Of course, almost any book in social science will be an example of categorization and generalization. It is difficult to find a source that specifically discusses the issues described in this chapter. Here are works that have especially influenced me in understanding social science, stereotyping, categorizing, and generalizing.

**Adorno, Theodor W., Else Frenkel-Brunswick, D. J. Levinson, and R. N. Sanford** 1950 *The Authoritarian Personality.* New York: Harper & Row.

**Allport, Gordon** 1980 *The Nature of Prejudice.* Reading, MA: Addison-Wesley.

**Anderson, Eric** 2005 *In the Game: Athletes and the Cult of Masculinity.* New York: State University of New York.

**Aronson, Elliot** 1995 *The Social Animal.* 7th ed. San Francisco: W. H. Freeman.

**Babbie, Earl M.** 1997 *The Practice of Social Research.* 8th ed. Belmont, CA: Wadsworth.

**Bailey, Kenneth D.** 1999 *Methods of Social Research.* 5th ed. New York: Free Press.

**Bainbridge, William Sims** 1997 *The Sociology of Religious Movements.* New York: Routledge.

**Barton, Bernadette** 2006 *Stripped: Inside the Lives of Exotic Dancers.* New York: New York University Press.

**Becker, Howard S.** 1973 *Outsiders.* Enlarged ed. New York: Free Press.

**Berger, Peter L., and Thomas Luckmann** 1966 *The Social Construction of Reality.* Garden City, NY: Doubleday.

**Best, Joel** 2001 *Damned Lies and Statistics: Untangling Numbers from the Media, Politicians, and Activists.* Berkeley: University of California Press.

**Blumer, Herbert** 1969 *Symbolic Interactionism: Perspective and Method.* Englewood Cliffs, NJ: Prentice Hall.

**Borchard, Kurt** 2005 *The Word on the Street: Homeless Men in Las Vegas.* Reno,NV: University of Nevada Press.

**Brown, Roger** 1965 *Social Psychology.* New York: Free Press.

**Cohen, Morris R., and Ernest Nagel** 1934 *An Introduction to Logic and Scientific Method.* New York: Harcourt Brace Jovanovich.

**Currie, Elliott** 2005 *The Road to Whatever: Middle-Class Culture and the Crisis of Adolescence*. New York: Henry Holt and Company.

**Denzin, Norman K., and Yvonna S. Lincoln (Eds.)** 2002 *The Qualitative Inquiry Reader.* Thousand Oaks, CA: Sage.

**Dooley, David** 2001 *Social Research Methods.* Upper Saddle River, NJ: Prentice Hall.

**Edin, Kathryn, and Maria Kefalas** 2007 *Promises I Can Keep: Why Poor Women Put Motherhood Before Marriage*. Berkeley, CA: University of California Press.

**Ehrlich, Howard J.** 1973 *The Social Psychology of Prejudice.* New York: Wiley Interscience.

**Emerson, Robert M. (Ed.)** 2001 *Contemporary Field Research: Perspectives and Formulations.* 2nd ed. Prospect Heights, IL: Waveland Press.

**Goffman, Erving** 1963 *Stigma: Notes on the Management of Spoiled Identity*. Englewood Cliffs, NJ: Prentice Hall.

**Goldblatt, David (Ed.)** 2000 *Knowledge and the Social Sciences: Theory, Method, Practice.* New York: Routledge/Open University.

**Goldhagen, Daniel Jonah** 1996 *Hitler's Willing Executioners: Ordinary Germans and the Holocaust.* New York: Alfred A. Knopf.

**Gusfield, Joseph R.** 2000 *Performing Action: Artistry in Human Behavior and Social Research*. New Brunswick, NJ: Transaction.

**Hamilton, Davis L.** 1981 *Cognitive Processes in Stereotyping and Intergroup Behavior.* Hillsdale, NJ: Lawrence Erlbaum.

**Hancock, Angie-Marie** 2004 *The Politics of Disgust: The Public Identity of the Welfare Queen.* New York: New York University Press.

**Hayes, Sharon** 2003 *Flat Broke with Children: Women in the Age of Welfare Reform.* New York: Oxford University Press.

**Helmreich, William B.** 1984 *The Things They Say behind Your Back: Stereotypes and the Myths behind Them.* New Brunswick, NJ: Transaction.

**Hertzler, Joyce O.** 1965 *A Sociology of Language.* New York: Random House.

**Hull, Kathleen E.** 2006 *Same-Sex Marriage: The Culture Politics of Love and Law.* Cambridge: Cambridge University Press.

**Iceland, John** 2003 *Poverty in America: A Handbook.* Berkeley, CA: University of California Press.

**Jacobs, Aton K.** 2006 "The New Right, Fundamentalism, and Nationalism in Postmodern America: The Marriage of Heat and Passion." In *Social Compass,* 53:(3): 357–366.

**Jacobs, Nancy R., Mark A. Segal, and Carol D. Foster** 1994 *Into the Third Century: A Social Profile of America.* Wylie, TX: Information Plus.

**Keister, Lisa A.** 2005 *Getting Rich: America's New Rich and How They Get That Way.* New York: Cambridge University Press.

**Kincaid, Harold** 1996 *Philosophical Foundations of the Social Sciences.* Cambridge, England: Cambridge University Press.

**Linden, Eugene** 1993 "Can Animals Think?" *Time,* 141:(12) (March 22): 54–61.

**McNamee, Stephen J., and Robert K. Miller**, Jr. 2004 *The Meritocracy Myth.* Landham,MD: Rowman & Littlefield Publishers, Inc.

**Mills, C. Wright** 1959 *The Sociological Imagination.* New York: Oxford University Press.

**Neuman, William Lawrence** 2000 *Social Research Methods: Qualitative and Quantitative Approaches.* Boston: Allyn & Bacon.

——— 2004 *Basics of Social Research: Qualitative and Quantitative Approaches.* Boston: Pearson.

**Newman, Katherine** 2004 *The Social Roots of School Shootings.* New York: Basic Books.

**Parenti, Michael** 1998 *America Besieged.* San Francisco: City Lights Books.

**Phillips, Bernard S.** 2001 *Beyond Sociology's Tower of Babel: Reconstructing the Scientific Method.* New York: Aldine de Gruyter.

**Rotenberg, Paula S.** 2002 *White Privilege: Essential Readings on the Other Side of Racism.* New York: Worth Publishers

**Ryan, William** 1976 *Blaming the Victim*. Rev. ed. New York: Vintage.

**Scott, Jacqueline L., Judith K. Treas, and Martin Richards (Eds.)** 2004 *The Blackwell Companion to the Sociology of Families*. Malden, MA: Blackwell.

**Shapiro, Thomas M.** 2003 *The Hidden Costs of Being AfricanAmerican*. New York: Oxford University Press.

**Shibutani, Tamotsu** 1970 "On the Personification of Adversaries." In *Human Nature and Collective Behavior*. Ed. Tamotsu Shibutani. Englewood Cliffs, NJ: Prentice Hall.

**Simpson, George E., and J. Milton Yinger** 1985 *Racial and Cultural Minorities: An Analysis of Prejudice and Discrimination*. 5th ed. New York: Harper & Row.

**Steger, Manfred** 2003 *Globalization: A Very Short Introduction*. New York: Oxford University Press.

**Sullivan, Maureen** 2004 *The Family of Woman: Lesbian Mothers, Their Children, and Undoing of Gender*. Berkeley, CA: University of California Press.

**U.S. Bureau of the Census** 1993 *Statistical Abstract of the United States*. Washington, DC: U.S. Government Printing Office.

**Waldinger, Roger David, and Michael I. Lichter**. 2003 *How the Other Half Works: Immigration and the Social Organization of Labor*. Berkeley, CA: University of California Press.

**Whorf, Benjamin Lee** 1956 *Language, Thought, and Reality*. New York: John Wiley.

**Witt, Griff** 2004 "As Income Gap Widens, Uncertainty Spreads." *The Washington Post* (September 20).

**Zetterberg, Hans L.** 2002 *Social Theory and Social Practice*. New Brunswick, NJ: Transaction.

**Zirakzadeh, Cyrus** 2006 *Social Movements in Politics: A Comparative Study*. New York: Palgrave McMillan.

# Glossary

**Age cohort** Generational position in society. Position in a social structure based on age.

**Alienation** Separation from other people, meaningful work, or ourselves as active beings.

**Authority** Legitimate power; a structure of power in organization that is relatively stable and in which people come to believe that others have a right to command and they feel an obligation to obey; a system of inequality regarded as legitimate and right. A relatively stable system of power built on people's loyalty to the community or society.

**Beliefs** What people believe; the ideas they hold to be true.

**Bureaucracy** A certain system of authority in which ranking is intentionally created in an organization for purposes of efficiency. Lines of authority are clearly spelled out, and obedience to authority becomes a central value. That system of organization that dominates the twentieth and twenty-first centuries.

**Capitalism** The most common definition of capitalism is an economic system in which most of the means of production and wealth are in private hands and in which private accumulation of profits is a primary incentive for owning and operating business. Prices are determined by supply and demand in an open market, and, theoretically at least, government intervention in the economy is discouraged, especially if it works against private ownership, individual profit, and the interests of those who have power in the economic system.

Often called laissez faire capitalism–does not exist in reality. Instead, the concept "capitalism" should be understood as one end of a continuum. Thus, we need to understand that some societies are more capitalistic than others and some periods of a society's development will be more capitalistic than another period. Capitalism is an economic system; democracy (also never pure) is an political system or a type of society.

To Karl Marx, capitalism was an economic system controlled by and for the capitalist class, those who owned the factories and banks, and who thus were able to control all of society. To Marx, capitalism was a corrupt, selfish system that exploited laborers and allowed owners to take profit for themselves without caring for the welfare of anyone else or society at large. Far from being a free system, Marx believed, capitalism's inequities enslaved workers and even dominated capitalists, who had no choice but to be concerned only with achieving more and more profit. Inevitably, capitalism would lead to communism, which Marx saw as a much more just and democratic economic system.

To Max Weber, capitalism was a certain approach to doing business. Capitalists typically created and organized their own business for the primary purpose of profit. Where capitalism prevailed in society, so did creative business entrepreneurs, who built and owned their own small businesses and sought meaning in work and success in accumulating profit. Weber saw capitalism as a stage in the history of society, a stage that ultimately would lead to a time when giant bureaucratic enterprise would come to dominate society.

**Category, categorization** The isolation of a chunk of reality, distinguishing that chunk from all other parts of reality, giving it a name, and associating ideas with it.

An isolated part of the environment we notice and generalize about. An important part of human socialization making it possible to understand the environment in complex and sophisticated ways; a part of the human ability to see important and subtle similarities and distinctions not available to those animals that do not categorize and generalize with words.

Creating useful and accurate categories about people is difficult to do unless we work hard at it. An inaccurate and sloppy category whose purpose is not to understand people but to condemn them is *stereotyping.* A isefi; categpr os desciptive, allows for exceptions, is recognized to be one of many categories used to describe a given object, changes with evidence, is carefully created, and facilitates understanding cause.

**Charismatic authority** To Max Weber, authority that arises from the perception that a certain individual is extraordinary and special, representative of a community of believers, and wrapped within a sacred source. Tends to arise among people who are dissatisfied with the traditional or legal-rational authority in place and therefore is a revolutionary force for social change. Also tends to arise within periods of history when the old world is collapsing, institutions no longer seem to work well, and old ideas and authority no longer seem to be legitimate.

**Civil religion** The United States is sometimes described as having a civil religion, a political philosophy taught to and practiced by its citizens that values democracy, pluralism, and diversity; functions like a religion; has become more important than any single organized religion; is sacred in its own right; and is thus important for bringing people together into community.

**Class society** Society in which some are advantaged based on the economic resources they are able to accumulate; a society where one is able to pass down advantages or disadvantages to children through educational opportunities, social contacts, or direct inheritance.

**Commitment to Society** Self-imposed emotional duty and loyalty by members of a society.

**Competition** A form of conflict that takes place within clearly specified and accepted rules.

**Conditional loyalty** Commitment to organization based on whether it benefits the actor or conforms to principles in which he or she believes.

**Constructive social conflict** Conflict based on disagreement; promotes constructive solution to problems; discourages destructive social conflict; conflict characterized by negotiation and compromise; conflict where all parties are able to achieve something, organization is able to change, people's interests are heard, and real problems are identified and dealt with.

**Critical** Habitual questioning of the ideas held and presented by others as well as ideas that one has learned.

**Cultural Theme or Cultural Trend** Used in this book to express the idea that there are three general cultural directions throughout the world, amking it very difficult to argue that the world is becoming one culture. The three themes are the modern/global/capitalist, the traditional-religious, and the human rights-democratic.

**Culture** One of three social patterns in society (see also *social institutions* and *social structure*); arises out of social interaction and taught in social interaction; made up of rules, beliefs, and values that are useful to people in social organization. Agreement among people in organization; a shared perspective.

Facilitates cooperation within organization.

A set of taken-for-granted truths; a set of assumptions accepted by people within an organization without serious questioning. A central part of what is taught by various institutions in society.

What people use to guide understanding and self-control. Culture is used to perceive both reality outside ourselves and within ourselves. A context within which experience is perceived and interpreted.

Makes it possible for people to understand rather than to respond; makes possible diversity among societies as well as the different directions people may take in life. Because of culture, humans are able to interpret the world they observe, and their lives are not fixed by biology.

**Democracy** A society characterized by four qualities: (1) Individuals are free, (2) government and other sources of power are effectively limited, (3) human differences are respected and protected, and (4) equal opportunity exists for all. Is a matter of degree rather than an absolute. To Alexis de Tocqueville, a society difficult to achieve; possible only with certain social conditions and social patterns that support its continued existence; easily lost.

**Destructive social conflict** Conflict governed by anger and hostility; threatens the continuation of organization. The other side is seen as the enemy, anger and a desire to hurt or destroy others are encouraged, and escalation to violence is common.

Ignores real issues that need to be dealt with for social organization to continue.

An important cause of human misery. Destructive of the victim, the perpetrator, and the legitimate order.

Often a product of social inequality or socialization; encourages violence and running away from serious problems that need to be addressed; also occurs when constructive conflict is repressed and negotiation and compromise are discouraged.

**Division of labor** Dividing activities among people in an organization. Diversity of occupations, division between employers and employees, division between leaders and followers. An important cause of social inequality.

**Empirical** Proof arising from careful observation of events in nature; proof that forms the basis for scientific discovery. Form of proof in both natural and social science, including sociology. An attempt to support or refute an idea on the basis of independent observation by many individuals.

Similar to rational proof because both are attempts to apply neutral measuring sticks to determine whether an idea is true.

**Ethnocentrism** The tendency for people in a social organization to regard their own culture as central to the universe and to believe their own ideas, values, and rules must be true; leads to judging other people according to how close they are to this organizational culture. Ethnocentrism involves little or no evidence but assumes that one's own socially created culture is right.

**Exploitation** Selfish use of other human beings as a means to one's own ends.

**Fascism** A set of beliefs, norms, and values developed in Germany and Italy after World War I that became a basic part of Nazi Germany under Hitler; a political ideology that includes a positive attraction to war and power, a belief in the natural inequality among categories of people, and a tendency to see democracy and freedom as human weakness. A political ideology that is used by some political leaders to try to justify their own policies through appealing to a people's fears as well as their strong feelings of nationalism.

**Flat world** A concept created by Thomas Friedman. The interdependence of the world creates a "flat world," one where it is not simply nations or corporations or giant organizations are competing, but increasingly individuals are. It is a world where anyone who knows the veloping communication technology can become competitive. Those individuals in the world who did not have the tools to compete with the United States and Western Europe, now have a much better competitive opportunity to succeed educationally and economically. Businesses, schools, news agencies, commodities, ideas, and work no longer are divided between various societies and nations. Almost everyone has access. This view of the world is what Friedman sees as the dominant trend. Giant corporations and other organizations continue to exist, but they have much greater efficiency, making it possible to find laborers, capital, professionals, buyers, knowledge, and information throughout the world.

**Free capital** Capital is wealth, especially wealth that is easily invested in creating private and public business. The term "free capital" is sometimes used to create a situaiton where capital can be easily invested anywhere in the world without government protection or regulaiton.

**Free trade** Trade within and among nations and societies without government resticion, encouraging trade without taxes, regulation, and tariffs on commodities. Like capitalism, freedom, and democracy, pure free trade does not exist in any pure way. Instead, it is a matter of degree. Free trade, like capitalism, limits the intervention of government, so that the free market place allows maximum profit to business.

**Freedom** The actor is in control of his or her own life; understanding and determining choices, directing one's self, taking an active stance in relation to the environment. Making a difference in one's own life.

Freedom has to do with both thinking and action. Free thinking is a prerequisite for free action, but it does not necessarily lead to free action. Free action is movement that is not interfered with by factors outside one's own control. Free action is restricted by limits on thinking, social institutions, socialization, social structure, and positions within social structure.

Freedom is always relative: It is never absolute, and there are always limits. If it exists at all, it arises from the use of language, self, and mind— all socially derived qualities.

One of the characteristics of a democratic society.

***Gemeinschaft*** A feeling of community; a sense of "we."

**Generalization, generalizing** An essential part of understanding. The application of what one learns in one situation to similar situations. A statement that describes a whole category of objects created by humans as they try to understand reality. Encourages the individual to see differences and similarities of objects and events in the universe. Often a generalization is an attempt to understand the cause and effects of various objects and events.

If not done well, generalizing can actually stand in the way of understanding. The purpose of science is to generalize carefully; the purpose of social science is to develop good generalizations and to question poor ones about the human being.

**Generalized other** The integration of one's significant others into a consistent whole whose rules become guides to the individual. Instead of understanding and using the rules of one individual at a time, the individual is able to generalize to the group or society as a guide to action.

**Globalization** A process, a direction, a trend, a cluster of certain activities throughout most of he world. The drection is toward a world system, an integrated interdependent world. Much of this is economic, but there are also technological, political, cultural, and social trends toward a world order. This trend goes back to at least the sixteenth century, but now it is accelerating greatly. Communication is international, corporations are multi-national, trade, immigration, outsourcing, and the spread of Western culture are some of the activities that make up globalization. Some believe it is a dramatic change; some believe it is simply a long term developing trend over centuries; some see it to be a good development; some are very critical.

**Homogenization of culture** Many believe that the world is becoming one culture, and thus the people of the world aare losing diversity and moving toward sameness. "Homogenization" is usually a term is critical of globalization, United States cultural domination, and shalloenss.

**Human beings** Mammals and primates who are characterized by being social and cultural.

To Mead, humans are unique in nature because of symbol use, selfhood, and mind—all socially created qualities; these allow humans to create, shape, and change society and the environment, as well as be shaped by it.

**Identity** Both the name one gives oneself and announces to others through action. Formed in social interaction; tied to position in social structure or commitment to the group itself.

**Ideology** That part of culture that works to defend society as it exists rather than to simply try to understand and explain. An attempt to justify the way society works; tends to be created and taught by those who are powerful in society; thus a large part of ideology is meant to protect the position of the powerful in society and to justify the system of inequality that prevails.

For some, "capitalism" and "globalization" are examples of ideology.

**Instinct** The biological origin of behavior; much of the behavior of most animals. A certain instinct is characteristic of a species or other grouping of animals. An instinct is usually a pattern of complex action rather than a simple reflex. An instinct allows the organism to do what it has to do in nature without needing to learn it. Humans are thought to have many simple reflexes, but have few if any instincts and thus must learn how to survive in their environment.

**Integration** Uniting individuals into a whole; bringing people together in an organization.

**Iron law of oligarchy** Michels's view that organization inevitably creates oligarchy when leaders are chosen. Oligarchy means that a few people end up dominating the organization, and the members of that organization end up following what the few expect or demand. According to Michels, democracy is a fiction as long as people organize themselves.

**Liberal arts education** An education that is meant to be liberating for the student; one whose purpose is to seriously challenge what the student takes for granted and to encourage the student to break out of the immediate world of his or her earlier socialization and to broaden his or her views and interests.

**Loyalty** An important way in which organization is made possible. Commitment to the whole; feeling positive about being a part of the whole social organization. Tendency to believe that what we are part of should be regarded as important and right;

brings a feeling of obligation to serve and to defend; linked to willingness to obey authority; often encourages ethnocentrism.

**Meaning** In sociology, meaning is used in two ways. When it is applied to social interaction and symbolic communication, meaning is that which symbols represent. Meaning is socially designated so that people can understand one another. In the sociology of religion, meaning is that which is important to life. One purpose of religion is to bring meaning to one's life. Religion attempts to explain that which is difficult for people to understand. It is an attempt to understand our place in the universe, and it attempts to give life purpose and fulfillment. Religion is a community's attempt to bring meaning to life above the immediate, physical, and profane.

**Mind** The action the individual takes toward himor herself that we call *thinking;* internal communication about the environment. Socially created because it depends on two other socially created qualities: symbols and self.

**Misery** A state of chronic suffering and unhappiness; a subjective feeling created by objective conditions, sometimes biological but often social in origin. In sociology, the objective social conditions are usually created by the social patterns that exist.

Although impossible to end, one of the aims of a democratic society is to lessen human misery and the social conditions that create it.

**Modernization** The social trend away from traditional society, where the present and future become more important than the past, where progress in this life is believed to be achievable, where science, rationalism, urbanization, industrialization, bureaucratic organization, social and geographic mobility, individualism, and perhaps secularization become increasingly important.

**Nation** A political organization of people; a political organization includes government, law, and political boundaries. Usually refers to a political organization that is distinguished from kingdoms, fiefdoms, city-states, and empires; a particularly modern political organization that is characteristic of much of the world today.

May include one society, a part of a society, or more than one society.

**Nationalism** A feeling by a people that it constitutes a society and has a right to establish a nation; a feeling by people of loyalty to a nation that represents their society.

**Natural cause** The assumption in science that natural events are caused by other natural events.

Natural cause is not easy to isolate and understand and is often probabilistic, emphasizing tendencies in nature rather than absolute links.

**Natural law** The assumption of science that nature is governed by predictable regularities; the belief that the past can be explained and the future predicted because of this regularity. Natural events are thought to be patterned rather than simply random or the result of supernatural forces.

**Objectivity** The attempt to observe the world as it really is—that is, as an "object" to be understood apart from our own subjective perception of it. Although impossible to have in any absolute sense, objectivity is a primary goal in science, including social science.

**Organized religion** A religion that exists within a community, with a structure, culture, and set of institutions. Organized religion contrasts with both less established informal religions and people who are individually religious. In sociology, organized religions are often divided into major religions, denominations, sects, and cults.

**Outsourcing** The activity by which businesses are able to hire workers anywhere in the world in order to lower costs, increase efficiency and productivity. Competition for jobs become worldwide rather within a single nation-state or society.

**Perspectives** An angle on reality; an approach to perceiving and understanding reality; a context within which an individual or organization interprets reality.

A socially constructed outlook. The culture of any social organization is a perspective; the positions we have in organization also become our perspectives.

The individual is encouraged to take on a perspective by being embedded in the social organization and through both continuous social interaction in the organization and social sanctions by the organization, as well as by the need to check out views with those around him or her. A perspective is believed by the individual as long as it is useful in dealing with his or her situation. Usefulness depends on commitment to the organization, the need to be successful in one's position, and the need to understand and handle situations as they arise.

A perspective influences perception, thinking, and action. Perspectives are often shaped by the powerful in organizations.

**Pluralism** In relation to religion, pluralism is a characteristic of society where a diversity of religions exist with the recognition that they are to be tolerated. In a pluralistic society, it is not one religion that unites society but tolerance of and belief in diversity.

**Positions** Locations in social structure; placement in organization. Each location has attached to it a role, a perspective, an identity, and a rank. Positions are ranked according to their attendant power, privilege, and prestige.

Positions act as angles from which one sees reality; they, along with culture, create the perspectives that influence the actor's thoughts.

A position influences how the actor is supposed to act in the world (role), the identity the actor assumes in the world, the perspective the actor uses to see reality, and the amount of power, privilege, and prestige the actor has in the organization. Because positions are so important for what actors become, when they change position, they will be influenced to change and are highly likely to do so.

**Poverty** An economic position in society that means dependence on others for one's own survival; it means lack of control over one's own existence and lack of power in the direction of society and its patterns. (See also *social power*.)

**Prestige** Honor accorded to a position in social structure. Honor that others accord to an individual actor because of his or her social position.

Prestige is one of three qualities that rank positions in social structure. It is usually but not always associated with the other two qualities (privilege and power), so there is a strong tendency for the amount of prestige to be associated with the amount of privilege and power in position. Indeed, the amount of prestige usually influences the amount of power and privilege; the amount of power and privilege usually influences the amount of prestige.

**Primary group** A small, relatively permanent, intimate, and unspecialized group that develops a sense of "we"; a face-to-face group that entails close emotional ties. A type of group that is less characteristic of groups in modern society where impersonality and individualism tend to dominate.

**Primary socialization** Earliest, most powerful, and most basic socialization that the individual experiences. Where we first experience affection, support, and the ability to exercise self-control.

A filter through which we view new approaches to reality and new ideas. It tends to ensure stability of belief over time.

**Private property** Ownership of whatever is both valued in a given society and can be captured and controlled by an individual; a source of power that allows those who have it to accumulate more.

**Privilege** Something in society that is valued and struggled over by actors; benefits that are unequally distributed to positions in social structure. Distribution of privilege favors those at the top of the social structure; it is difficult for others to achieve and thus is a source of power and prestige for those in higher positions.

Privilege is one of three qualities that rank positions in social structure. Usually but not always associated with the other two qualities (prestige and social power), so there is a strong tendency for the privilege granted in a position to be associated with the amount of prestige and social power. Indeed, the privilege obtained through position usually influences the amount of social power and prestige; the amount of social power and prestige usually influences the privilege one gets.

**Proof** The use of rational or empirical evidence to establish an idea's truth or falsehood. Using a neutral measuring stick rather than the beliefs one already accepts to determine the truth or falsehood of an idea.

**Rational proof** Proof governed by the rules of logic, considered by many to be the basis of good thinking. Testing ideas through the use of continuous questioning and thoughtful examination.

**Rationalization of life** To Max Weber, the dominant trend in modern society; a way of approaching all areas of life that relies on calculation, efficiency, problem solving, and goal-directed behavior; decreasing reliance on tradition, values, and feeling for guiding action.

**Religion** Émile Durkheim defined religion as a set of beliefs and practices that divides the universe into the sacred and the profane. Ultimately, what is designated as sacred represents the community itself. Religious beliefs and practices toward the sacred are really aimed at the community and help bring the community together. Durkheim believes that religion is universal. It is necessary for society to exist.

To Max Weber, religion is part of culture, a belief system that people hold toward the universe that affects much of what they do. It is important to a people's norms, values, and view of reality.

To Peter Berger, religion is a belief system that helps people make sense of their existence and the events they encounter.

To Karl Marx, religion is an important part of society meant to control individuals through giving them relief or escape from their oppressed state as well as protect the ruling class—that is, the wealthy and powerful.

In sociology, religion is a socially constructed view of the universe, a central part of a people's culture, necessary for the functioning of society. It is a recognition by a people that there is more to existence than the profane, immediate, and physical. It is important for both human action and the continuation of society. There is a strong tendency for sociology to regard religion as an organized community affair rather than simply as an individual's spiritual beliefs.

**Religious fundamentalism** A certain way of looking at reality found among certain religious individuals, communities, and social movements. Emphasis is placed on tradition, acceptance of a revealed truth, a disciplined rigid code, and a sharp distinction between believers and nonbelievers. Religious fundamentalism often attracts those who oppose several trends in modern life, especially secularization, individualism, and rationalism.

**Rituals** Traditional action in an organization whose primary function is to bind people together, to facilitate loyalty to the organization. Action whose primary function is to symbolize the organization rather than fulfill a practical purpose. Often, a ritual does in fact fulfill a practical purpose secondarily.

**Role** A script handed to the actor in the position that tells the actor how to act and think. A set of expectations that focus on the actor in a position. To act out a role is to act according to these expectations. Actors are influenced to act roles in positions, and others are constrained to accept performances that conform to the role.

**Rules** Part of culture; includes taboos, customs, morals, procedures, and informal expectations.

Part of positions in social structure; a set of rules—formal and informal—that make up one's role in one's position.

**Ruling class** Term used by Karl Marx to describe those who own the means of production in society and, as a consequence, dominate the economy, the government, and all other institutional areas.

**Sacred** The sacred, according to Durkheim, is that part of the universe that stands in contrast with the world of the profane (the everyday). It is socially created. The sacred is regarded as special, above the physical, beyond our senses, beyond the immediate, universal, and to be honored, respected, and held in awe. What is determined to be sacred is treated in a special way by believers. What is sacred is symbolic, especially of the community. The belief in the sacred is a recognition that there is more to life than the profane, physical, and immediate. The sacred contributes to the idea that life is meaningful; it is more important than simple pleasure.

**Sacred objects** Symbols of the community that encourage loyalty; the value of sacred objects is not found in their physical nature but in the worship or respect they are given because they symbolize the community. Objects that are valued more than for their everyday usage, but are thought to be above the ordinary and used to symbolize the special nature and rightness of the community.

**Science** An approach to understanding objects and events in nature that uses empirical proof as evidence. An approach to proof that relies on careful measurement based on our senses, primarily by observation. Similar to mathematics and philosophy in that it attempts to measure truth or falsehood through a neutral measuring stick rather than by what people have been taught by culture, others, or personal experience.

An approach to understanding that involves tentative conclusions rather than absolute final ones; an attempt to understand the natural universe objectively and carefully without allowing moral judgments to enter into the actual findings; an attempt to understand by recognizing that conclusions are open to further challenge.

An attempt to understand nature that assumes the existence of natural law and natural cause. The acceptance of natural law makes possible generalization; the acceptance of natural cause means that events in the natural world happen because of other identifiable natural events.

An attempt to examine without bias; a critical approach to understanding; a refusal to accept authority; an attempt to encourage open debate using empirical evidence and rational analysis; the testing of ideas democratically in the sense that it is not what someone says that makes something true but the evidence provided because it can be checked out by anyone who uses scientific procedures.

**Secondary socialization** Socialization arising from those with whom we interact after primary socialization. Often secondary socialization reaffirms what we learn in our primary socialization; in modern society, it becomes increasingly important, often changing our directions in significant ways.

**Secularization** A trend that some philosophers and social scientists believe occurs because of the modernization of society. Secularization decreases the importance of religion in society, especially traditional organized religion. Some thinkers believe that this trend is inevitable; others believe that religion changes in modern society but does not decline in importance. Some believe that secularization is a good trend; others are critical of this trend; still others simply see it as important but without being good or bad.

**Self** Socially created object that is integral to and makes possible actions by the actor directed inwardly. The self is the actor's internal environment to which he or she acts in response.

Makes possible self-awareness—that is, the ability to see one's own actions objectively. Makes possible the ability to judge and control ourselves.

Possession of self has consequences for action, including the ability of the actor to direct his or her own action in a situation.

**Self-control** A product of socialization; if not achieved, the individual will act impulsively without considering consequences and how action will affect others.

**Significant others** Individuals with whom we interact, who are important to us, and who socialize us. The origin of many of the beliefs, values, and rules the individual comes to accept.

**Social** Living around others, doing things in relation to others, cooperating with others, engaging in social conflict with others, needing others, being socialized by others, surviving through others, becoming human through others, existing in social organizations.

**Social action** Action that takes other people into account.

**Social change** An inherent part of all social organization; results from the acts of individuals, groups, and social movements, as well as social conflict and social trends. Inevitable but not necessarily progressive or regressive.

Real social change is change in social patterns.

**Social conflict** A struggle between actors in which some win and some lose; struggle in which scarcity exists; struggle in which some dominate and others are denied what is valued.

Often contributes to the creation of an unequal social structure because those who win in social conflict are able to establish patterns that protect their interests.

Often contributes to social change in organization. To Karl Marx, social conflict is the central force for social change.

An inherent part of all social life; arises through differences or scarcity; arises from and encourages ethnocentrism.

Can be both constructive and destructive in organization. On the one hand, can help identify and deal with important social problems; on the other hand, can create actions that are destructive of people and social order.

**Social construction of reality** The creation by people in social interaction of what they regard as true. What people come to understand as they interact. Whatever reality may exist objectively, part of what humans always do is to see that reality through a social construction to some extent.

**Social control** In sociology, social control includes all the various ways a society attempts to control the individual. Religion is said to exercise social control through making morality sacred, creating a view of a just universe in which people come to believe that goodness is rewarded and evil punished, and teaching that working for the community is right.

**Social controls** All the various attempts to control the population in order to protect society's social patterns, both through rewards for conformity and punishments and threats for nonconformity.

**Social institutions** A term used in many different ways to describe what exists in society (see also *culture* and *social structure*). The three most important ways it is applied are to (1) an organization of people that is given special importance, (2) a sphere of society in which a set of patterns exists that deals with universal problems in that particular society (e.g., the economy, religion, government, and education), and (3) the particular ways a society deals with ongoing universal problems (the Supreme Court, public education, separation of church and state, computerization). In this book, the term is applied to the third usage. As Berger and Luckmann argue, institutions are "typifications," typical ways that society develops to deal with various situations.

Established ways of dealing with ongoing situations in society; established grooves that people follow in organization; accepted ways that help ensure smooth continuous action; the ways that are set up to deal with society's basic problems.

Every society establishes its own social institutions; alternatives are possible, but there is always an attempt to legitimate its particular ways. Berger and Luckmann describe social institutions as having historicity (a long history), objectivity (real forces), and legitimacy (a perception that they are right).

Like social structure and culture, a third set of patterns in society. Institutions not only aid the smooth continuation of society, but they are also used by the powerful to protect their favored position in society. Normally, institutions protect and expand inequality unless there is a conscious effort in society to create greater equality.

Created through a competition of people who make efforts to create institutions to serve their interests; social institutions tend to be successfully created, sponsored, supported, and defended by people with social power in society. Few institutions are meant to solve the problems of human misery except when misery touches the lives of those who are powerful or when democratic principles that find misery to be unacceptable are important in society.

**Social interaction** Action built up back and forth among actors acting around one another. Actors taking account of one another as they act; actors acting with one another in mind. Mutual social action among actors; mutual social influence. Symbolic interaction: actors meaning to communicate to one another and trying to understand one another.

The building block of society; creates and confirms social patterns (culture, social structure, and social institutions); facilitates cooperation and negotiation; the way people are socialized into organization.

Tends to isolate actors from those outside the interaction; outsiders often appear different, strange, maybe deviant. Encourages ethnocentrism. Creates differences among all social organizations, including societies.

**Social movements** Loose organizations of large numbers of actors who can be effectively mobilized around leaders to march, protest, boycott, strike, and actively confront oppositional groups; many people working together in defending or, more usually, opposing established social patterns.

**Social organization** An organization of people that arises out of and is maintained in social interaction and that possesses the social patterns of social structure, culture, and social institutions. Developed over time, social organizations include dyads, groups, formal organizations, communities, and societies.

**Social patterns** One of two qualities that make up all social organization. Developed out of social interaction. Make social interaction regularized, regulated, stable, predictable, routinized. Make possible a cooperative organized order. Guides to people's actions in organizations.

The longer and more intense the social interaction, the more important and established become such patterns. Social patterns tend to hang on, and they are historical forces that seem right. They are defended by those who like them and have power in society.

Durkheim describes social patterns as taking on a life of their own, possessing an objective existence and being invisible, real, external, and controlling.

**Social power** Ability to achieve goals in relation to other people; ability to achieve one's will in social relationships. Over time, social power creates a social structure of inequality, and those who continue to have social power protect that structure and their position in that structure.

Social power is one of three qualities that rank positions in social structure. It is usually but not always associated with the other two qualities, so there is a strong tendency for the amount of social power to be associated with the amount of privilege and prestige in position. Indeed, the amount of social power usually influences the amount of prestige and privilege; the amount of privilege and prestige usually influences the amount of social power.

Inequality of any kind can be translated into social power.

Social power makes possible an individual's or group's ability to change organization; usually in the hands of those in high position in organization who are not motivated to significantly change organization; to determine one actor's power in organization, one must take into account the power of others.

**Social science** Attempt to understand, categorize, and generalize about people without stereotyping. A highly disciplined process of investigation whose purpose is to question many of our uncritically accepted stereotypes and generalizations. An attempt to understand without being judgmental about categories of people. An attempt to create categories and generalizations that are tentative, probabilistic, and developed through carefully gathered evidence; and a willingness to change with new evidence. An attempt to understand not only categories of people, but also why those categories exist and what they cause.

**Social solidarity** The degree to which a community is integrated.

**Social structure** One of three patterns (see also *culture* and *social institutions*) that make up all organizations, including society; a set of relationships, positions, and ranks in organization.

A set of positions (sometimes called *statuses, social locations, locations, status positions*) that arise in social interaction. Almost always a system of inequality.

The sorting of people in society, often resulting from something other than qualifications; contributes to control, socialization, cooperation, and interdependence of people in organization.

Social structure, once established, protects the favored position of the powerful, increases their competitive edge, and allows them to exploit others.

**Social trend** Change that arises from the actions of many individuals who act in a similar direction and produce a cumulative effect on society; partly a result of individuals attempting to deal with their everyday situations without intending to change society; usually the result of even larger and more abstract trends or forces in society.

Long-lasting, far-reaching, general developments that have important impacts on social patterns in society. In the long run, the most important forces for social change, setting up an almost irreversible direction for society; once begun, trends have an inertia and take on a life of their own.

**Socialization** The process by which representatives of society teach people the ways of society and, in so doing, form their basic qualities. Through socialization, people learn the ways of society and internalize these ways. Socialization is also the process by which individual qualities and directions people take in life are created.

Socialization is the lifelong process by which people learn how to act and think from significant others, perceive what opportunities exist for them, and use others as models for what they become and do.

The process of learning in two stages: learning from individual significant others and then uniting these others into a whole, a generalized other.

The process by which the individual is taught to control himor herself according to the rules of the group, and through this control is able to take part in cooperative actions. The process by which actors are taught in society to accept the system of inequality and their own position in that system. The process by which actors learn the role, identity, perspective, power, privilege, and prestige associated with their position in social structure. The process by which the culture of an organization is learned. The process by which people come to feel a part of a community and come to feel an obligation to obey people who represent the community.

The process by which society creates the language, self, and mind of the individual, thereby creating much of the essence of the human being.

Inadequate socialization can result in misery for many people because it may bring too little love and support to the individual or too little selfcontrol necessary for successful problem solving. Even if there is adequate love, support, and teaching of self-control, socialization may lead the individual in directions that are destructive and bring misery to oneself and others. Socialization can also bring misery through creating unreasonable expectations for the actor.

**Society** A social organization of people who share a history, a culture, a structure, a set of social institutions, usually a language, and an identity. The largest social organization with which one identifies and is socialized within. All other social organizations—groups, formal organizations, and communities—are similar to society in their basic qualities, and all of these other organizations exist within society.

Durkheim sees society as a unity, a larger whole, a reality that is more than the sum of the individuals who make it up.

**Sociology** An academic discipline that focuses on society, social patterns, social interaction, socialization, social conditions, social causes, and social problems.

To Peter Berger, a type of consciousness, a perspective that is profound, unusual, critical, and humanistic in its concerns. Liberating, questions taken-for-granted ideas, and makes people aware of the power of society.

The science of society; the attempt to analyze the nature of society objectively through careful observation; an attempt to arrive at ideas about society supported by careful observation rather than cultural bias. A critical and realistic approach to understanding society, including the degree to which an individual can have an impact on social change.

An attempt to encourage people to wonder, investigate, and carefully examine their lives; an attempt to carefully and systematically examine what most people only casually and occasionally think critically about; an attempt to examine the power of culture, social structure, and social institutions in human life; an attempt to understand and appreciate the power of society.

A perspective that tries to enhance a liberal arts education; an attempt to liberate the individual through understanding society; one way to influence individuals to confront their own ideas, acts, and being. An attempt to scrutinize life; an attempt to make truth about human life more tentative for the individual.

**Spiritual** Religious in the sense that an individual believes that there is more to life than the profane. This would include those who accept an organized religion as well as those who hold individualistic beliefs and practices that focus on the sacred. Belief in God is not necessary for one to be spiritual, but one believes in more than just the physical aspects of the universe.

**Stereotype** An attempt to categorize people for the principal purpose of making judgments about others, usually negative, rather than carefully understanding them. An attempt to blame, condemn, or praise a category of people; to carelessly create sharp, absolute, and rigid categories; to make people unidimensional; to refuse to accept evidence that questions the categories and generalizations created; and to ignore attempts at carefully understanding the reason that the category exists. Stereotypes are categories not carefully created in the first place, and they usually allow few exceptions.

Highly oversimplified and exaggerated view of reality; quick to condemn; used to justify oppression of others; arises from hearsay and culture; stands in the way of accurate understanding.

Stereotypes of people act as a contributing factor to what they do and are. Often, stereotypes have a negative effect on those who are stereotyped and place them at a disadvantage in society.

**Symbols, symbolic** A form of socially created communication and thinking that is understood and intended by the one who communicates; makes possible much of human uniqueness.

**Thinking** Talking to oneself; using language to understand reality; and giving meaning to one's own actions and the actions of others. Allows for reflection and perhaps some freedom.

**Traditional society** Traditional society is a societal type that social science contrasts with *modern*society. Both are regarded as ideal types—that is, they are understood to be a matter of degree. Traditional societies tend to be characterized as societies dominated by the past, where progress in this life is neither assumed to be real nor necessarily good, where commitment to community rather than the individual is valued, where traditional religious beliefs and practices are all-encompassing and powerful, and where a single organized religion is dominant.

**Values** Part of culture; commitments based on our image of what is good and not good in life and what people regard as important; influences action; organizes cooperative action. Standards that people use to judge their own acts and the acts of others. What we use when we make statements using the word *should.*Humans tend to believe their own values are true when, in fact, they are usually matters of cultural preference.

**World society** A single world where social interaction, social structure, culture, and social institutions. Instead of independent nations, societies, and local communities are the social organizations that are important to people, the world itself becomes the dominant organization.

**World system** A term developed by Emmanuel Wallerstein who argues that the world has become an interdependent system of nation-states, where "core nations"–wealthy, industrial, highly diversifed, and stable–become central economic centers, and are able to dominate other nation-states called "peripheral" and "semi-peripheral." This is an example of a worldwide social structure.

# Index

## D

## G

## H

## M



## N



## O



## P

**T**



**U**



**V**



**W**